W9-BBO-760

TREE DISEASE CONCEPTS

Second Edition

PAUL D. MANION

State University of New York,
College of Environmental Science and Forestry
Syracuse, New York

PRENTICE HALL CAREER & TECHNOLOGY, New Jersey 07632

Library of Congress Cataloging-in-Publication Data

Manion, Paul D.
    Tree disease concepts / Paul D. Manion. — 2nd ed.
        p.   cm.
    Includes bibliographical references and index.
    ISBN 0-13-929423-6
    1. Trees — Diseases and pests.   2. Trees — Wounds and injuries.
I. Title.
SB761.M22   1991
634.9′63 — dc20

                                                90–44249
                                                CIP

TO MY WIFE, NANCY, AND MY SONS, WILL AND ED

Editorial/production supervision and interior
design by **Marcia Krefetz**
Cover design by **Lungren Graphics, Ltd.**
Cover Photo by **Herb Levart/SuperStock**
Manufacturing buyers: **Mary McCartney**
    and **Ed O'Dougherty**

© 1991, 1981 by Prentice Hall Career & Technology
Prentice-Hall, Inc.
A Paramount Communications Company
Englewood Cliffs, New Jersey 07632

Printed in the United States of America

10 9 8 7 6 5 4 3

ISBN   0-13-929423-6

Prentice-Hall International (UK) Limited, *London*
Prentice-Hall of Australia Pty. Limited, *Sydney*
Prentice-Hall Canada Inc., *Toronto*
Prentice-Hall Hispanoamericana, S.A., *Mexico*
Prentice-Hall of India Private Limited, *New Delhi*
Prentice-Hall of Japan, Inc., *Tokyo*
Simon & Schuster Asia Pte. Ltd., *Singapore*
Editora Prentice-Hall do Brasil, Ltda., *Rio de Janeiro*

CONTENTS

Contents

PREFACE

The subject of tree diseases has finally come of age. From a topic taught by a few peculiar people called forest pathologists to forestry students we have come to the point where young schoolchildren are made aware of the damages to our forest by "acid rain".

The news media splash attention-getting pictures of dead and dying trees before the public as profits of doom extoll their special-interest positions. Political and economic groups establish and or defend their positions. It is as though we have recycled the clock backward by hundreds of years. "Wrath of the gods" on our crops has been substituted by "wrath of man (industrialized society)" on our forests. There is no question that industrialized society causes changes in our forests, but no one is properly served when simple insect defoliations, diseases, and climatically induced injuries are used by scientists and politicians as examples of unusual recent forest destruction.

Scientists of many disciplines as well as the general public need to have a framework for understanding natural and unnatural forces that are causing problems with trees and forests. The need for a better understanding will be even more important in the future as the topic of the "greenhouse effect" builds. Climatic changes caused by increased carbon dioxide and other gases in our atmosphere will induce subtle effects that will be difficult to differentiate from natural effects. We can leave it up to the political forces and special-interest groups to lump everything under the wrath of industrialized society or we can understand the natural forces sufficiently for a proper recognition of unnatural forces.

Changes and additions to this revised edition of *Tree Disease Concepts* result both from additions to our scientific base of information and additions to my understanding of concepts that have been well established.

I continue to be indebted to many people within the profession of forest pathology for ideas, information, and reviews. Individual chapters were reviewed by Drs. William Smith, Dale Bergdahl, William Merrill, Everett Hansen, Harry Powers, William McDonald, Robert Blanchette, William Jacobi, Frank Hawksworth, David Houston, and Wayne Sinclair. My colleagues in the faculty of Environmental and Forest Biology, SUNY College of Environmental Science and Forestry, Drs. John Castello, James Nakas, Chun Wang, Hugh Wilcox, James Worrall, David Griffin, and Robert Zabel, have also been an invaluable resource of diverse ideas, concepts, and reviews. I have been particularly influenced in recent years by my interactions with a number of European scientists. But in the final analysis it is my students, both undergraduate and graduate, that give me a reason to try to understand and to organize my understanding into this book.

PREFACE TO THE FIRST EDITION

The forest pathology profession began about a century ago when Robert Hartig published his observations on the interrelationship of fruit bodies or conks on the outside of trees and fungus hyphae causing decay in the wood of the tree. Up until the late 1940s, there were a very limited number of forest pathologists in the world, but they contributed the major share of our present understanding of the causes of various diseases of trees. The forest pathology text written by John Shaw Boyce in 1938, and updated in 1948 and 1961, is a testimonial to the industriousness of early forest pathologists.

The tone for forest pathology set by the early workers centered on the cause and control of specific disease problems. Teaching of forest pathology and texts in forest pathology have emphasized these aspects. The student is generally exposed to a massive array of diseases. He or she is expected to identify the disease, causal agent, and be able to remember the control measure. If one looks at the list of control measures, it becomes obvious that we do not have control measures for diseased plants but rather recommendations on how to prevent the problem in the first place. Therefore, the student of forest pathology has been traditionally trained to write proper epitaphs for diseased trees with little emphasis on the basic biological understanding of representative disease systems, the ecological role of disease, and the economic interaction of disease in forest and urban management systems.

Rather than write an updated treatise cataloging diseases of trees for the specialist, I have attempted to write an introductory book which emphasizes the biological understanding, ecological considerations, and interactions of diseases of forest and urban trees with management practices for a limited number of selected diseases. These are intended to represent the types of tree disease problems that one might encounter.

The thrust of the book is to present tree diseases in such a way that beginners can understand and categorize things they see around them. If the majority of beginners can be stimulated to be more critically observant of the role of disease in management practices as they relate to both urban and forest trees, my task will have been successfully completed. To the few who will eventually become professionals in forest pathology or related areas, I hope the initial stimulus will provide direction for the vast array of fascinating and potentially extremely important interactions among trees, microorganisms, and environments.

Although specific references are not cited in the text, the reference lists should make it readily evident where many of my concepts originate. These lists are by no means complete but will provide the interested reader with a place to begin a more comprehensive search of the literature.

The synthesis of a book of this type requires more than can be acquired through reading the literature. I am very much in debt to many of my associates for ideas, concepts, criticism, suggestions, and encouragement. If I were to properly acknowledge the many people who have contributed ideas that I have freely drawn upon, the list would begin with the professors and graduate students at the University of Minnesota, where I began my training in forestry and phytopathology. The list would continue through many of the participants of the national and northeastern regional meetings of the American Phytopathological Society and Northeastern Forest Pathology Workshops, and further through many colleagues in various parts of the United States and Canada who have graciously hosted me as a visitor to their region.

I especially wish to acknowledge Dr. Harrison Morton, Dr. Robert Zabel, Dr. Edson Setliff, Dr. Wayne Sinclair, Dr. William MacDonald, and Dr. Gary Lahey for their assistance in reviewing drafts of this manuscript. I am also grateful for specific comments and criticism from my graduate students, in particular Dr. Robert Bruck, Dr. Patricia Gowen, Miss Anne Mycek, and Miss Barbara Schultz, who on many occasions have had to interpret what I have written for the undergraduate students in my courses. I would also like to gratefully acknowledge the typing assistance of Mrs. Julia Thomas, Mrs. Regina Carlin, and Miss Penny Weiman, and the photographic assistance of Mr. George Snyder.

Paul D. Manion

1

INTRODUCTION TO TREE DISEASE CONCEPTS

- *HISTORICAL PERSPECTIVE OF PLANT DISEASES*
- *FOREST PATHOLOGY IN RELATION TO PLANT PATHOLOGY*
- *ROLE OF TREE DISEASES IN NATURAL ECOSYSTEMS*
- *IMPORTANCE OF TREE DISEASES*
- *DISEASE IN RELATION TO OTHER DISORDERS OF PLANTS*
- *SYMPTOMS OF TREE DISEASES AS A REFLECTION OF DISTURBED PHYSIOLOGICAL FUNCTION*
- *PROOF OF PATHOGENICITY*
- *CATEGORIZING TYPES OF TREE DISEASES*
- *RECOGNITION OF BIOTIC, ABIOTIC, AND DECLINE DISEASES*
- *BRIEF OUTLINE OF THE BOOK*

Plants interact with their environment and other organisms in a wide range of ways. The plants most fit to survive are in balance with their environment. In the short run, imbalance caused by the presence of disease agents may produce serious economic and ecologic effects. These are the concerns of the plant pathologist.

HISTORICAL PERSPECTIVE OF PLANT DISEASES

To properly understand plant pathology today, it is helpful to look back and see how plant pathology developed. The perspective of the past should improve our capacity to comprehend the present and may sharpen our ability to predict and influence the future.

Your immediate reaction to the idea of "predicting and influencing the future" may be that all you want is a little understanding of the topic of tree diseases. But a little knowledge is dangerous. You will find or may already have found that people expect you to be conversant on a wide array of topics related to trees and plants. It is difficult to separate the casual conversation from real questions, but misinformation in either case may be costly. You will be asked to make judgments on problems and make suggestions on how to "control" the problem, which is really influencing the future.

Instead of starting this historical account at the beginning, let us start at the present and work backward.

1990–1900: Modern plant pathology. In the United States today, there are approximately 3000 professional plant pathologists. What role do they play? They are engaged in a cooperative effort with other professionals in the agricultural and forestry fields to provide a stable supply of food and fiber for our modern industrialized society. Agricultural technology in the United States allows an average farmer to supply food for 80 people rather than just himself and his immediate family. One can appropriately question specific environment, societal, and economical aspects of the agricultural technology, but there should be no doubt in anyone's mind that the majority of the people in this country and a large number of people in other countries of the world are highly dependent upon the success of each year's food crop.

What role do plant pathologists play in the success of each year's crop? It is difficult to single out the role of plant pathologists when, in reality, an array of disciplines are integrated into a successful system. Plant pathologists are involved in the development of disease resistance, fungicidal controls, monitoring disease buildups, making recommendations on when and what to spray, predicting yields and losses, and a host of other aspects of agricultural production.

Rather than continue to elaborate in general terms, let us look back just a few years to 1970, when a major disease epidemic reduced corn production 15% nationwide. Losses due to the corn leaf blight caused by *Helminthosporium maydis* were as high as 50% in some states. Even more severe losses were predicted for 1971, but because of dryer weather conditions, the fungus disease epidemic never materialized. By 1972, new resistant varieties were available, thereby reducing the threat of major losses, at least for the time being.

The 1980s have brought another challenge to plant pathologists. The news media have focused public attention on the "acid rain" threat. Initial concern was with lakes and rivers, but concern has spread to crops and forests. The hypothesis is that pollutants are affecting current and future agricultural and forest resources. Governments interact and respond to public pressure to do something about industrial pollution. Plant pathologists and forest pathologists are engaged in scientific inquiries to properly validate or invalidate many hypotheses that have emerged from this emotional issue. More recently the emphasis has been shifting to another emotional issue, global warming. Pathologists will again be involved in characterizing and modeling impacts.

The involvement of plant pathologists in acid rain illustrates some of the breadth of topics of modern plant pathology. Plant pathology is a discipline concerned with plant health problems caused by an array of biotic and nonbiotic factors.

Throughout most of this century, plant pathologists have been highly effective in reducing losses due to major epidemics. Today we rarely have a major disease epidemic. Twentieth-century plant pathologists can be credited with eliminating from agricultural production the merciless ravages of disease epidemics that have plagued humankind throughout recorded and unrecorded history.

Over the past 50 years the number of professional plant pathologists has increased from about 300 to 3000. During this period, viruses and mycoplasmas have become recognized as agents of plant disease. A wealth of information on diagnoses and control of plant diseases is available today to anyone who is interested, through extension specialists associated with state agencies and agricultural colleges.

1900–1850: Beginning of plant pathology.   During the latter half of the nineteenth century, modern plant pathology began. It was spawned out of the activities of three great scientists: a botanist, Anton deBary; a chemist, Justis Freiherr von Liebig; and a chemist-bacteriologist, Louis Pasteur.

From 1667, when Robert Hooke introduced the microscope, a dogma of spontaneous generation had become firmly entrenched in the thinking of the major intellectual figures of the scientific community. This dogma asserted that the structures seen in the microscope emerged spontaneously from diseased or decomposing matter.

Ignorant farmers had a more pragmatic outlook on disease. They recognized, for example, that wheat stem rust was caused by barberries. They controlled this disease of wheat by eliminating barberries. Present-day understanding of this disease demonstrates that wheat stem rust is caused by a fungus with a complex life cycle involving two hosts and a number of different spores (see Chapter 11). Elimination of barberries interrupts the life cycle of the pathogen.

A few independent thinkers had established that spores germinated and that chemical treatments that prevent germination prevented disease. They were on the right track but not in the main stream of scientific activity of the time.

Obviously, plant pathology, medical pathology, and microbiology could not emerge until it was recognized that fungi and bacteria were the cause rather than the product of the disease. It was a major uphill battle to disprove spontaneous generation and explain in chemical and biological terms what was really taking place during fermentation and decomposition of organic matter. Research conducted independently by deBary, von Liebig, Pasteur, and others eventually set the foundation for our present concepts of these processes.

In 1853, deBary demonstrated the causal nature of rust and smut fungi in diseases of cereals by carefully demonstrating the roles of the various spores in the life cycles of these fungi. He later identified the spores of the particular fungus that caused late blight of potatoes but could not work out the total life cycle. Based on his contributions, deBary is credited with being the father of modern plant pathology.

Julius Kuhn, in 1858, published a text on plant pathology that incorporated the concepts of deBary on the causal nature of fungi and mycological concepts of early nineteenth-century mycologists.

1847–1845: Late blight epidemic. Moving back just a bit further, to 1845–1847, we recognize the potato famine of Europe as a major stimulus for botanists and mycologists to apply themselves to problems associated with the economic welfare of mankind. The death by malnutrition of a million people, and the emigration of another million and a half, reduced the population of Ireland by one-half in a period of 5 years. Many of us with Irish ancestors can trace our roots back to this devastating epidemic. The need for plant pathologists was clear. It took almost a century to develop the concepts and trained professionals to fill the need.

1800–1700: Classification period. A botanist, Carolus Linnaeus, published a two-volume work, *Species Plantarum*, in 1853. These volumes were characteristic of the eighteenth century, a period of classification and taxonomy for botanical sciences. A number of other authors attempted disease classification. Diseases were named and classified more for the sake of classification than for their value to agriculture.

During the eighteenth century, occult influences that presumed diseases to be the wrath of angry gods still persisted, although there was some recognition of the effects of external environmental factors. During this period, the first pruning wound dressing was developed for fruit trees. This appears to be the origin of the idea that wound dressings do something for trees. This occult-fostered idea persists to this day.

1700–1600: Renaissance of earlier philosophical writings. For plant pathology, the seventeenth century was a period of renaissance or revival of interest in the writings of early philosophers. This followed a period, A.D. 500 to 1600, the dark or middle ages, during which science and learning appeared to slumber.

The introduction of the microscope by Robert Hooke in 1667 opened up a new world for amateurs and professionals to explore. This new world was teeming with shapes, colors, and movement. Reasonable thought processes recognized that people, mice, and trees came from other people, mice, and trees. Like begot like. Nevertheless, a mystic spontaneous generation process was assumed to produce things in the microscopic world. It took about 200 years to recognize that the fundamental biological properties of the macro- and microscopic world were the same.

A.D. 500–300 B.C.: Philosophers. The writings of such philosophers as

Theophrastus and Pliny during the period 300 B.C. to A.D. 500 were the source of concepts for the renaissance plant pathologists. These ancient philosophers observed, described, and speculated on the nature of diseases. It is interesting to note that Theophrastus (about 300 B.C.) recognized that wild trees were not liable to the ravages of disease, whereas cultivated plants were subject to an array of devastating diseases.

Beginning of recorded history. Moving back to the beginning of recorded history, we find references to plant diseases in Greek and Hebrew writings. Various crop maladies, such as blighting, blasting, rust, mildews, and smuts, were assumed to result from the wrath of gods. Biblical references can be found in Gen. 41:23; 1 Kings 8:37; Deut. 28:22; Amos 4:9; Hag. 2:16, 17; and 2 Chron. 6:28. There is every reason to assume that plant diseases have plagued human beings from the very beginnings of plant cultivation.

To summarize the development of plant pathology, we see a series of transitions from philosophical and occult interpretation to descriptive and taxonomic classification, to the application of scientific investigation to plant-related economic and social problems. The latter half of the nineteenth century saw the expansion of plant pathologists to most of the developed countries of Europe. Plant pathology was introduced into North America early in the twentieth century. Today modern agriculture, producing food for a hungry world, is sustained and advanced through the efforts of plant pathologists of many nations.

FOREST PATHOLOGY IN RELATION TO PLANT PATHOLOGY

Forest pathology is the branch of plant pathology that deals with diseases of woody plants growing in natural forests, in plantations, and in urban environments. For historical and other reasons, the deterioration of forest products is often included in forest pathology. Some would exclude from forest pathology the woody plant diseases of commercial fruit and nut crops. The individual importance of these crops has resulted in a series of subdisciplines associated with the individual crops. In my opinion, it is important not to exclude aspects of fruit crops from the topic of forest pathology.

Although there are about 3000 plant pathologists in the United States, there are just a few hundred forest pathologists in the world. Obviously, by number alone, their impact on societal needs is somewhat less apparent than the impact of plant pathologists on agriculture. The major disease epidemics of this century such as chestnut blight, white pine blister rust, and Dutch elm disease have had a stimulating effect on the profession much as the late blight epidemic of potatoes had on agriculture over a century ago. Just like the agricultural pathologists of the nineteenth century, forest pathologists have a well-developed taxonomic classification of diseases but are just beginning to understand how to manipulate conditions to reduce the impact of disease. Forestry is presently transitioning from a gathering to a resource management profession. As forests are cultivated and managed, applications of pathological understanding will have greater utility.

The forest pathology profession is still plagued with wound dressings as a disease-preventive measure. Some persons continue to utilize "bloodletting" activities such as the selective removal of diseased individuals for canker and decay problems, and "sugar-pill" therapy, such as the use of fertilizer and water for an array of diseases. The value of such occult-based activities is difficult to disprove; indeed, such methods may be warranted in certain cases. But a major task of present-day forest pathologists is to separate the witchcraft from the proven, scientifically based therapy.

ROLE OF TREE DISEASES IN NATURAL ECOSYSTEMS

The organisms of this planet have evolved relationships one with another. Plants are the predominant autotrophs capable of photosynthetically capturing the sun's energy for biochemical synthesis. Heterotrophic organisms are dependent upon the complex carbon compounds synthesized by the autotrophs. In turn, autotrophs are dependent upon heterotrophs to break down the structure and release the elements locked by previous populations of both heterotrophs and autotrophs. All the chemical elements of biological synthesis are cycled again and again through the various organisms and recycled back through the soil and air to be reused in biological synthesis.

It is appropriate to keep this balance of nature in mind when considering biological agents of disease. Disease may not be an imbalance in nature but rather a very normal part of the cycling and recycling of elements.

The ancient philosopher Theophrastus recognized the difference between natural ecosystems and cultivated plants. Natural ecosystems were not subject to the destructive effects of diseases as were cultivated crops. I would suspect that something may have been lost in the translation and interpretation of the works of this observant philosopher, because if one critically observes natural ecosystems, one sees the destructive effects of disease. The same disease-causing organisms are present in the natural ecosystem, culling out less-fit individuals. The spread rate and overall visual impact of these organisms are tempered by the buffering effect of genetic and/or species diversity.

In the natural ecosystem, the pathogen and host populations have evolved a balanced relationship. The natural population of plants may have genetic diversity, age diversity, and species diversity. There is little or no selection pressure on pathogens to increase rapidly. Under this type of system, diseases play a role in eliminating less vigorous plants and in facilitating succession. Widespread lethal diseases are unknown because there is no way for a large population of susceptible plants to develop. The pathogen population would quickly shift to eliminate the development of such a population.

Diversity is the key to long-term survival. In contrast, uniformity is often the key to short-term success. Uniformity establishes a selection pressure, often leading to uniformity within the pathogen population and an epidemic of disease which subsequently reestablishes the diversity and the population balance.

A superficial look or the misinterpretation of a concept might cause an observer to assume that diseases are not as significant to a natural forest as they are to a cultivated crop. But in actual practice, disease-causing factors are a dominant part of the ecological balance.

Natural ecosystems are not immune to imbalances introduced from the outside. The introduction of the Asian fungus *Cryphonectria parasitica (Endothia parasitica)* to North America produced a major imbalance, resulting in catastrophic destruction of the American chestnut during this century. This is a very unusual epidemic event of a natural population of plants. Pollen records document the chestnut blight destruction of the chestnuts. An event of similar magnitude involving the eastern hemlock is documented in the pollen record for the eastern United States about 4,800 years ago. The principles of stability in natural populations are not invalidated by such infrequent events. We should recognize the potential utility of some of the diversity features of natural population stability in developing our future managed forests.

IMPORTANCE OF TREE DISEASES

Evaluation of the importance of tree diseases is often based on evaluation of the lethal effects of diseases. A number of major diseases have produced extensive losses of this type. Syracuse, New York, which in 1950 had 53,000 elms along its streets, now has fewer than 300 as a result of the ravages of Dutch elm disease. The chestnut, once the major hardwood timber species in the eastern United States, has been reduced to a useless brush species by chestnut blight. Western white pine, one of the most valuable timber species of the northwestern United States, is so adversely affected by white pine blister rust that management objectives in high-hazard areas suggest discrimination against white pine in favor of any other species. The factors affecting infection hazard are discussed in Chapter 11.

Although dramatic, these few examples of the importance of diseases do not begin to represent the total picture. In 1958, the U.S. Forest Service published *Timber Resources for America's Future*, in which they attempted, in the forest protection section, to evaluate the roles of disease, insects, fire, animals, and weather on losses. They separated losses into mortality and growth loss, which added together were called growth impact. Table 1–1 shows that diseases rank as the most destructive agent. Approximately 45% of all losses in saw timber are attributable to diseases. The actual loss due to diseases is equivalent to about one-half of the annual amount cut.

One may question the specifics and applicability of these figures today. Times have changed. We no longer harvest vast quantities of old-growth western timber, so it may be that heart rots have been reduced. Possibly other factors, such as root rots, have increased. Management of southern pines has changed the picture there also.

No one has properly summarized the impact of disease of trees in recent years. The U.S. Forest Service recently estimated the yearly unsalvaged insect and disease losses in the United States at 67.9 million cubic meters (2.4 billion cubic feet), which is equivalent to about one-fourth of the annual timber harvest. Canadian annual losses to tree diseases were recently estimated at 65 million cubic meters. This is

TABLE 1-1 ANNUAL PERCENT GROWTH LOSS, MORTALITY, AND
TOTAL GROWTH IMPACT CAUSED BY VARIOUS AGENTS ON THE
SUPPLY OF SAW TIMER IN THE UNITED STATES[a]

	Growth loss	Mortality	Growth impact
Disease	40	5	45
Insects	8	11	19
Fire	15	2	17
Weather	1	8	9
Other	7	3	10
Total	71	29	100

Source: Data are for 1952 from *Timber Resources for America's Future,* U.S.
Forest Service, 1958.

[a]Percent of total loss (43.8 billion board feet).

equivalent to about 42% of the annual harvest in Canada. Insect-induced losses in
Canada are about 63 million cubic meters, so combined insect and disease losses in
Canada are equivalent to more than 80% of the annual cut. It is difficult to
understand the differences in the losses between the two countries, but the primary
point of these figures is to indicate that disease losses are major factors in the forest.
Diseases "harvest" a large volume of wood.

DISEASE IN RELATION TO OTHER DISORDERS OF PLANTS

Disease involves disturbance in the normal physiologic functioning of a plant, has
many causes, and exhibits an array of appearances. Plant pathologists do not agree
on a precise definition of plant disease. A useful concept of disease should distinguish it from plant injury, from disease symptoms, and from disease incitants
(pathogens). There is a major difference between the pathogen–host interaction of a
rust fungus and a white pine tree compared to the interaction of a camper-wielded
hatchet and the same tree. One is considered disease, the other injury. These are two
extremes in a continuum of plant–incitant interactions that are damaging to the
plant.

Disease will be defined here as any deviation in the normal functioning of a
plant caused by some type of persistent agent. How long must an agent persist in its
interaction with a plant to cause disease? This is where the continuum comes in. The
hatchet blow is a very short interaction. An air pollutant such as fluoride released
suddenly in large amounts as a result of an industrial accident also causes injury.
The same pollutant continuously released in small quantities as the result of an
ongoing industrial process causes disease. The boundary line between when something causes injury only or results in disease is not particularly important, since the
problems that fall into this area can be handled as specific cases. The more important point is to recognize that disease is generally caused by a persistent biotic or
abiotic agent.

Any agent that causes disease is called a pathogen. As we shall see, pathogens may be either biotic agents such as fungi or abiotic agents such as air pollution. Some pathogens are parasites, but not all parasites are pathogens. Any organism that lives on and derives nutrients from another organism is a parasite. Only those parasites that cause a disruption in the normal physiological function of the host are called pathogens.

SYMPTOMS OF TREE DISEASES AS A REFLECTION OF DISTURBED PHYSIOLOGICAL FUNCTION

Disease symptoms resulting from the interaction of specific pathogens and hosts are characteristic signatures of the pathogen and host. The plant pathologist can often readily recognize the presence of a specific pathogen based on symptoms alone. Why are symptoms so characteristic?

The symptoms of diseases are expressions of disturbed or abnormal physiology of the host plant. The woody plant has evolved a complex structure to separate and yet tie together various functions necessary for competitive survival. In Fig. 1–1, an elementary understanding of the structure and function of the woody plant is superimposed on the diagrammatic tree. There is a division of function, and therefore a limit to the range of expression, which various parts of trees can produce in response to invasion by pathogens.

Along the right and left columns of the diagram are listed the various abiotic and biotic agents of disease. At the center top a third category of diseases, declines, is tied to both biotic and abiotic agents. Decline diseases are characterized later in this chapter and discussed more fully in Chapter 18. As we shall see, they are complex diseases involving interacting biotic and abiotic agents. The center portion of the diagram relates the physiological functions of various parts of trees with the general categories of disease.

The biotic pathogens have evolved to fit into specific niches. A fungus that has evolved the capacity to survive by competing with soil microorganisms and the responses of tree roots is most likely to be found causing a disease of the roots. If we recognize the function of roots, we see why root problems produce characteristic symptoms of decay and necrosis of the root system and an overall appearance of mineral deficiency in the rest of the tree. Pathogens that disrupt DNA-directed meristematic cell division result in cancerous-like growths called galls.

Pathogens that parasitize the cambium, phloem, and sapwood xylem cells for available sugars and other nutrients result in the death of the invaded area. Death of a localized stem area prevents secondary growth in the affected area. The bark may change color. A depressed area on the stem results from the lack of stem enlargement in the diseased area. As the stem is being completely girdled by the invasion of an aggressive canker fungus, the roots are the first remote part of the tree to deteriorate. The canker interrupts the production and maintenance of functional phloem.

Heart rot pathogens have evolved the capacity to utilize the cell-wall materials of woody plants (i.e., cellulose and lignin). They differ from saprobic decay fungi in that the heart rot fungus is able to tolerate the dynamic chemical and morphological

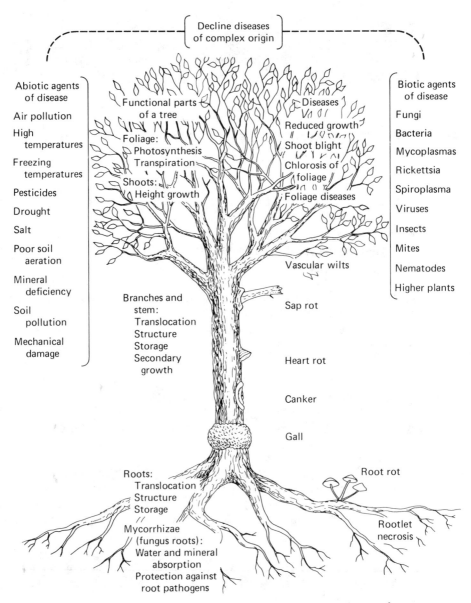

Abiotic agents
of disease

Air pollution

High
temperatures

Freezing
temperatures

Pesticides

Drought

Salt

Poor soil
aeration

Mineral
deficiency

Soil
pollution

Mechanical
damage

Decline diseases
of complex origin

Functional parts
of a tree

Foliage:
Photosynthesis
Transpiration

Shoots:
Height growth

Branches and
stem:
Translocation
Structure
Storage
Secondary
growth

Roots:
Translocation
Structure
Storage

Mycorrhizae
(fungus roots):
Water and mineral
absorption
Protection against
root pathogens

Diseases

Reduced growth

Shoot blight

Chlorosis of
foliage

Foliage diseases

Vascular wilts

Sap rot

Heart rot

Canker

Gall

Root rot

Rootlet
necrosis

Biotic agents
of disease

Fungi

Bacteria

Mycoplasmas

Rickettsia

Spiroplasma

Viruses

Insects

Mites

Nematodes

Higher plants

Figure 1–1. Summation of abiotic and biotic agents involved in diseases of trees, types of diseases, and functional parts of the tree. Decline diseases are caused by a combination of biotic and abiotic agents.

defense mechanisms of the living stem. The saprobic decay fungi do not compete well against the chemical and morphological defenses of a vigorous living stem. Therefore, they successfully invade dead or dying branches, large wounds, and eventually the dead or dying tree. The effects of decay or rot fungi are to weaken the structural integrity of the stem, roots, or branches.

Vascular wilt pathogens are adapted for survival in vessels of the sapwood xylem. Disruption of xylem vessels by wilt pathogens reduces the capacity of the vessels to translocate water from the roots to the top of the transpiring tree. During hot, dry periods, insufficient water is translocated to the leaves, causing them to wilt and die.

Foliage diseases affect the photosynthetic activity of trees. Viruses induce subtle color changes such as mottling and chlorosis, as well as other morphological and metabolic abnormalities. Obligate parasites such as rust and mildew fungi disrupt photosynthetic activity without causing serious mortality of leaves. Other fungi and bacteria cause necrosis of invaded portions, thereby reducing the effective area of the leaf. Abiotic toxicants, including salt, pesticides, and air pollutants, accumulate in leaves, disrupting or reducing photosynthetic activity.

Chlorosis (yellowing) of foliage may result from the direct effects of biotic and abiotic factors on leaves or the indirect effects of biotic and abiotic factors on roots. The most common symptom of mineral deficiency in plants is chlorosis.

Shoot blight is caused by microorganisms that aggressively parasitize succulent, rapidly growing shoots. These fungi and bacteria are also foliage and canker pathogens. They may gain access to shoots through infection of foliage, flowers, or succulent shoots, and may persist as stem cankers at the base of the infected shoot. The effect of shoot blight on young seedlings is more pronounced because killing the terminal shoot may destroy a great deal of the aboveground portion of the plant. As trees get larger, the killing of a shoot or shoots induces lateral buds to take over and compete for dominance as the new leader. A bushy-crowned tree may result from the inability of one lateral to gain dominance, or from the successive deaths of new leaders.

Reduced growth may occur as a consequence of the effects of any one or a combination of the problems discussed above. Reduced growth is also a characteristic symptom of decline diseases. Reduced growth may be the only aboveground symptom of some destruction of the root system. But reduced growth is a very subjective symptom, which may not be caused by disease agents. One must keep in mind that the capacity to grow is a combination of the age, genetic makeup of the tree, environmental effects on those genes, and possible pathogens.

Although the profile of a tree has been emphasized in developing this introduction to disease symptoms as a reflection of disturbed physiological functions, it is appropriate also to think of the functions of a tree in cross section. The tree stem is a complex structure consisting of (1) inner xylem (heartwood), functioning basically for structure; (2) outer xylem (sapwood), for storage and translocation of water; (3) cambium, as the meristematic layer of cells which, by mitotic division, produces xylem cells on one side and phloem cells on the other; (4) phloem, as a region where photosynthetic products, produced in the leaves, are translocated down to the stem and roots; and (5) bark, as a protective envelope of dead cells surrounding the living cells and providing a physical as well as a chemical barrier to invasion by microorganisms. Bark cells are produced from a cork cambium (phellogen) layer between the phloem and the bark. As the tree enlarges the cork cambium becomes interrupted. The interrupted cork cambium produces the rough or platy bark characteristic of older trees. This is a rather simplified characterization of the stem cross section, but it gives a framework on which one can impose the activity and effects of various diseases of the stem and branches.

This quick survey of disease as a reflection of disturbed physiologic function is meant simply as an overview. Subsequent chapters provide details regarding the interaction of pathogens and hosts.

PROOF OF PATHOGENICITY

Proof of the pathogenicity of specific biotic agents has generally been accomplished by the following set of procedures originally proposed by Robert Koch (1843–1901). The modified procedures are as follows:

Koch's postulates

1. There must be constant association of the suspected causal agent and the disease.
2. The suspected causal agent must be isolated and grown in a pure culture.
3. When inoculated into healthy plants, the agent that has been isolated must induce the disease.
4. Re-isolation from the disease-induced plants must yield the same causal agent.

Certain modifications of the procedures are necessary for specific types of disease agents that cannot be cultured — viruses, nematodes, mycoplasms, and some fungi. These are covered later in the appropriate sections.

CATEGORIZING TYPES OF TREE DISEASES

Biotic Plant Diseases

Biotic plant disease is the product of the plant, the pathogen, and the environment (Fig. 1-2) interacting over time. It is important to recognize that all three factors interact to produce diseases and that we may therefore prevent or control diseases by manipulating any one of the three. It is also important to recognize the time factor. Some diseases develop quickly within a plant; others develop slowly. There is also a time factor related to the spread and increase of the pathogen population within the host population.

By tradition, plant pathology is concerned with all diseases of plants except those caused by insects. This is a very artificial separation of an important group of pathogens. In actual practice, a useful plant pathologist must also recognize and understand insect–plant interactions.

Abiotic Plant Diseases

Diseases can also be caused by abiotic agents such as high or low temperature, phytotoxic gases, nutritional imbalance, soil-oxygen deficiency, moisture stress, and

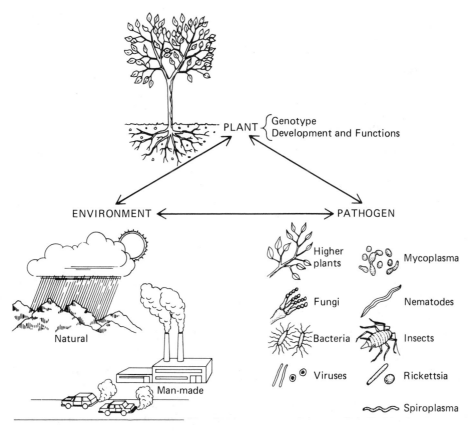

PLANT { Genotype / Development and Functions

ENVIRONMENT ⟷ PATHOGEN

Higher plants
Fungi
Bacteria
Viruses

Mycoplasma
Nematodes
Insects
Rickettsia
Spiroplasma

Natural

Man-made

Figure 1–2 Biotic plant disease is the product of three interacting factors over time.

other abiotic factors. Abiotic diseases are sometimes very similar to injury, so that separation of disease from injury is often more academic than practical.

Decline Plant Diseases

Major emphasis in plant pathology has been directed toward single biotic or abiotic primary-causal-agent diseases. There is a third category of diseases, called declines, which result not from a single causal agent but from an interacting set of factors (Figure 1–3). Terms that denote the symptom syndrome, such as dieback and blight, are commonly used to identify these diseases. One must be cautious, though, because these are not terms used exclusively for declines.

RECOGNITION OF BIOTIC, ABIOTIC, AND DECLINE DISEASES

The reasons for separating disease into three groups will become more evident once an understanding of each of the three types has been developed, but a few generaliz-

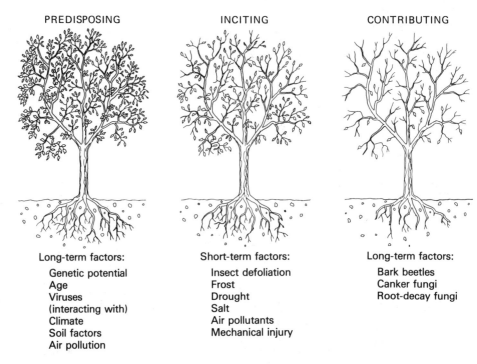

PREDISPOSING	INCITING	CONTRIBUTING
Long-term factors:	Short-term factors:	Long-term factors:
Genetic potential	Insect defoliation	Bark beetles
Age	Frost	Canker fungi
Viruses	Drought	Root-decay fungi
(interacting with)	Salt	
Climate	Air pollutants	
Soil factors	Mechanical injury	
Air pollution		

Figure 1-3   Factors influencing declines.

ations may be helpful at this point. If we shift from the theoretical concept that three types of disease exist to the practical question of how a person involved in the management of trees recognizes the three types, the concept should be more understandable.

One must look at both the diseases of individual trees and the diseased trees in a population to recognize differences among the three categories of disease. The biotic, abiotic, and decline diseases will be compared using four criteria: symptoms, signs, host specificity of disease, and spatial distribution.

Symptoms

Biotic agents of disease produce symptoms on specific plant parts. The affected parts are usually not randomly distributed on the plant; more often, only a portion of a plant is affected. Environmental or inoculum dispersal factors may account for the unevenness of disease symptoms. A progressive invasion of tissues is also a good symptom of biotic-induced disease.

Abiotic disease may or may not be plant-part-specific but is usually uniform in its symptom expression on the plant. There may be exceptions to a general distribution throughout the plant, caused by portions of the plant not being exposed. For example, the lower branches of a tree may be covered with snow and therefore not exposed to salt spray or the desiccating effects of winter winds. Abiotic disease does not generally occur as a progressive invasion like biotic infection, so that evidence of

Introduction to Tree Disease Concepts   Chap. 1

callous ridges on stem infection, or necrosis of leaves surrounded by chlorotic or other colored tissue, is usually not seen.

The most characteristic symptom of decline disease is the progression of symptom expression on individual plants and between plants. Another characteristic symptom is a reduction in growth. Some trees show very slight symptoms, others are dead, and others are intermediate in condition. A range of symptom expression may also occur with biotic diseases as a result of genetic variation and the spread patterns of the pathogen.

Signs

Signs are fruiting or other structures of biotic causal agents of diseases. Signs are most useful with fungal-induced diseases. The fungus can be identified and recognized as a causal agent of disease by the presence of specific fruiting structures.

With abiotic diseases, a confusing array of fungal structures may be seen. These can be identified and recognized as known saprobic organisms and are therefore not signs of tree pathogens.

Signs associated with decline diseases are not uncommon either. Some of these may be the confusing saprobes such as those often found on trees suffering from abiotic diseases, and others are facultative or weak parasites that contribute to the decline. Identification of a specific organism as a known contributor to declines is a good indication of decline disease.

Host Specificity

Biotic diseases are usually host-specific or occur on limited numbers of related or unrelated hosts. This concept fits best for fungal and bacterial pathogens. As you will see later, other pathogens may be more general in their hosts.

The most characteristic feature of abiotic disease is the occurrence of similar symptoms on two or more totally unrelated hosts. Decline diseases are host-specific problems, but more than one tree species in a region may have its own specific decline syndrome.

Spatial Distribution

Biotic diseases, because they are caused by infectious agents, usually show a clumping distribution pattern of diseased individuals. Inoculum produced by diseased individuals is most concentrated around the diseased individuals, thereby contributing to a higher incidence of disease in localized areas. Only with initial infection caused by inoculum dispersed from a distance does the distribution of disease approach randomness. Topographic features that produce moisture or temperature conditions favorable for inoculum production, dispersal, and infection may contribute to clumped disease distribution patterns typical of biotic disease.

Abiotic disease is usually random in a population except when the agent is distributed in a nonrandom fashion. For example, a point source of pollution will produce a progressive intensification of symptoms as one nears the source. Over distance, the distribution of symptoms is progressive, but at a given distance the individuals affected will be randomly distributed. Decline diseases have a random symptom distribution pattern within a given location.

All three types of diseases occur nonrandomly if one looks at a region as a whole. Thus, differences from one stand to another can be caused by many factors — site, environmental, and genetic factors of the host, to name just a few.

BRIEF OUTLINE OF THE BOOK

Each of the three types of diseases will be discussed at length in subsequent chapters. The abiotic diseases will be discussed first. The biotic agents of disease will then be developed, and finally the decline diseases will be presented as a complex interaction of at least three factors from the biotic and abiotic groups.

A series of overview chapters discussing plant disease epidemics, genetic control of resistance, diseases of seedlings in the nursery, pathological considerations of urban tree management, and pathological considerations of intensively managed forest plantations will draw upon earlier chapters to develop some of the concepts applied to forest management systems.

REFERENCES

ANONYMOUS. 1972. Genetic vulnerability of major crops. National Academy of Sciences, Washington, D.C. 307 pp.

BURDON, J.J., and R.C. SHATTOCK. 1980. Diseases in plant communities. *In* Applied biology, Vol. V, ed. T. Coaker. Academic Press, Inc., New York, pp. 145–219.

DAVIS, M.B. 1981. Outbreaks of forest pathogens in Quaternary history. Proc. 4th Int. Palynol. Conf. Lucknow (1976–77) *3*:216–227.

DINOR, A., and N. ESHED. 1984. The role and importance of pathogens in natural plant communities. Annu. Rev. Phytopathol. *22*: 443–466.

HEPTING, G.H. 1961. Forest pathology in forest management in the United States. *In* Recent advances in botany. University of Toronto Press, Toronto, Canada, pp. 1565–1569.

HEPTING, G.H. 1970. The case for forest pathology. J. For. *68:* 78–81.

HEPTING, G.H., and G.M. JEMISON. 1958. Forest protection. *Section of* Timber resources for America's future. USDA For. Serv. For. Resour. Rep. 14, pp. 184–220.

HORSFALL, J.G., and E.B. COWLING, EDS. 1977. Plant disease: an advanced treatise. Vol I. How disease is managed. Academic Press, Inc., New York. 465 pp.

HORSFALL, J.C., and S. WILHELM. 1982. Heinrich Anton DeBary: Nach einhundertfünfzig Jahren. Annu. Rev. Phytopathol. *20*: 27–32.

SINCLAIR, W.A., H.H. LYON, and W.T. JOHNSON. 1987. Diseases of Trees and Shrubs. Cornell University Press, Ithaca, N.Y. 574 pp.

STAKMAN, E.C., and J.G. HARRAR. 1957. Principles of plant pathology. The Ronald Press Company, New York. 581 pp.

STARK, R.W. 1987. Impacts of forest insects and diseases: significance and measurement. Crit. Rev. Plant Sci. *5*: 161–203.

STEWART, J.L. 1985. Current use and potential for implementing forest management practices in intensive forest management in the USA. For. Chron. *61*: 240–242.

WHETZEL, H.H. 1918. An outline of the history of phytopathology. W.B. Saunders Company, Philadelphia.

WHITNEY, R.D., R.S. HUNT, and J.A. MUNRO. 1983. Impact and control of forest diseases in Canada. For. Chron. *59*: 223–228.

2

SOIL CONDITIONS AFFECTING TREE HEALTH

- *MINERAL NUTRITION*
- *MOISTURE*
- *SALT*
- *SOIL AERATION*

The quantity and quality of the soil has a major impact on the health and vigor of trees. This chapter will highlight some of the chemical and physical features of soil that cause abiotic tree diseases.

The interaction of various physical factors of the soil, such as moisture, oxygen, mineral content, structure, and profile, with tree health is complex enough to make separation of single factors very difficult. For example, a heavy clay topsoil with impeded drainage will have a serious oxygen-deficiency effect, particularly during seasons of excessive rainfall. Anaerobic bacteria tie up nitrogen and sulfur under these conditions (see Chapter 7). Tree root development is impeded in this type of soil, resulting in mineral-deficiency symptoms in the tree crown. Application of fertilizer may temporarily alleviate the crown symptoms but will not solve the long-run imbalance between root regeneration and crown demands on the root system. Another complicating factor of soil environment on tree health involves the indirect effects of physical factors on the mycorrhizae (Chapter 9), pathogens (Chapter 16), and many interacting species of soil microorganisms. Finally, the impact of soil factors as stress-inducing agents in tree decline disease syndromes (Chapter 18) further exemplifies the importance and interaction of soil in plant health and disease.

Plant species differ in their tolerance of deficient soil environments. Those able to tolerate low oxygen resulting from high moisture are able to avoid competition from other, less-tolerant species in bog or wet sites. Those able to tolerate excessive vapor-pressure deficits in droughty sites avoid competition by growing in arid or highly drained sandy soils. A point to consider is that species that tolerate deficiencies do not necessarily have an obligate requirement for the deficient environment and may actually do well in a better soil if competition is removed.

MINERAL NUTRITION

An ideal environment should supply a balance of all necessary major nutrients, such as nitrogen (N), phosphorus (P), potassium (K), calcium (Ca), magnesium (Mg), and sulfur (S). Micronutrients, such as iron (Fe), manganese (Mn), zinc (Zn), boron (B), copper (Cu), and molybdenum (Mo) are needed in smaller quantities. The tree roots compete with microorganisms and the chemical attraction of soil structure bonds for these elements, so that availability is dependent upon more than just total concentration.

Serious deficiencies in chemical nutrients are uncommon in the "natural" forest because the species have evolved and occupy the sites that supply the required nutrients. Exotic trees planted on sites not normally occupied by the species may show deficiencies. Short chlorotic needles typical of potassium-deficiency symptoms in plantations of white spruce in central New York provide an example of this type of problem (Fig. 2–1).

Figure 2–1 The white spruce trees in the center of the photograph have short chlorotic needles typical of a mineral deficiency. The problem is most evident in the poorly drained sections of the plantation. Scots pines on the left do not show mineral-deficiency symptoms.

Good examples of mineral deficiency associated with plantations are reported from Germany during the 1980s. Second- and third-generation Norway spruce plantings, on sites that were formerly occupied by beech forest, are showing chlorosis of older needles typical of magnesium and or calcium deficiency. These elements are very mobile in plants and can be translocated out of older tissues to younger tissues. The parent rock material of some sites supplies a limited quantity of these elements. As the biomass of the forest grows, it ties up most of the available magnesium and calcium in plant tissues. The specific deficiency can be diagnosed by selective fertilization. Recovery of normal color following application of magnesium, for example, would demonstrate a magnesium deficiency.

The example above only scratches the surface of a very interesting spruce decline topic with far-reaching environmental and political ramifications. The reader is encouraged to read widely on this topic to understand the workings of science and politics.

Trees grown in urban environments or on highly disturbed sites also will often develop mineral-deficiency symptoms. Regular fertilization of ornamental plantings may improve their development on such imbalanced sites.

One should be cautious in recommending fertilization. Most trees, even forest trees growing on good sites, will respond to nitrogen fertilization if moisture conditions are sufficient. Cosmetic improvement by fertilization may not be the most appropriate solution to a problem with a different cause.

Forest fertilization should only be considered if results of foliar chemical analysis demonstrate a potential benefit of fertilization. For example, with loblolly pine in the south, a phosphorus content of needles less than 0.10% would indicate a phosphorus deficiency. Application of 50 lb/acre (56 kg/ha) of phosphorus could increase height at age 25 by 10 to 15 feet (3–4.6 m) on soils of the Lower Coastal Plain. Nitrogen fertilization in the same area should show a high response if the foliar content is less than 1.2% and if the site has adequate moisture.

MOISTURE

Soil moisture is divided into water held in the soil because of impeded drainage to lower layers, water held in the macrocapillary structure, water held in the microcapillary structure of small soil particles, and water chemically bound (Fig. 2-2). Macrocapillary water is utilized by tree roots and soil microorganisms. Microcapillary water is generally not available to plants, nor is chemically bound water. The ideal soil has a large expanse of macrocapillary structure to hold water for extended periods. Humus or decomposing organic matter is the major source of macrocapillary water holding capacity; therefore, the amount of organic matter in the soil is very important to tree health. Excess water held in the soil, filling air voids, is detrimental to most tree roots.

The availability of moisture to plants is measured as soil matric potential. A well-watered, yet aerated soil may have a matric potential of -0.1 megapascals (MPa). At -1.5 MPa most plants begin to wilt permanently.

The physical structure of the soil, as described above, is the major factor

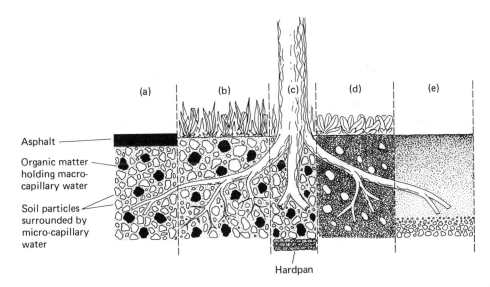

Asphalt

Organic matter
holding macro-
capillary water

Soil particles
surrounded by
micro-capillary
water

Hardpan

Figure 2-2 Interactions between tree roots and soil water with aeration, drainage, and soil structure: (a) degenerate roots in a good soil that is poorly aerated because of asphalt or concrete covering the surface; (b) healthy fibrous roots in a well-drained soil with extensive macrocapillary water; (c) degenerate roots in a poorly aerated soil because a hardpan impedes free water drainage; (d) degenerate roots in a soil with little macrocapillary water but much microcapillary water; (e) degenerate roots in a highly compacted soil with little macrocapillary water and aeration.

affecting matric potential in most soils. A second factor affecting matric potential results from the effects of ions dissolved in water. Dissolved ions decrease matric potential to a greater or lesser extent depending upon the concentration and chemical characteristics of the ions.

Trees respond to drought initially by closing their stomates to reduce transpiration. If moisture deficit persists, the leaves and new shoots will droop. Norway maple and other broadleaf species may develop a marginal necrosis to the leaves in midsummer. Early fall coloration and early leaf drop may occur. Conifers may show yellowing of the needle tips. Unfortunately, neither the symptoms on hardwoods nor conifers are particularly diagnostic since they can be produced by a variety of other agents.

Most established trees can withstand drought conditions and respond only by reduced growth, but extreme or extended drought conditions may cause twig dieback in the outer crown. Extreme extended drought conditions, over a period of years, may contribute to the death of trees.

SALT

Salt applied to de-ice highways generally does not accumulate in soil because it is very water-soluble and is washed out with spring runoff. In low areas where water accumulates or near intersections where salt application is excessive, salt will accumulate and be taken up by plants. Distribution of salt from high-speed highways to the foliage of conifers or buds of hardwoods via small droplets of salt spray also occurs. This salt can be taken up by the foliage and accumulate in plants.

Along Interstate 95 (Connecticut Turnpike) during the winter of 1969–1970, sodium chloride was applied at approximately 35 tons per two-lane mile (about 20 metric tons/km) and calcium chloride was applied at 1.5 tons per two-lane mile (about 1 metric ton/km). Under these conditions, samples of foliage of white pine had chloride concentrations greater than 1% (dry-weight basis). A typical highway in New England receives 20 tons of salt per mile annually (about 11 metric tons/km).

The effects of salt on plants are threefold. One is a direct internal effect resulting from sodium and chloride uptake and accumulation. Chloride translocation to points of active growth concentrates the toxic chloride ion in leaves and growing shoots. Sodium displaces the uptake of potassium by the root system.

The second effect of salt on plants results from changes in soil structure. Accumulation of sodium cations in the soil causes a displacement of other ions absorbed to clay particles, resulting in a change in interparticle forces. Air spaces are reduced, and displaced essential mineral elements are no longer available to plant roots.

The third effect of salt is on moisture availability. Salt decreases the matric potential of the soil. As the matric potential of the soil decreases, the tree roots have more and more difficulty drawing water from the soil.

Symptoms of salt damage to plants are seen in the spring and vary depending upon the plant and the method of salt uptake. In conifers, tip chlorosis and browning of needles occur with low concentrations. With higher concentrations the entire needle turns brown. Usually, the bud is not killed, so that new growth in the spring maintains the photosynthetic leaf surface. By late summer, brown needles of the previous winter have fallen off and the trees look reasonably healthy. A somewhat thinner crown, caused by the loss of needles, may be apparent. Growth reduction and predisposition to infection by root, shoot, and foliage pathogens may complicate the damage due to salt.

Usually, the damage to conifers is most apparent on the road side of trees. Salt-spray droplets, resulting from high-speed driving on the deiced surface, are wind-blown to the trees. Foliage on the road side of the tree intercepts most of the salt spray. Symptoms expressed on foliage are related to the concentration of chloride directly taken up by the foliage.

On hardwoods, the symptoms of salt damage are expressed as a marginal chlorosis and browning of foliage. Twig dieback is common. The crown of a hardwood tree damaged by salt usually has a concentration of foliage along large branches and the stem. A scattering of small dead twigs over the outer surface of the crown gives a pincushion appearance to the tree (Fig. 2–3)

Hardwood trees may accumulate salt by root uptake from soil or from droplets deposited on small twigs. The chlorides translocated to growing points generate the typical twig dieback and marginal leaf chlorosis symptoms.

Plant species vary in their tolerance to salt. Limited evaluation of salt damage has shown white pine and sugar maple to be very sensitive to salt, whereas Austrian pine and Norway maple are somewhat more tolerant to salt.

Figure 2–3 Salt-spray damage on red maples along Interstate 81 in Syracuse, New York.

Salt is a pollution problem involving more than just plant damage. Rusting of automobiles, deterioration of concrete, and increasing salt concentrations in fresh-water lakes are some of the concerns caused by salt in our environment. Damage to plants is probably well down the list in order of concern. Therefore, the best present methods for avoiding damage are to place trees well back from high-speed highways (at least 35 m was suggested for the Connecticut study), and to avoid planting trees in places where water runoff from roads accumulates salt in the soil.

SOIL AERATION

Oxygen is as essential to the root system of plants as it is to the rest of the plant. Soil oxygen and carbon dioxide in the ideal soil should be at about the same level as in the atmosphere (approximately 20% O_2, < 1% CO_2). Decreased levels of O_2 are more significant than increased levels of CO_2 to the health of plant roots. Factors such as heavy clay soils, soil compaction, filling over root systems with asphalt or concrete, or saturation of soil because of excessive rainfall or impeded drainage reduce the exchange of gases in the soil with the atmosphere and thereby reduce O_2. Increased metabolism of soil microorganisms increases CO_2 and consumes O_2, which needs to be exchanged with the atmosphere. Anything that interrupts air exchange is a potential problem.

In soil deficient in oxygen, root growth is retarded, amino acid leakage from roots to soil presumably increases, mycorrhizal development is reduced, and water and mineral absorption is consequently reduced. Avoiding conditions that reduce O_2 in the soil, as well as planting species that tolerate low O_2 levels, are ways to handle the oxygen problem.

Species that tolerate low oxygen also tolerate excesses of water, so as a general rule, bog species such as American elm, sycamore, hackberry, and honeylocust are preferred hardwood species for low-O_2 environments. You may note that these are species commonly used for urban plantings. Soil-oxygen deficiency is very common in urban areas. It is evident why these species have proven to be the best for use in the urban environment.

REFERENCES

HACSKAYLO, J., R.F. FINN, and J.P. VIMMERSTEDT. 1969. Deficiency symptoms of some forest trees. Ohio Agric. Res. Dev. Cent. Res. Bull. 1015. 68 pp.

HOFSTRA, G., and R. HALL. 1971 Injury on roadside trees: leaf injury in relation to foliar levels of sodium and chloride. Can. J. Bot. *49*: 613–622.

KOZLOWSKI, T.T. 1985. Soil aeration, flooding and tree growth. J. Arboric. *11*: 85–96.

PARKER, J. 1969. Further studies of drought resistance in woody plants. Bot. Rev. *35*: 317–371.

SCHARPF, R.F., and M. SRAGO. 1974. Conifer damage and death associated with the use of highway deicing salt in the Lake Tahoe basin of California and Nevada. USDA For. Serv. For. Pest Control Tech. Rep. 1. 16 pp.

SMITH, W.H. 1970. Salt contamination of white pine planted adjacent to an interstate highway. Plant Dis. Rep. *54*: 1021–1025.

WELLS, C., and L. ALLEN. 1985. When and where to apply fertilizer: a loblolly pine management guide. USDA For. Serv. Gen. Tech. Rep. SE–36. 23 pp.

WESTING, A.H. 1969. Plants and salt in the roadside environment. Phytopathology *59*: 1174–1181.

YELENOSKY, G. 1964. Tolerance of trees to deficiencies of soil aeration. Int. Shade Tree Conf. Proc. *40*: 127–147.

ZAK, B. 1961. Aeration and other soil factors affecting southern pines as related to little leaf disease. USDA For. Serv. Tech. Bull. 1248. 30 pp.

3

WINTER DAMAGE
TO TREES

- *MECHANISMS TO AVOID WINTER INJURY*
- *PHYSIOLOGICAL CHLOROSIS*
- *DESICCATION*
- *RAPID TEMPERATURE CHANGES*
- *LOW TEMPERATURE*
- *LATE SPRING AND EARLY FALL FROSTS*

Winter damage to trees is a topic of concern in climates where subfreezing temperatures occur. Native vegetation has evolved methods for advanced recognition of seasons when freezing conditions will occur. The triggering or recognition signals are probably decreasing temperatures and shortening of the photoperiod. Regional adaptation of vegetation for recognition of specific dates when freezing temperatures occur and the photoperiod shortens generally avoids damage to natural vegetation.

When people move plants about, regional adaptation for one site may not fit the new site, and under these conditions, winter damage will often occur. Therefore, as a general rule, we may say that winter damage is relatively unimportant in locally adapted vegetation but is often a problem with any introduced plant, even plants of the same species as native vegetation. In eastern North America, movement of plants more than 160 km (100 miles) north or south of the seed origin can potentially result in winter damage. In the mountainous regions of the west, elevation changes produce major climatic differences over much shorter distances.

Nurseries supplying planting stock should attempt to coordinate seed sources and planting locations. With exotics, one can only hope that some of the population of plants will by chance have the proper timing to the environment in which the plants are to be grown. Those that do not will not survive or will be injured year after year. Tulip poplars planted in central New York sometimes survive, but often the plants are severely injured by winter conditions. There are pecan trees growing in Syracuse, New York, but these are just chance selections from a species normally adapted for warmer climates.

MECHANISMS TO AVOID WINTER INJURY

Freezing temperatures generate a number of problems for plants. Those that have evolved a capacity to survive freezing conditions utilize a number of approaches to specifically avoid intracellular freezing. Ice formation within cells is lethal to both freezing-tolerant and freezing-intolerant plants. Intracellular ice formation disrupts vital functions of membranes, resulting in rapid cell death.

Mechanisms to avoid intracellular freezing involve, at the simplest level, a capacity of the plasma membrane surrounding the cell to move water out of the cell as freezing temperatures develop. Reduced internal water reduces the freezing point of the protoplasm of the cell. At a more advanced level, intracellular freezing of dehydrated cells is prevented at the -5 to $-10°$ C temperature range through additional accumulation of sugars and other antifreeze-like chemicals in the protoplasm.

Although intracellular freezing can be prevented by reducing the water and accumulating sugars, this does not necessarily prevent injury to the plant. Extracellular freezing does occur in plants. Extracellular ice produces a water potential gradient that draws additional water from cells, producing a freeze-dehydration condition. Very low temperature hardy plants must maintain the ion pumping capacity of the plasma membrane at temperatures below $-10°$ C. Without proper adjustments the plasma membrane would stop functioning at these temperatures. The lipids of the membrane would no longer be liquid, the proteins of the membrane would begin to denature, and shrinkage would change the spatial features of the functional sites on the membrane. Those plants that survive freeze-dehydration can avoid these detrimental effects to the membrane.

The maintenance of cell membrane function at temperatures below $-10°$ C is theorized to involve protection of membrane protein SH groups, avoidance of lipid peroxidation, and synthesis of additional membrane protein and lipids to avoid membrane separation from the cell wall. As the temperature drops, quaternary and tertiary structural changes occur in proteins. The quaternary structure or aggregation of two or more proteins into functionally active enzymes breaks down. The tertiary structure or folding within proteins is controlled largely by SS bonds between the SH groups of amino acid cysteine as it occurs at various points in the amino acid chain making up the protein. At lower temperatures, the folding bonds open up, exposing SH groups that are free to react with SH groups on other proteins, resulting in precipitation and denaturation of the proteins. Plants able to survive very low freezing conditions must protect the SH groups of the membrane proteins from this type of precipitation and denaturation. The lipids of the membrane are sensitive to oxidative changes at low temperatures. These reactions must be avoided by increasing the reductive capacity of the cells.

Hardy plants must also be able to tolerate the increased concentrations of various compounds in the protoplasm and avoid mechanical injury associated with shrinkage of the protoplasm. How plant cells cope with increased concentrations and shrinkage is not well understood. One concept suggests that the protoplasm becomes a gel matrix that holds various organells and reactive compounds in appropriate spatial arrangement. Indiscriminate mixing and contact would have lethal consequences.

The mechanisms described above for coping with freezing-dehydration, increasing concentration of compounds within the cell, and mechanical aspects of shrinking are a conceptual model that trees of the boreal forest would utilize to tolerate low temperature extremes. Some species of spruce, fir, poplar, birch, and willow are examples of trees with extreme-low-temperature tolerance. Most eastern deciduous forest trees, such as maples, beech, ash, and oaks, cannot survive the temperatures of the boreal forest. As described above, intracellular freezing is not tolerated by any plant, but these eastern deciduous trees cannot tolerate intercellular freezing of buds and parenchyma tissues. To avoid ice formation, these trees have a different mechanism that allows deep supercooling of tissue solutions. The mechanisms are not fully understood, but aspects of cell size, structure, changes in water content, and lack of ice nucleation centers can allow tissue solutions to drop to close to $-40°$ C before ice formation occurs. Once ice nucleation starts, the ice expands and rapidly kills the tissues.

The two mechanisms of tolerance or avoidance of ice formation in plants as described above represent ways that plants survive the consequences of low-temperature stresses. Other plants avoid low temperatures by staying small and covered by protective snow.

Plants go through a winter-hardening process each fall as the cells prepare for freezing stress. In the spring the physiological processes leading to winter hardening are reversed and plants become more and more susceptible to low-temperature stress. At the point of bud flush the succulent parts of plants have no tolerance for low-temperature stress and are rapidly killed by intracellular freezing caused by a light frost.

PHYSIOLOGICAL CHLOROSIS

Although conifers and other evergreen plants do not go through the major physiological change of deciduous plants, they do produce changes in chlorophyll content, structure, and function during the dormant season. These changes may appear as chlorosis, as in some varieties of Scots pine, or may appear as red to bronze color changes, as in some junipers.

DESICCATION

Many southern plants grown in cooler climates suffer damage due to desiccation. Water uptake capacity is decreased between 5 and 0°C, and therefore a southern plant such as rhododendron suffers desiccation even without having the root or stem frozen. In contrast, freeze resistant trees require extended continuous freezing conditions of roots and/or stem to prevent water uptake.

Plant species differ in their resistance to dehydration. Scots, jack, and red pines are examples of dehydration-resistant plants. Injury is enhanced by wind speed, sunshine, and low humidity. Symptoms of desiccation are yellow-brown to red-brown foliage and stems. In rhododendron, desiccation produces darkened brown or black zones that fade to brown or tan near leaf margins. Freezing damage in rhododendron produces a contrasting blackening of leaves, with no tan or brown outer edges. Symptom differences between desiccation and freezing damage in other plants are not well documented. It should be possible to reduce desiccation damage to plants with antidesiccant sprays, but a better way to avoid the problem is by selection of tolerant plants.

Some have suggested that desiccation injury occurs when a frozen root system cannot supply water lost by the crown of the tree. Although this may seem like a logical explanation, it is difficult to demonstrate that this is what really occurs in the field. Usually, there is cell damage caused by rapid temperature change and then loss of water from damaged cells. Under these circumstances, the condition of the root system does not play a role.

RAPID TEMPERATURE CHANGES

Winter damage from rapid temperature changes is very common even among native trees. Foliage of conifers or bark of young sugar maples exposed to heating by the sun can be damaged by the rapid temperature drop at sunset or when intermittently shaded by clouds. A drop from 2°C to −8°C in 1 minute and to −12°C in 6 minutes was measured for foliage of arborvitae (northern white cedar) at sunset with an air temperature of −12°C. Intracellular ice forms upon rapid freezing of plant tissue. The ice formation inside the living cells results in disruption of cell membranes, denaturation of some proteins, and other less-understood cellular dysfunctions. Winter damage due to rapid temperature change is recognized as dead foliage or dead patches of bark, always on the south-facing side of the tree (Fig. 3–1).

Rapid-temperature-change freezing has been a major part of the red spruce "decline" problem in the northeast. The winters of 1981 and 1984 had an extended warm period in February followed by a very cold period. The first visual evidence of damage occurred in early spring as a browning of needles on the outer branches. Closer examination revealed that the cambium of the twigs had been killed and that the needle browning was a response to this injury.

Another, better recognized example of rapid temperature change is the red belt phenomenon of western conifers. This situation occurs in areas and years when warm chinook winds buffet south and west mountain slopes. Lodgepole pine, very

Figure 3-1 Winter sun scald on sugar maple 1 year later.

severely affected during the winters of 1948 and 1971 in Alberta, Canada, appeared as red-brown bands on the mountains when viewed from a distance. Some have suggested that the needles dry out because the frozen root system cannot supply water, but it is more probable that the warm winds caused the trees to loose winter hardiness and that they were subsequently damaged by a rapid drop in temperature.

One can avoid the damage with ornamentals by wrapping the stems of young maples with paper or by shading the southern exposure of arborvitae. One can also avoid damage by recognizing that physical factors of the site, such as buildings, produce shadows that abruptly block out the sun. Planting in the zone of intermittent shadow of buildings should be avoided. Recognition of the potential problem of southern exposures and grouping different types of plants to produce a diffusion of sunlight rather than exposing individual plants to the direct effects of the sun provide a long-run solution to winter injury of this type.

LOW TEMPERATURE

Low temperatures generally do not produce damage to plants that can survive freezing conditions. Jack pines exposed to −195°C were not injured if they were hardened for 12 days at subfreezing temperatures. If the drop in temperature is slow enough, cells accommodate to the lower temperatures by concentrating solutes in the cytoplasm to prevent death due to intracellular freezing.

The most pronounced effect of low temperatures on trees of northern climates that have evolved mechanisms of tolerance for the extracellular freezing is a frost cracking or physical splitting of the woody stem. Dehydration of outer xylem cells by extracellular ice produces tangential contraction without radial contraction. The forces are such that when the outer xylem splits, a sound like a rifle shot can be heard. Additional splitting or cracking is enhanced by water which gets into the crack and speeds up the extracellular ice formation of cells on the margin of the crack as well as physical expansion of ice forming within the crack. Frost cracks may result because of low temperatures and/or may interact with wood decay and wound defects to produce a nonhealing vertical seam (Fig. 3-2).

Prevention of low-temperature damage involves recognition of sites where extremely low temperatures are expected to occur. Within cold regions, valleys or basins in the topography will generally have measurably lower temperatures than those in the surrounding upland country. Planting should be avoided in such areas. When planting exotics it is important to recognize the tolerance or lack of tolerance for freezing conditions of the planting stock. Match the plant to the climate of the site.

Another point to recognize is that extreme low temperatures are what separate the northern hardwood from the boreal forest. At Old Forge in the Adirondack Mountains of New York a temperature of $-47°C$ was recorded for February 18, 1979 (NOAA Climatological Data, National Climatic Center, Asheville, N.C.). This lowest temperature ever recorded for New York state occurred at the end of a 10-day period when the daily temperature lows ranged from -23 to $-44°C$. Although never properly documented, this low temperature, characteristic of the boreal

Figure 3-2 Seam in beech.

Winter Damage to Trees   Chap. 3

forest, undoubtedly caused injury to maples of the region, which in subsequent years became part of the "acid rain/forest decline political issue" (see Chapter 18).

LATE SPRING AND EARLY FALL FROSTS

Damage to plants by freezing temperatures during periods when plants are not dormant is related to hardiness. Hardiness is a nebulous, variable, physiologically based, state of plant development. During active growth, hardiness, the ability to withstand freezing conditions, is nonexistent. The effects of extracellular freezing are disruptive to actively growing plants.

Any factor that reduces or slows growth seems to induce hardiness. Decreased temperature and shortening of the photoperiod are examples. Fertilization increases growth and therefore should be avoided in late summer.

Chemically induced protection from spring frosts in apples has been accomplished experimentally with application of 2-chloroethyl trimethylammonium chloride (CCC), *N*-dimethylaminosuccinamic acid (Alar or B-nine), or malic hydrazide (Alar). How practical such chemicals may be for ornamentals is not yet known.

Late spring frost injuries are quite common even on native vegetation. On conifers, the curling downward and death of the succulent shoot is very obvious for a few weeks after the event. By late summer the dead tips are shed and growth from buds that had been dormant during the freezing conditions restores the crown to a relative normal appearance.

Spring frost injury on hardwoods causes blackened shoots and defoliation. A second flush of growth may restore the normal crown, but in many instances there is twig dieback and epicormic branching. These symptoms and reductions in radial growth increment may persist for years. A follow-up study on a frost injury in 1977 to Appalachian hardwoods of Virginia and West Virginia found that twig dieback of cherry increased for 4 years and was still evident 6 years after the event. In beech, the twig dieback peaked at 2 years but the trees were almost normal at 6 years. Sugar maple initially responded with an 88% reduction in normal increment but had little evidence of twig dieback at 6 years.

It is interesting to note that in this example the injury was most pronounced on the dominant and codominant trees of the stand. The most vigorous trees should theoretically be able to recover, but as discussed in Chapter 18, these specific trees may have been predisposed by some combination of events and therefore may never fully recover.

REFERENCES

ALDEN, J., and R.K. HERMANN. 1971. Aspects of cold-hardiness mechanism in plants. Bot. Rev. *37*: 37–142.

BARNARD, J.E., and W.W. WARD. 1965. Low temperature and bole canker of sugar maple. For. Sci. *11*: 59–65.

FRIEDLAND, A.J., R.A. GREGORY, L. KARENLAMPI, and A.H. JOHNSON. 1984. Winter damage to foliage as a factor in red spruce decline. Can. J. For. Res. *14*: 963–965.

GERHOLD, H.D. 1959. Seasonal discoloration of Scotch pine in relation to microclimatic factors. For. Sci. *5*: 33–343.

HERRINGTON, L., J. PARKER, and E.B. COWLING. 1964. The coefficient of expansion of wood in relation to frost cracks. Phytopathology *54*: 128.

JONES, J.K. 1971. Seasonal recovery of chlorotic needles in Scotch pine. USDA For. Serv. Res. Pap. NE 184. 9 pp.

LEVITT, J. 1980. Responses of plants to environmental stress, 2nd ed. Vol. 1, 497 pp.; Vol. 2. 607 pp. Academic Press, Inc., New York.

MAZUR, P. 1969. Freezing injury in plants. Annu. Rev. Plant Physiol. *20*: 419–448.

RAST, E.D., and R.L. BRISBIN. 1987. Six-year effects of two late spring frosts on Appalachian hardwoods. North. J. Appl. For. *4*: 26–28.

ROBINS, J.K., and J.P. SUSUT. 1974. Red belt in Alberta. Northern Forest Research Centre, Edmonton, Alberta, Inf. Rep. NOR-X-99. 6 pp.

SAKAI, A. 1970. Mechanism of desiccation damage of conifers wintering in soil-frozen areas. Ecology *51*: 657–664.

SCHUBERT, G.H. 1975. Silviculturist's point of view on use of nonlocal trees. USDA For. Serv. Gen. Tech. Rep. RM-11. 12 pp.

ZALASKY, H. 1975. Low-temperature-induced cankers and burls in test conifers and hardwoods. Can. J. Bot. *53*: 2526–2535.

4

TREE DISEASES CAUSED BY AIR POLLUTION

- **PLANT PATHOGENIC AIR POLLUTANTS AND SOURCES**
- **TOXICITY AND SYMPTOMS OF POLLUTION ON PLANTS**
- **WEATHER CONDITIONS AFFECTING AIR POLLUTION PROBLEMS**
- **REAL-WORLD INTERACTION OF POLLUTANTS, WEATHER SYSTEMS, AND PLANTS**
- **ACIDIC DEPOSITION**

Increased urbanization and industrialization, together with increased demands for mobility and the use of electricity, have placed excessive demands upon our environment to supply the needed raw materials and reabsorb the waste products. The problems produced fall into ecological, social, economic, and political areas of specialization which are interrelated into what seems like an unsolvable dilemma. I do not propose to develop or express the needed course of action out of this dilemma, because there is no single easy solution. My approach will be to describe the problems of air pollutants as they affect trees. You can develop the answers to our pollution problems. What this country needs is someone with the "right answer" who can convince others that he or she is right and who has enough money and political influence to instigate the solutions. In the meantime, those of us in biology who do not have the answer can strive to develop the background information that will help in the development of acceptable solutions.

Air pollutants, like many other abiotic factors, cause both disease and injury. If the pollutant is a persistent agent, it results in disease. If the agent interaction with the plant is of short duration, it produces injury. Let us not quibble over the separation of disease and injury, but rather consider phytotoxic gases of the atmosphere as a unified topic.

PLANT PATHOGENIC AIR POLLUTANTS AND SOURCES

In the past, pollution problems related to vegetation usually involved recognition of symptoms, proving that the injury was caused by an airborne gas, and locating the source of the phytotoxicant. Air pollution of this type centered around smelting (Fig. 4–1), chemical processing, and coal-burning electrical-generating installations. Since the source of the problem was relatively obvious, the answer was also obvious. Clean up the problem at the source. Political processes eventually resulted in legislation that forced reductions in ground-level concentrations of pollutants around point sources.

Ground-level pollutant concentrations were reduced through reduction in emissions or by dispersion of the pollutant over larger areas. Changes in fuel materials, changes in the combustion process, and installation of pollutant scrubbers on exhaust stacks were used to reduce emissions. These efforts resulted in 10% reduction in sulfur oxide emissions from 1970 to 1980. Increased dispersion was achieved by building higher smokestacks and increasing the exit velocity and temperature of the material being exhausted.

The combined effects of reduced emissions and increased dispersion have been dramatically successful in reducing plant damage around point sources of pollution. For example, the Sudbury region shown in Fig. 4–1 is no longer totally devoid of vegetation. Increased stack height and effective exhaust recovery procedures have dramatically reduced pollutants. Vegetation is returning to previously denuded areas.

As the point source pollution problem was being solved, a more difficult problem was emerging. The more dispersed pollutants of point sources in combination with pollutants coming from increasing numbers of mobile point sources (the

Figure 4–1 Extreme effects of air pollution associated with smelting, Sudbury, Ontario.

automobile) were generating more subtle region-wide impacts. The problems and political solutions have become more complex.

Many of the conveniences associated with modern civilization emanate from the combustion of fossil fuels. Fossil fuels are a source of entrapped, life-giving energy of the sun but also a source of many life-destroying toxic gases when combusted.

Because combustion of fossil fuels is central to the pollution problem, it is important to understand the products of combustion. In Fig. 4-2, the combustion process is summarized. Ideally, hydrocarbons are oxidized to produce CO_2 and water. In actual practice the common contaminant sulfur is also oxidized to SO_2, and additives, particularly in gasoline, are also released to the environment. The hydrocarbons are seldom totally combusted, so some are released into the atmosphere as carbon monoxide and unburned hydrocarbons.

One can improve the combustion and reduce the amount of unburned hydrocarbons by increasing the temperature and pressure of combustion, but this is not a solution without problems. Other pollutants, nitrogen oxides (NO_x), are produced under high temperature and pressure by chemical fusion of nitrogen and oxygen.

The array of primary pollutants from fossil fuels are carbon monoxide (CO), sulfur dioxide (SO_2), nitrogen oxides (NO) and (NO_2), lead (Pb), and hydrocarbons (HC). Of these, SO_2 and NO_2 are presently recognized as plant toxicants. These pollutants have an average lifetime of few days in the atmosphere. Dry and/or wet deposition bring them back to the earth's surface. Wind velocity and direction determine where they are deposited.

Lead is being investigated and may represent additional plant problems. Annual lead deposition in 1978 in a remote hardwood forest of New Hampshire was estimated at 266 g/ha. Today, lead-free fuels have reduced the inputs substantially. The accumulated deposits of many years of leaded fuels remains in the litter because

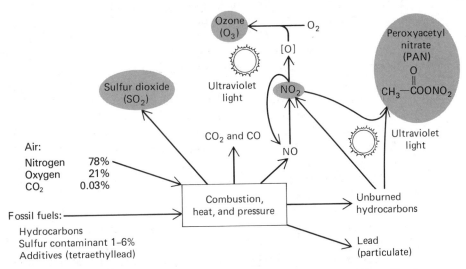

Figure 4-2   Plant-pathogenic air pollutants resulting from the combustion of fossil fuels.

of the relative insolubility of lead. The possible impact of these lead deposits on litter-decomposing microorganisms and symbiotic microorganisms is not fully understood, but it is reasonable to assume that there should be some effects and that these will be with us for many years to come.

The picture regarding CO is not clear, but some plants appear to incorporate CO in the same manner as they do carbon dioxide. Most of this CO is broken down by soil microorganisms. No primary plant toxicant role is known for hydrocarbons.

Two secondary plant toxicants are produced by photochemical reactions of the primary pollutants. Ultraviolet light from the sun supplies energy for the chemical oxidation of NO to NO_2. Hydrocarbons play a role as reductants during this oxidation. The NO_2 can further react with hydrocarbons in the presence of ultraviolet light to form peroxyacetyl nitrate (PAN). It can, also with ultraviolet light, form ozone (O_3). The ozone formation again produces NO. Nitrogen oxide can be reoxidized with ultraviolet light and hydrocarbons to form NO_2. In this way, a limited quantity of nitrogen oxide, acting as a catalyst, can produce a very large amount of ozone.

An important point to recognize with secondary pollutants is that they do not generally cause a problem in areas with high emissions. In the northeast, automobiles and industrial emissions of nitrogen oxides in urban areas generate about a 20% increase in ozone levels 40 to 50 km downwind. The San Bernardino Mountains region of California is an extensively studied pollution injury problem associated with secondary air pollutants generated downwind from a primary pollutant source, the Los Angeles area.

In these situations, the relationships among sunlight, primary pollutant sources, and secondary pollutant buildup is evident. The reaction that generates ozone from nitrogen oxide is reversible. Ozone will also react with sulfur oxides. Therefore, high ozone concentrations do not develop in areas with high emissions of primary pollutants. Peak ground-level concentrations of ozone occur in the noon to 4:00 P.M. period on sunny days in areas downwind from the primary pollutant source.

Another group of plant-toxic air pollutants are by-products of industrial processes rather than fuel combustion. Of these, fluoride produced in the smelting of copper and other ores is the most significant plant toxicant.

The major sources of primary pollutants are presented in Table 4-1. The secondary pollutant, ozone, is not listed in the table, but it should be recognized that ozone alone or in combination with other pollutants accounts for 90% of the air pollution damage to crops in the United States. Note that transportation is the largest source of nitrogen oxides, the primary reaction products associated with the production of ozone.

Most of the discussion of this chapter has concentrated on the contribution of modern industrialized society to air pollution. For perspective, it is important to recognize that natural sources of pollutants are a part of the problem. Sulfur- and nitrogen-containing gases are generated by forests. Hydrocarbons emitted by natural sources provide some of the distinctive smells of the forest but may also react with nitrogen oxide to produce ozone. In the eastern United States, reductions in hydrocarbon emissions from automobiles would have little or no effect on the ozone levels because of natural background.

TABLE 4-1 RELATIVE AMOUNTS OF PLANT PATHOGENIC AIR POLLUTANTS EMITTED INTO THE AIR BY VARIOUS SOURCES ($\times 10^6$ TONS/YR = 907×10^6 KG/YR)

Pollutant	Transportation	Industry	Generation of electricity	Space heating	Refuse disposal
Sulfur oxides	1	9	12a3	3	<1
Hydrocarbons	12	4	<1	1	1
Nitrogen oxides	6	2	3	1	<1
Fluorides		<1			
Particulates	1	6	3	1	1
Miscellaneous others	<1	2	<1	<1	<1
Total	<21	<24	<20	<7	<5
Percent	28	30	26	9	7

Source: Wood (1968).

Ozone is produced "naturally" during electrical storms. It can also be circulated down from the upper atmosphere, where it is not a pollutant but a very important ultraviolet-light protective screen. Ozone episodes at Whiteface Mountain in New York appear to be associated with downward circulation from the upper atmosphere. The relative contributions of "natural" and "unnatural" pollutants is difficult to estimate. These factors and other uncertainties contribute to the difficulties in making appropriate political decisions on the pollution problem.

TOXICITY AND SYMPTOMS OF POLLUTION ON PLANTS

Air pollutants vary in their toxicity to plants. Table 4-2 summarizes the concentrations and times necessary to produce injury as well as the most characteristic symptoms of injury; see also Figs. 4-3 to 4-9. One must also recognize that pollutants can produce growth reductions in the absence of obvious foliage symptoms.

TABLE 4.2 CONCENTRATIONS OF POLLUTANTS CAPABLE OF CAUSING INJURY WITH ONE FOUR-HOUR[a] Exposure and Typical Symptoms of Damage to Trees

Pollutant	Dose (ppm/4 hr)[a]	Typical symptoms	
		Hardwoods	Conifers
SO_2	0.5	Interveinal necrosis	Tip burning
NO_2	10.0	Interveinal necrosis	Tip burning
O_3	0.2[b]	Upper surface flecking	Tip burning, flecking, and banding
PAN	0.2	Lower surface bronzing	Chlorosis and early senescence
Fl	?[c]	Marginal chlorosis	Tip burning

[a]Parts per million (PPM). 1 ppm = 2600 μg/m³ of SO_2 = 1900 μg/m³ of NO_z = 1960 μg/m³ of O_3 = 4945 μg/m³ of PAN.

[b]Some conifers are sensitive at 0.07 ppm/4 hr dose.

[c]Fluorides accumulate in the plant from small and large doses. Symptoms occurs when the accumulative level reaches 100 μg/g dry weight of tissue.

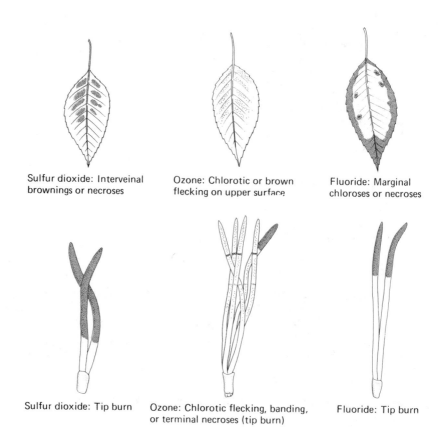

Sulfur dioxide: Interveinal
brownings or necroses

Ozone: Chlorotic or brown
flecking on upper surface

Fluoride: Marginal
chloroses or necroses

Sulfur dioxide: Tip burn

Ozone: Chlorotic flecking, banding,
or terminal necroses (tip burn)

Fluoride: Tip burn

Figure 4–3 Typical air-pollutant injury of hardwood leaves and conifer needles
associated with various air pollutants.

    The table is presented as the concentration of pollutant for a 4-hour exposure.
A 4-hour exposure would correspond to a typical afternoon exposure in the field.
These data are based on sample information from the literature. One should
recognize that plant injury is affected by pollutant concentration and exposure time.
At higher concentrations, it takes less time to cause equivalent injury. At lower
concentrations, it takes more time to cause injury. Unfortunately, the relationship is
not necessarily a simple linear dose response of pollutant concentration and exposure time. A number of other variables, such as plant species, genotype, age,
cultural conditions, developmental phase, and unknown factors, also affect a plant's
susceptibility to a particular pollutant dose.

    The information on PAN pollution injury to plants is known only from
growth chamber tests. It is very difficult to monitor PAN accurately in the field and
also difficult to find the characteristic metallic bronzing symptoms.

    Fluoride is a serious pollutant even at low levels because it accumulates in the
plant. PAN and O_3 are toxic in concentrations of less than 0.2 parts per million. It is
difficult for most people to visualize concentrations in parts per million (ppm), so I
will describe ppm in more detail. If you were to make a box to hold 1 m³ of air, the
box would be 100 cm on each side and therefore contain 1 million cm³. If 0.2 of one

(a)

(b)

Figure 4-4 Ozone injury (0.5 ppm for 7.5 hours) on (a) honeylocust and (b) "Bloodgood" London plane. (Photographs compliments of Dr. David Karnosky.)

(a)

(b)

Figure 4-5 SO$_2$ injury (1.0 ppm for 7.5 hours) on (a) honeylocust and (b) "Bloodgood" London plane (Photographs compliments of Dr. David Karnosky.)

Toxicity and Symptoms of Pollution on Plants

Figure 4-6 Fumigation chamber for testing air pollution sensitivity of ornamental plants. (Photographs compliments of Dr. David Karnosky.)

Figure 4-7 Chlorotic dwarf eastern white pine in Pennsylvania associated with air pollution. The small tree at the bottom of the picture is about 1 m tall. Its nonsensitive neighbors of the same age are 5 m tall.

Figure 4–8 Probably ozone injury on eastern white pine in Wisconsin. (Photographs compliments of Dr. David Karnosky.)

(a)   (b)

Figure 4–9 Leaf symptoms caused by (a) herbicide applied to the soil of pin oak or (b) flecking on sugar maple from leafhopper feeding may be confused with air pollution.

Toxicity and Symptoms of Pollution on Plants

of these cubic centimeters was ozone, this would be 0.2 ppm. In liquid measures the 1 m³ tank would hold about 265 gal. If 5 drops of liquid were added to this 265 gal., this would be approximately 0.2 ppm. Sulfur dioxide is slightly less toxic than ozone or PAN. Nitrogen dioxide is much less toxic and therefore is more important as a reactant in the photochemical production of O_3 and PAN than as a primary pollutant.

If one integrates Tables 4–1 and 4–2, it is obvious that transportation is the single most important source of air pollution. Because we all contribute to this type of pollution, change brought about in this area will be slow. It is much easier but less effective to condemn industry or electrical-power generators and force them to clean up than it is to reduce our own personal contribution to air pollution.

WEATHER CONDITIONS AFFECTING AIR POLLUTION PROBLEMS

Air pollution buildup is usually directly related to weather inversions that prevent dispersal of pollutants. Therefore, it is important to understand this type of weather condition.

An inversion is a condition in which warmer air sits on top of a layer of cooler air (Fig. 4–10). In the absence of an inversion, the temperature of the atmosphere decreases with altitude at a rate of 1°C per 100 m. This is called the adiabatic lapse rate.

Figure 4–10 Valley inversion over Syracuse, New York. The exhaust from the industrial complex in the background is visible just at the top of the inversion; the buildings are lost in the haze. The water vapor and pollutants from the stack in the foreground exhaust above the inversion and are therefore dispersed away from the city.

The reason air temperature normally decreases with increasing altitude is because of a decrease in pressure. Recall the relationship between temperature, pressure, and volume of a gas from an introductory physics course. The general gas law is $PV = MRT$ (pressure × volume = mass × a constant × temperature).

If a volume of polluted air emitted from a smokestack is warmer than the air around the top of the stack, it will rise because it is lighter than the surrounding air. It will lose temperature as it rises at a rate of 1°C per 100 m and will theoretically continue to rise indefinitely. Loss of some heat as it mixes with the surrounding air increases the heat loss with altitude, so that eventually the polluted air equilibrates in temperature with the surrounding air and is widely dispersed.

If an inversion occurs in the atmosphere, the parcel of polluted air from the smokestack may not be as warm as the air above the inversion and therefore will not rise through the warmer air. It is held below the inversion, which acts like a lid on a boiling kettle.

Inversions also prevent vertical dispersion of secondary pollutants such as ozone that are lighter than air. Thermal activity, wind turbulence, and the physical laws that govern dispersion of one gas within another provide thorough mixing of the pollutant in the air below the inversion.

There are many types of inversions. Two are rather regional in occurrence, because specific land formations contribute to their prevalence. Others are more general in occurrence.

Geographic Inversions

The most notorious inversion problems in North America are associated with geographically induced inversions along the California coast (Fig. 4–11). A general westward movement of moisture-laden air is often trapped by the coastal range of mountains. To rise over the mountains, the air must condense the moisture it holds.

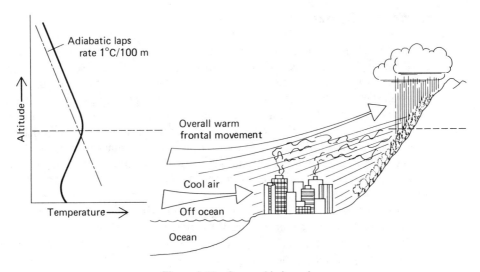

Figure 4-11 Geographic inversion.

Heat is released as moisture condenses, warming the air, so that eventually it will pass over the mountains. At the same time, a lower-level offshore wind brings inland cool air and fog underneath the warmer upper air that is attempting to get over the mountains. The final product is a mass of warm air on top of a mass of cool air.

Pollutants released into the cooler mass of air rise but are not warm enough to rise through the warmer upper layer and therefore are not dispersed from the coast. The fog and pollutants combine to produce a very unpleasant smog, which may remain over the metropolitan areas and surrounding basin for days.

Another type of pollution problem of the San Bernardino Mountains west of Los Angeles is associated with dry polluted air that is funneled like a chimney each day from the urban areas of the coast to the mountains. The ozone pollution builds to damaging levels in late afternoon in this type of system.

Valley Inversions

A second type of regional inversion is the valley type of inversion. Radiation of heat from the surface of the earth on clear nights causes cooling at the surface. The cool air, being heavier than the warm air, drains down into low-lying areas. If the topographic relief between hilltops and valleys is a few hundred meters, a mass of cool air may be trapped below a warm mass. A significant pollution problem occurs if there is a source of pollutant in the valley.

A town in a valley that expands its industrial or electrical-generating capacity must seriously consider the dispersion of pollutants out through the valley inversion. The engineering question of how to disperse the pollutants in this situation is easily handled. All that is required is to place a stack high enough so that the heated pollutant gases penetrate the inversion. This may be a local solution but not a regional solution, because the pollutants may be trapped by inversions covering larger areas.

Radiation Inversions

Radiation inversions are rather generally occurring inversions which when associated with a point source of air pollution, may produce a serious but localized problem. No highly specific topographic features are necessary, so that these can occur almost anywhere. The radiation inversion results from a cooling of the surface during clear nights. Radiation of heat into the atmosphere causes a cool mass of air to occur near the surface. Above 250 m the temperature of the air more normally approximates the adiabatic lapse rate, so that a cool air mass develops under a warm air mass. The inversion dissipates daily during midmorning because of heating of the surface by the sun. By midafternoon the surface temperature is above the normal adiabatic lapse rate.

Primary pollutants released at night and during the early morning may build up underneath the inversion and cause problems. Some may be brought down in localized areas during the up and down air movements associated with thermal activity during midday. These localized air-pollution injury pockets may be remote enough from the source to make it almost impossible to determine the actual source of the problem.

Anticyclone Inversions

The most serious type of pollution inversion on the eastern half of the North American continent occurs when an anticyclone or high-pressure system settles in for a few days (Fig. 4–12). Usually, highs and lows move across the United States from west to east rather quickly, but sometimes one gets held up by a major weather system off the Atlantic coast.

An anticyclone produces a subsidence inversion. The air temperature with altitude of a high-pressure system is slightly above what a normal adiabatic lapse rate would predict for various altitudes. The air in an anticyclone is descending and spreading out as winds from the center outward. The descending air is warmed along adiabatic-lapse-rate lines because of increasing pressure with decreasing altitude. Some of the increased heat is lost during mixing with cooler air, but the overall effect of subsidence is to increase the temperature of the descending air. Air near the top of the 5000 m anticyclone can descend for 4 to 5 days at 100 m per day and can, theoretically, be warmed by up to 50°C. Air in the anticyclone at 1000 m can only be warmed 10°C. Obviously, the air closer to the ground can descend very little and therefore is warmed little above the original temperature. The product of descending and differential warming is a warmer upper atmosphere on top of a cooler lower atmosphere—an inversion.

The anticyclone inversion is very large and may cover most of the northeast

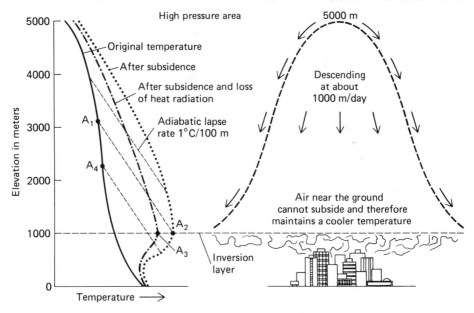

Figure 4–12 Anitcyclone inversion.

under a single blanket. Within the lower air mass, primary and secondary pollutants build up and cause problems over vast areas.

Frontal Inversions

A frontal inversion is very much like one side of an anticyclone inversion. Occluded or stationary fronts can more easily concentrate pollutants along the front to phytotoxic levels, but fast-moving fronts also concentrate as well as generate pollutants. Electrical storms generate ozone. The frontal inversion concentrates the electrical-discharged ozone and other, man-made pollutants to produce plant injury.

Turbulence Inversions

A turbulence inversion is a low-altitude inversion of air in the zone of turbulence caused by wind passing over irregular topographic features. The movement of air up and down by turbulence causes it to be warmed or cooled along the adiabatic lapse rate. If one averages all the adiabatic-lapse-rate lines for the turbulence zone, the interface of the upper point of the turbulence zone is cooler than the nonmixed air.

Turbulence inversion may concentrate and funnel polluted air from sources to remote areas along prevailing wind directions. It may also produce intermittent pollution problems due to irregular wind directions.

REAL-WORLD INTERACTION OF POLLUTANTS, WEATHER SYSTEMS, AND PLANTS

Although it is easier to separate the individual phytotoxic gases into specific concentrations and types of symptoms produced, it is more realistic to think of pollution as being caused by a mixture of pollutants. It is also easier to separate the inversions into types, but realistically they may superimpose and negate or enhance the concentrating effects. It is the synergistic and negating interactions of individual pollutants and types of weather systems that make real-world diagnosis of pollution problems much more difficult than one may initially be led to believe.

Ozone, the most important air pollutant, has a complex relationship with other pollutants. When the concentration of ozone and sulfur dioxide are too low to produce symptoms individually, the combination of the two pollutants may function synergistically, producing plant responses. At moderate concentrations of the two pollutants the symptom expression is approximately additive. At high concentrations the two pollutants interact antagonistically, thereby reducing the impact on plants.

The diagnosis of air pollutant injury to plants in the field is quite difficult (see Fig. 4–9). It is important to utilize a combination of features of pollution injury to arrive at a diagnosis. The first feature is the presence of acute symptoms, as described previously, on more than one plant species in the area. If one plant is affected, there is a very good chance that other species of plants are also affected. Since individual trees and forest stands are thoroughly immersed in the pollution mixture, pollution symptoms will occur uniformly over the susceptible plant, affecting all tissues of similar physiological and developmental state. The same uniform distribution occurs in the forest stand. Variation from one tree to another, caused by genetic variation and physiological differences, will occur, but if these differences are randomly distributed in the population, the pollution injury will also be randomly distributed. Pollution injury, associated with primary pollutants, may occur in localized areas associated with the pollutant source, but secondary pollutant injury will cover vast areas.

In the absence of acute symptoms, it is very difficult to identify air pollution problems without extensive growth and sensitivity testing using chambers (Fig. 4-6), in which the air is controlled through proper filtering. Comparisons are made with nonfiltered air and air in which natural levels and artificial levels of pollutants have been added.

The procedures and interpretations are sometimes confusing. Sometimes there is a stimulation in growth at low levels of pollutants. Interactions of pollutants and other stress-inducing agents may affect the interpretations. These and other problems limit our understanding of the affects of low levels of pollutants. At this time the "proper" methods for recognizing pollutant affects, in the absence of visible symptoms, are still in the developmental stages.

ACIDIC DEPOSITION

Acid rain has captured public attention in recent years. The news media have popularized the topic such that even young school children recognize that acid rain is killing fish and trees. Unfortunately, the public awareness is running far ahead of the documented real-world situation. Anyone who seriously looks into the topic quickly finds that untested and poorly tested hypotheses are the basis of many of the "facts" on acid rain. The topic is so emotional and political that it is difficult to objectively question some of these facts. Scientists and educators need to objectively question facts and to pass on this objectivity to others. Therefore, I will depart from the popularized dogma on acid rain and attempt to present the topic with a degree of objectivity.

It is important to recognize that serious interest and research on effects of acid rain on trees and forests are of very recent origin. Although the roots of the acid rain topic go back many years, current research output on acid rain and lakes began early in the 1970s. This lakes research fostered the beginnings of terrestrial vegetation effects research about 5 years later. Today there are tremendous volumes of literature being generated that expand the topic to include both wet and dry deposition of acidic substances, gaseous pollutants, and heavy metal pollutants as individually, or in part, contributing to decline and death of forests of North America and Europe.

Before we expand the topic to decline and death of forests, let us go back and understand acid rain. Acid rain literally refers to any rain at a pH of less than pure distilled water, pH 7. The average pH of the rain in the northeast is around 4.2 to 4.5. Rain in most places in the world is less than pH 7. In fact, distilled water, if left in "clean air," would not be pH 7. Distilled water is usually around pH 5.6, because dissolved CO_2 forms carbonic acid with water. Analysis of pH 4.5 rainwater finds other dilute acids, primarily sulfuric and nitric acids. Initially it was assumed that SO_2 and NO_X, from the combustion of fossil fuels and smelting of ores, were oxidized by ozone, hydrogen peroxide, and superoxides in the atmosphere to acids. Unfortunately, there are other natural sources of these acid precursors. One of the major issues today is how much of the acidity is due to natural sources and how much is due to anthropogenic, nonnatural, sources. One current estimate suggests that up to 25% of the total wet sulfur and nitrogen deposition in the northeast is from natural sources. Others suggest that up to 50% of the sulfur deposition is from natural origin.

There is no question that human activity is releasing large amounts of pollutants into the atmosphere. One can readily observe the obvious pollution sources and distribution by looking out the window of an airplane on a clear morning. Total emissions can be calculated from fuel consumption and similar data. Emission densities for sulfur in eastern North America is estimated at 1.9 g/m^2 per year. Nitrogen emission density in the Great Lakes region is about 3 g/m^2 per year. The issue becomes a bit more complex, though, when it is estimated that only about 25% of these emissions come back down as wet deposition. The fate of the remaining emissions and the relative contribution of natural versus nonnatural pollutants make it very difficult to estimate possible changes in acid rain that would occur if large amounts of money were spent to reduce emissions. These are some of the considerations being debated in the political arena.

A second area of considerable debate revolves around the impact of deposition of these and other pollutants in the environment. Laboratory tests demonstrate that very low pH solutions will etch the plant cuticle and leach nutrients from leaves. Field and laboratory tests demonstrate that low pH solutions affect availability and possible leaching of essential plant nutrients in the soil. The effects on growth and survival of plants from these activities is not clear. Another effect of acid precipitation is on the interaction of plants and pathogens. Although there has been limited work in this area, we can speculate that future work will demonstrate both positive and negative effects of acid precipitation on plant disease systems.

One should recognize that nitrogen and sulfur are essential plant nutrients and that acid rain is, in essence, fertilization to many plants. Acidification can increase the solubility and therefore the availability of other essential plant nutrients, such as calcium and magnesium. On the other hand, the increased solubility may result in increased leaching from the soil and a subsequent nutrient deficiency. The positive or negative effects of acid deposition depend on the concentration of the acid and the soil properties. Negative effects are difficult to demonstrate unless treatments utilize pH levels significantly below those found in the real world.

Acidic deposition has very little effect on most soils because of the large amount of buffering capacity. This buffering capacity is very high in deep fertile soils developed from parent material rich in limestone. The buffering capacity is lowest in shallow course soils developed from granitic parent material. Organic matter contributes to buffering capacity, but thin alpine soils composed of poorly decomposed litter material are already acidic and have little basic material to act as a buffer against additional acidic inputs.

If the pH of the soil drops much below 4.0, these conditions change the solubility of aluminum compounds, which in turn generate a toxic environment for many plants. One of the prominent hypotheses on the effects of acidic deposition suggests that there has been a lowering of the pH such that aluminum toxicity now affects root development in some forests. One of the conflicts with this hypothesis is that similar-looking forest problems occur in areas in which the pH is out of the range necessary for aluminum toxicity.

A second problem with demonstrating acid deposition effects relates to the documentation of changes in soil pH and the possible causes of these changes. Changes in the pH of soil can be caused by aging of the forest or by changing the forest composition. Climatic variables such as moisture and temperature affect the soil pH. These and many other factors make it very difficult to demonstrate that aluminum toxicity is a major part of the acid rain forest decline problem.

Growth reductions, tree declines (see Chapter 18), tree mortality, and an array of abnormalities have become entangled in the acid rain topic. The news media and many scientists are not very careful in selecting subjects to display as examples of acid rain damage to trees. It is my judgment, having visited many of the areas where acid rain damage is supposed to occur, having talked to many of the people directly involved, and having read many of the critical research papers on this topic, that no one has very convincingly demonstrated that acid rain is a problem for forest trees. The aspects of "normal" growth reductions and tree deterioration associated with cultural practices, stand dynamics, and age have not been adequately addressed. Symptoms and signs of "normal" diseases and insects have generally been neglected in assigning fault to acid rain. The impacts of climatic variables such as drought, excess rain, frost, and winter injury have been considered only superficially. The bottom line is to recognize that there may be problems in the forest associated with acid rain and other pollutants, but we will not be able to recognize them until we can properly recognize and quantify the impacts of the many "normal" factors affecting our forests.

REFERENCES

BENOIT, L.F., J.M. SKELLY, L.D. MOORE, and L.S. DOCHINGER. 1982. Radial growth reductions of *Pinus strobus* L. correlated with foliar ozone sensitivity as an indicator of ozone-induced losses in eastern forests. Can. J. For. Res. *12*: 673–678.

CLEVELAND, W.S., and T.E. GRAEDEL. 1979. Photochemical air pollution in the northeast United States. Science *204*: 1273–1278.

DAVIS, D.D., and F.A. WOOD. 1972. The relative susceptibility of eighteen coniferous species to ozone. Phytopathology *62*: 14–19.

DOCHINGER, L.S., AND C.E. SELISKAR. 1970. Air pollution and chlorotic dwarf disease of eastern white pine. For. Sci. *16*:46–55.

FISHER, B.E.A. 1986. The effect of tall stacks on long range transport of air pollutants. J. Air Pollut. Control Assoc. *36*: 399–400.

GARSED, S.G., and A.J. RUTHER. 1982. Relative performance of conifer populations in various tests for sensitivity to SO_2 and implications for selecting trees for planting in polluted areas. New Phytol. *92*: 349–367.

GSCHWANDTNER, G., K. GSCHWANDTNER, K. ELDRIDGE, C. MANN, and D. MOBLEY. 1986. Historic emissions of sulfur and nitrogen oxides in the United States from 1900 to 1980. J. Air Pollut. Control Assoc. *36*: 139–149.

HEPTING, G.H. 1968. Diseases of forest and tree crops caused by air pollution. Phytopathology. *58*: 1098–1101.

HUSAR, R.R. 1985. Chemical climate of North America relevant to forests. In Air pollutants effects on forest ecosystems. The Acid Rain Foundation, St. Paul, Minn., pp. 5–38.

JACOBSON, J.S., and A.C. HILL, EDS. 1970. Recognition of air pollution injury to vegetation: a pictorial atlas. Air Pollution Control Association, Pittsburgh, Pa.

KOZLOWSKI, T.T. 1980. Impacts of air pollution on forest ecosystems. BioScience *30*: 88–93.

LOOMIS, R.C., and W.H. PADGETT. 1973. Air pollution and trees in the east. USDA For. Serv. State Private For., Northeast Area and Southeast Area. 28 pp.

MANION, P.D. 1985. Effects of air pollution on forests: prepared discussion. J. Air Pollut. Control Assoc. *35*: 919–922.

MANION, P.D. 1987. Decline as a phenomenon in forests: pathological and ecological considerations. In Effects of atmospheric pollutants on forests, wetlands and agricultural ecosystems, ed. T.C. Hutchinson and K.M. Meema. NATO ASI series, Vol. G16. Springer-Verlag, Berlin, pp. 267–275.

MANION, P.D. 1988. Pollution and forest ecosystems. In Proc. 14th Int. Botan. Congr., Berlin, Federal Republic of Germany. ed. W. Greuter and B. Zimmer. Koeltz Scientific Books, Königstein. pp. 405–421.

McBRIDE, J.R., AND P.R. MILLER. 1985. Responses of American forests to photochemical oxidants. In Effects of atmospheric pollutants on forests, wetlands and agricultural ecosystems, ed. T.C. Hutchinson and K.M. Meema. NATO ASI Series, Vol. G16. Springer-Verlag, Berlin, pp. 217–228.

McLAUGHLIN, S.B. 1985. Effects of air pollution on forests: a critical review. J. Air Pollut. Control Assoc. *35*: 512–534.

MILLER, P.R., G.J. LONGBOTHAM, and C.R. LONGBOTHAM. 1983. Sensitivity of selected western conifers to ozone. Plant Dis. *67*: 1113–1115.

PETTERSSEN, S. 1958. Introduction to meteorology, 2nd ed. McGraw-Hill Book Company, New York. 327 pp.

REINERT, R.A. 1984. Plant responses to air pollutant mixtures. Annu. Rev. Phytopathol. *22*: 421–442.

SCHUTT, P., and E.B. COWLING. 1985. "Waldsterben"—a general decline of forests in central Europe: symptoms, development, and possible causes. Plant Dis. *69*: 548–558.

SCORER, R. 1968. Air pollution. Pergamon Press Ltd. Oxford. 151 pp.

SHRINER, D.S. 1978. Effects of simulated acidic rain on host–parasite interactions in plant diseases. Phytopathology *68*: 213–218.

SMITH, W.H. 1971. Lead contamination of roadside white pine. For. Sci. *17*: 195–198.

SMITH, W.H. 1981. Air pollution and forests. Springer-Verlag, New York. 379 pp.

SMITH, W.H. 1985. Forest quality and air quality. J. For. *83*: 82–92.

SMITH, W.H., and T.G. SICCAMA. 1981. The Hubbard Brook ecosystem study: biogeo-chemistry of lead in the northern hardwood forest. J. Environ. Qual. *10*: 323–333.

WALTHER, E.G. 1972. A rating of the major air pollutants and their sources by effect. J. Air Pollut. Control Assoc. *22*: 352–355.

WOOD, F.A. 1968. Sources of plant-pathogenic air pollutants. Phytopathology *58*: 1075–1084.

5

NEMATODES
AS PLANT PARASITES
AND AGENTS
OF TREE DISEASES

- *TYPES OF NEMATODES*
- *MODE OF ACTION OF PLANT-PARASITIC NEMATODES*
- *NEMATODE DISEASE CYCLE*
- *SYMPTOMS OF NEMATODE DISEASE AND INJURY*
- *METHODS FOR DIAGNOSIS OF NEMATODE PROBLEMS*
- *EXAMPLES OF NEMATODE PROBLEMS OF TREES*
- *CONTROL OF NEMATODE DISEASES*

Nematodes are one group of plant-parasitic animals studied by plant pathologists. Other plant-parasitic animals, such as insects and mites, are usually left to the entomologists.

At the present time, nematodes are not considered highly significant limiting factors in the growth of ornamental or forest trees. In fact, nematode problems occur only where there are nematologists (people who study nematodes), and there are only a few of these people in forest pathology. One current exception to the limited significance of nematodes is the pine wood nematode (*Bursaphelenchus xylophilus*). This nematode causes significant mortality of conifers in Japan and has received considerable attention because it was first thought to be a new introduction to North America. Pine wood nematode is now recognized as being native to North America, but now other countries, particularly the Scandinavian countries, are concerned about its introduction to Europe. They have placed an embargo on the importation of certain kinds of conifer wood products that costs the North American forest industry millions of dollars annually.

Why devote a chapter to such a limited subject? If we can learn from the successes and mistakes of plant pathologists with regard to agricultural crops, we can see that there was the same skepticism regarding nematode importance when there were very few nematologists in that field. Intensive agricultural practices introduced new and different problems, one of which was nematodes.

We can expect and already see similar problems in some forest nurseries and seed orchards. Nematodes will also become a significant factor in plantations where the same species are used in second and third rotations.

TYPES OF NEMATODES

Nematodes are unsegmented round worms, in the phylum Nemathelmenthes, with well-developed digestive and reproductive systems (Fig. 5–1).

Nematodes can be placed into four groups. The majority fall into a group of small, free-living saprophytic nematodes characterized by a hollow cavity like a mouth for picking up bacteria and/or organic matter and feeding it into the esophagus (Fig. 5–2). A second group of variable-sized nematodes are animal parasites with mouth parts adapted for suction attachment to animal cells. A third group are predators on small invertebrates in the soil and have mouth parts with modified teeth-like projections. The group we are interested in is a fourth group, 0.5

Figure 5–1 Plant-parasitic and free-living nematodes as seen with the 100× magnification compound microscope.

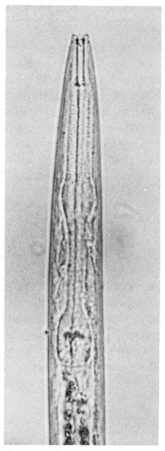

Figure 5-2 Free-living nematode (bacterial feeding) with a cavity-type mouth part.

to 2.5 mm in length. They are plant parasites characterized by a stylet or spear-like feeding apparatus used to penetrate cell walls of plants and fungi.

There are two types of plant parasitic nematode stylets. The stomatostylet is the most common (Fig.5-3). The odontostylet is a characteristic feature of some of the larger species of nematodes.

Plant parasitic nematodes are split into two groups based on feeding habits (Fig. 5-4). Some have a browsing feeding habit. They pierce a cell, remove the nutrients, retract the stylet, and move on to another location to pierce another cell. Another group has a sedentary feeding habit, which involves staying in the general location of their first feeding. These nematodes may nurse the cells along and remove only a small amount of nutrient, or they may continue to penetrate deeper and deeper into the root as cells die. The former group are called ectoparasites and the latter group endoparasites.

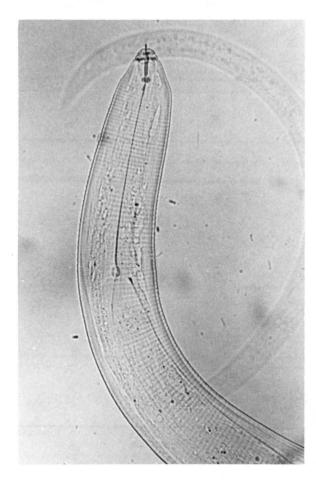

Figure 5-3 Plant-parasitic nematode with a stomatostylet used to penetrate cell walls.

MODE OF ACTION OF PLANT-PARASITIC NEMATODES

The stylet is a hollow pin-like structure used to pierce plant cell walls. Salivary juices are injected into the cell and soluble nutrients removed by means of the stylet. Plant-parasitic nematodes usually feed on the root systems of plants, but some feed on upper portions of plants. They may cause disease or injury directly by their feeding or by toxins injected into plants.

A second role of nematodes is to produce wounds for infection by soil fungi and bacteria. The destructive effects of the root-knot nematode (*Meloidogyne* spp.) and the *Fusarium* wilt pathogen on mimosa trees are more dramatic than either alone.

A third role of nematodes is as vectors of viruses. Tobacco ringspot virus of ash is transmitted by the dagger nematode (*Xiphinema americanum*). The odon-

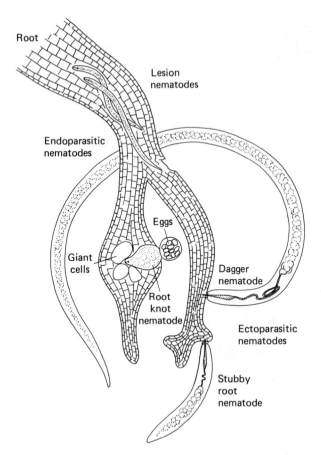

Figure 5-4 Types of plant-parasitic nematodes.

tostylet of this large nematode has a central opening large enough to allow uptake and transmission of polyhedron-shaped virus particles.

A fourth role of nematodes is the parasitism of mycorrhizal fungi. (Mycorrhizae are discussed in detail in Chapter 9.) The protective barrier and enhanced uptake of mycorrhizal roots is destroyed by the feeding activity of the lance nematode (*Hoplolaimus galeutus*).

The fifth role of nematodes is as a contributing factor in declines. (Declines are discussed in Chapter 18.) The stubby root nematode (*Trichodorus christiei*) is associated with littleleaf disease of southern pines.

NEMATODE DISEASE CYCLE

The typical plant–parasitic–nematode life cycle involves an egg, one larval stage within the egg, three larval plant-feeding stages, and an adult plant-feeding stage. Some species produce male and female individuals. Others are hermaphrodites with both male and female organs in the same individual. A third type reproduces parthenogenetically or in the absence of fertilization of eggs.

Those nematodes with odontostylets shed the stylet with each larval molt. One or two embryonic stylets to replace the shed stylet can often be seen in second- and third-stage larvae. Nematodes with stomatostylets maintain their stylets through the successive molts.

Under ideal conditions the life cycle from egg to adult and reproduction of up to 500 eggs is completed in about 1 month, so more than one generation per growing season is possible, even in northern climates.

Nematodes move in the film of free water surrounding soil particles. On their own, they can move only a few centimeters per year. One factor in parasitic nematode buildup in agricultural or intensively managed forest soils is increased potential for spread of the parasite on cultivation equipment and on transplants.

One unusual nematode, the pine wood nematode, moves from one tree to another by infesting the respiratory organs of wood-boring sawyer beetles (*Monochamus* spp.). The transmission stage of this nematode is called dauerlarvae.

SYMPTOMS OF NEMATODE DISEASE AND INJURY

Root destruction by nematodes produces symptoms in plants similar to nutrient deficiencies. Symptoms of nematodes on plants include growth reduction, imbalance of root to shoot ratio, sparse yellow foliage, reduced size of foliage, premature leaf drop, and abnormal wilting during hot, dry periods. Root symptoms include root lesions, impaired root growth, root rot (caused by fungi), excessive root branching, and galls.

The pine wood nematode invades the resin canals of the stem and branches of many species of conifers. The symptoms associated with this nematode are decreased resin production followed by rapid wilting and death.

METHODS FOR DIAGNOSIS OF NEMATODE PROBLEMS

Nematode diseases are detected initially by the symptoms produced on aboveground portions of plants. Examination of roots often shows characteristic symptoms of nematode activity, such as lesions or galls. To finally characterize a problem as being caused by nematodes requires extraction and identification of nematodes from the soil, roots, or stems.

The pine wood nematode can be recovered from recently killed branches or stems by placing chips of wood in a test tube of water. In a few days the nematodes can be seen in the water. It is interesting to note that this procedure will very often recover nematodes even though tree death was caused by some other agent, such as root rot or fungus infection of the shoots. The pine wood nematode can actively parasitize living trees if it is introduced into the tree through feeding wounds of sawyer beetles, but it can also be introduced into dying trees through oviposition slits made by the same beetles. In this instance the nematode feeds on blue stain fungi, which also invade the dying tissue.

Isolation of nematodes from soil is generally done by mixing a sample of soil in water. The slurry of soil and water is passed through a number of screens to remove the larger soil particles. Most plant parasitic nematodes are not small enough to pass through a 320-mesh (44-micron) screen, so nematodes and some soil particles in that size range are trapped on a 320-mesh screen. The concentrated nematode sample is washed from the 320-mesh screen and examined directly, or after further concentration by use of a centrifuge or Baermann funnel.

The Baermann funnel consists of a beaker in which the nematode slurry is placed (Fig. 5–5). A layer of cotton cloth or facial tissue is placed over the top of the beaker, which is then inverted into a water-filled funnel. Nematodes in the beaker wiggle their way through the tissue and settle to the neck of the funnel, where they are collected in a small volume of water a few hours later.

Any nematode with a stylet is potentially a plant parasite but the question arises as to which plants are parasitized. Some nematodes have very narrow host ranges and others very large. The species of the genus *Meloidogyne* (root-knot nematodes) are parasitic on at least 1500 species of plants, including 1,000 forest trees. Some stylet-bearing nematodes feed on fungi exclusively. Others may feed on both higher plant tissues and fungi.

Demonstration of parasitism of a particular plant by a nematode species requires controlled experimentation. The suspected parasite is introduced into the soil of a potential host grown in pots in a greenhouse. Extraction of nematodes from the pot at a later date should recover the introduced nematode. Evidence of larval stages indicates that the nematode can complete its life cycle. Evidence that the population increased is used to postulate parasitism of the particular host.

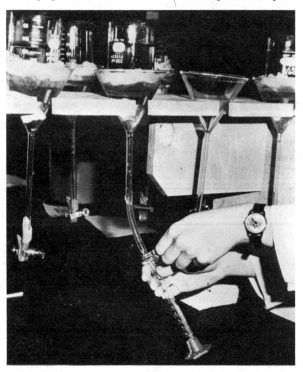

Figure 5–5 Baermann funnel for concentrating nematodes.

You can readily see the major difficulty in demonstrating the parasitism of nematodes of large trees based on this type of test. Because of this, very little information on the role of nematodes on mature trees is available.

An example of the circumstantial evidence for nematode interaction in mature trees is a study of maple decline in Connecticut. Larger populations of plant-parasitic nematodes were found in soil beneath declining trees than in soil beneath healthy trees. However, it is not possible to predict a cause-and-effect relationship on the basis of this evidence alone.

Proof of parasitism is not proof of pathogenicity. It is even more difficult to demonstrate pathogenicity, or the capacity to cause disease. Modification of Koch's rules of pathogenicity (discussed in Chapter 1) are needed. The first requirement of constant association can be demonstrated by extraction of a large population of parasitic nematodes from the soil or tissue of diseased plants. Because plant parasites can be extracted from almost any soil, we need some quantification of the term "large." As a general rule, populations of fewer than 200 nematodes per half liter of soil from the root zone do not cause problems, but populations in excess of 1000 per half liter of soil may result in a problem.

The second step, isolation and growth in pure culture, is a bit more difficult. The nematode is a complex organism with other microorganisms maintained internally and externally. Surface sterilization of the outer skin does not affect the internal microorganisms. So it is almost impossible to isolate the nematode in pure culture. Growth in pure culture is equally difficult because nematodes need plant cells to feed upon. Nematodes may be grown on tissue culture of a plant or various fungi, so that we can approach but not actually achieve pure cultures.

The third step, production of disease symptoms following inoculation with a "pure culture" of nematodes, and the fourth step, re-isolation, follow the procedures for proof of pathogenicity already described. Plant damage should be correlated with an increase in the nematode population.

EXAMPLES OF NEMATODE PROBLEMS OF TREES

Plant-parasitic nematodes are serious pathogens of agricultural crops and forest nursery crops. We also can expect, even though there is little evidence, that they play a role in reducing productivity of forest tree populations. The effects of nematodes on the diseases of natural populations of plants will be difficult to demonstrate. We find plant parasites present in almost any soil, but interactions of soil microorganisms and intermixing of roots of various plants retard the development of large populations of specific plant parasites.

Root-Knot Nematode

Over 1500 species of plants, including 1000 species of trees, both conifers and hardwoods, are parasitized by the root-knot nematode (*Meloidogyne* spp.). Feeding by the female causes giant cells to form in the root. Cell walls between adjacent root cells dissolve, nuclear division increases, and cells enlarge. The female *Meloidogyne* nematode is an endoparasite and spends her entire life in one location within the root gall, feeding upon the few giant cells formed. Eggs form parthenogenetically in the adult female and she swells to a pear shape. The eggs extruded from the female remain together in a sac-like membrane on the surface of the root gall. The young that hatch often penetrate the root in the same general area as the parent and further accentuate gall formation.

Dagger Nematode

Dagger nematodes (*Xiphinema* spp.) are very large plant-parasitic nematodes, at least 10 times larger than most others. The stylet is of the odontostylet type, with a central opening large enough to allow virus particles to be taken up during feeding. The stomato stylet of other plant-parasitic nematodes has too small an opening to allow the uptake of virus particles. Dagger nematodes are ectoparasites which in sufficient numbers cause extensive destruction of roots. Many ringspot viruses are known to be transmitted by this nematode, so that their role as vectors of virus may be as significant as their direct parasitism.

Stubby-Root Nematode

The stubby-root nematode (*Trichodorus christiei*) is associated with many agricultural crops as well as short leaf pine showing littleleaf symptoms. It was found in 19 of 35 southern pine tree nurseries. The effect of this ectoparasite is to produce short stubby roots and an overall reduction in both root and top weights of southern pine seedlings.

Lesion Nematode

Lesion nematodes (*Pratylenchus* spp.) are endoparasitic nematodes of agricultural and forest crops. Populations of nematodes burrow within the host root, causing necrotic areas of root. The nematode does not feed on just a few cells as with *Meloidogyne*, but continues to expand the size of the lesion by feeding on living cells on the periphery of the lesion.

Pine Wood Nematode

The pine wood nematode (*Bursaphelenchus xylophilus*) has been recognized as causing pine wilt disease in Japan for most of this century. It continues to destroy ornamental plantings and large areas of pine plantations in Japan and other Asian countries. When it was first recognized as a problem on pines in the United States in 1979 there was a major concern that it was an introduced pest that could pose a major threat to our pine forests. We now recognize that the nematode is probably native to North America and that, more than likely, it was introduced into Japan. The initial concern over an introduced pathogen produced a flurry of research. However, the pace of research slackened as people began to recognize the widespread occurrence of the nematode in North America as a secondary invader of trees killed by other agents. There has been limited evidence of dramatic mortality characteristic of the Japanese experience.

A second flurry of concern developed following the imposition of an embargo by Finland on the importation of all raw softwood products from North America. Other European countries followed suit to avoid introduction of the pine wood nematode into Europe. The economic consequences of an embargo on softwood products from Canada and the United States is very significant. For example, shipment of southern pine wood chips to Scandinavian countries amounts to revenues to the southern forest industry of $20 million annually. The Canadian forest industry reported a loss of $140 million in 1987 as a result of these embargoes.

Although the nematode does not represent a serious threat of mass destruction to our North American conifer forests, it can be a problem in seed orchards and plantations, particularly where the trees are growing under some type of environmental stress. The pine wood nematode has an interesting and complex relationship with wood-boring insects, bark beetles, blue stain fungi, and conifers. This relationship, summarized in Fig. 5–6, shows the Cerambycid beetles in the center because these wood-boring insects, called sawyer beetles (*Monochamus* spp.), are the only means for dispersal of the nematode. The trachea (respiratory system) of the pupal stage of the beetles becomes infested with up to 79,000 dauerlarvae of the nematode. These dispersal stage dauerlarvae can be introduced into branches of healthy conifers during cambium feeding by the adult insect. They can also be introduced into trees dying from other causes during the oviposition (egg-laying) activities of the adult insects.

Within the tree the nematodes multiply and move about in the resin canals while feeding on the adjacent epithelial cells. The disruption of the resin canal system affects the tree's ability to produce its defensive resins. The weakened trees are unable to defend themselves properly against *Dendroctonus* and *Ips* species of Scolytid bark beetles. These bark beetles introduce blue stain fungi (see Chapter 15) that invade and kill parenchyma cells. The pine wood nematode can also feed on the blue stain fungi. The dying trees are attractive for egg laying by the female sawyer beetles and development of the beetle larvae. Nematodes attracted to the pupal stage of the beetles complete the disease cycle.

Because there are so many participants in the pine wilt disease complex, there is some difficulty in identifying the specific cause of death. The complex system is also rather difficult to manage. Although a number of conifers (pines are the most common) can be affected, the insects and nematodes can be species specific, so there may be some limits on the movement of the pest from one tree species to another. Insect control is the logical place to break into this disease cycle, and in Japan this is the major thrust of their control program. There has been some work with parasites and pathogens of the nematodes and a recognition that stressed and dying trees set the stage for initiating attack by both sawyer beetles and bark beetles.

The concerns with introduction of the pine wood nematode into Europe are very real but need to be placed into proper perspective. It appears extremely unlikely that the nematode could be introduced into European forest ecosystems via wood chips shipped from North America. Although the nematodes can survive in the chips, the sawyer beetles do not, so unless there is some other vector that would go

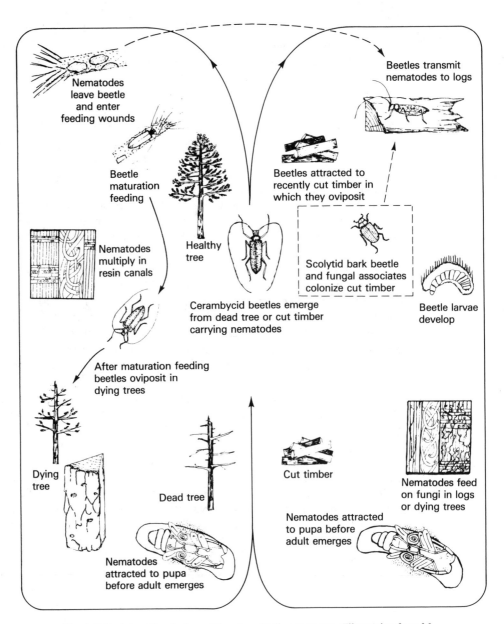

Figure 5-6 Interacting factors of the pine wilt disease system. (Illustration from M. J. Wingfield et al. J. For. 82:234.)

from wood chip piles to trees there should be no problem. Although wood chips are the primary form of raw material shipped for pulp production, the shipment of conifer logs certainly could provide a very logical means for introduction of the nematodes. Because of the complexity of the pine wood nematode system, there are many biological, economic, and political questions that have no immediate answers.

Consequently, the embargo problem experienced by North American forest products companies may continue for some time to come.

Many additional nematodes are presently known to parasitize trees, but those described above represent the range of interaction between nematodes and trees.

CONTROL OF NEMATODE DISEASES

Control of nematodes is generally through application of chemical nematocides and sterilants to the soil. Crop rotations, interplanting nonhost cover crops, fallow periods between crops, organic amendments, and nematode resistance are some of the control methods that will be used in the future. None of the present or future techniques will totally eliminate nematode problems, so we must learn to live with some nematode-induced losses.

REFERENCES

BERGDAHL, D.R. 1988. Impact of pine wood nematode in North America: present and future. J. Nematol. *20*: 260-265.

CHRISTIE, J.R. 1959. Plant nematodes, their bionomics and control. Univ. Fla. Agric. Exp. Stn., Gainesville, Fla. 256 pp.

DWINELL, L.D., and W.R. NICKLE. 1989. An overview of the pine wood nematode ban in North America. USDA For. Serv. Gen. Tech. Rept. SE–55. 13 pp.

FRECKMAN, D.W., and E.P. CASWELL. 1985. The ecology of nematodes in agroecosystems. Annu. Rev. Phytopathol. *23*: 275–296.

GILL, D.L. 1958. Effect of root-knot nematodes on Fusarium wilt of mimosa. Plant Dis. Rep. *42*: 587–590.

HOLLIS, K.P. 1963. Action of plant-parasitic nematodes on their hosts. Nematologica *9*: 475–494.

KIYOHARA, T.,and K. SUZUKI. 1978. Nematode population growth and disease development in the pine wilting disease. Eur. J. For. Pathol. *8*: 285–292.

MAI, W.F., and H.H. LYON. 1975. Pictorial key to genera of plant-parasitic nematodes. Cornell University Press, Ithaca, N.Y. 219 pp.

MAI, W.F., J.R. BLOOM, and T.A. CHEN. 1977. Biology and ecology of the plant-parasitic nematode *Pratylenchus penetrans*. Pa. State Univ. Coll. Agric. Bull. 815. 64 pp.

MAMIYA, Y. 1983. Pathology of pine wilt disease caused by *Bursaphelenchus xylophilus*. Annu. Rev. Phytopathol. *21*: 201–220.

NICKLE, W.R. 1972. Nematode parasites of insects. Proc. Annu. Tall Timbers Conf. Ecol. Anim. Control Habitat Manage., pp. 145–163.

PETERSON, G.W. 1962. Root lesion nematode infestation and control in a plains forest tree nursery. USDA For. Serv. Rocky Mount. For. Range Exp. Stn. Res. Note 75. 2 pp.

RIFFLE, J.W. 1971. Effect of nematodes on root-inhabiting fungi. *In* Mycorrhizae, ed., E. Hacskaylo. First North Am. Conf. Mycorrhizae, Proc. Apr. 1969. USDA Misc. Publ. 1189, pp. 97–113.

RIFFLE, J.W. 1973. Effect of two mycophagous nematodes on *Armillaria mellea* root rot of *Pinus ponderosa* seedlings. Plant Dis. Rep. *57*: 355–357.

RUEHLE, J.L. 1964. Nematodes, the overlooked enemies of tree roots. Int. Shade Tree Conf. Proc. *40*: 60–67.

RUEHLE, J.L. 1964. Plant-parasitic nematodes associated with pine species in southern forests. Plant Dis. Rep. *48*: 60–61.

RUEHLE, J.L. 1966. Nematodes parasitic on forest trees. I. Reproduction of ectoparasites on pines. Nematologica *12*: 443–447.

RUEHLE, J.L. 1969. Forest nematology: a new field of biological research. J. For. *67*: 316–320.

RUEHLE, J.L. 1969. Influence of stubby-root nematode on growth of southern pine seedlings. For. Sci. *15*:130–134.

RUEHLE, J.L. 1971. Nematodes parasitic on forest trees. III. Reproduction on selected hardwoods. Nematologica *3*: 170–173.

RUEHLE, J.L. 1973. Nematodes and forest trees: types of damage to tree roots. Annu. Rev. Phytopathol. *11*: 99–118.

RUEHLE, J.L., and D.H. MARX. 1971. Parasitism of ectomycorrhizae of pine by lance nematode. For. Sci. *17*: 31–34.

VIGLIERCHIO, D.R. 1971. Nematodes and other pathogens in auxin-related plant-growth disorders. Bot. Rev. *37*: 1–21.

WINGFIELD, M.J., ed. 1987. Pathogenicity of the pine wood nematode. The American Phytopathological Society, St. Paul, Minn. 122 pp.

WINGFIELD, M.J., R.A. BLANCHETTE, and T.H. NICHOLS. 1984. Is the pine wood nematode an important new forest pathogen in the United States? J. For. *82*: 232–235.

6

VIRUSES AS AGENTS OF TREE DISEASES

- *TYPES OF VIRUSES*
- *MODE OF ACTION OF PLANT-PARASITIC VIRUSES*
- *VIRUS DISEASE CYCLE*
- *SYMPTOMS OF VIRUS DISEASES*
- *METHODS FOR RECOGNITION AND VERIFICATION OF VIRUS DISEASES*
- *EXAMPLES OF VIRUS DISEASES OF TREES*
- *CONTROL OF VIRUS DISEASES*

The topic of virus diseases of forest trees is still just getting off the ground. In a recent review by Nienhaus and Castello they list 108 viruses on 23 genera of forest trees. Many have been poorly characterized. Very few have been studied enough to demonstrate an impact.

This is a very difficult topic with a limited number of participants. The forest decline issues of the 1980s have caused some to ask: What is the role of viruses in the forest? We have a limited foundation of information to answer this question since most of our work with viruses comes from cultivated crops. Some would suggest that viruses are rather uncommon in forest trees and use the absence of information in the literature as their evidence. Others would speculate that viruses are very common based on detailed studies of a few species where a number of different viruses have been found. They would also cite references where viruses are readily recovered from forest soils or from water draining forest land.

The cataloging of viruses from trees is still a major activity. As to the question of the role of viruses in declines, we have a long way to go. It is fascinating to speculate on the role of these agents that modify the genetic information of the plant. There could be modifications that affect aging and senescence. There could be modifications that redirect needed metabolic activity. There could be modifications that affect sensitivity to environmental stresses, air pollutants, or infection by other pathogens. There may even be "positive" effects of virus on plant development. Research with agricultural crop viruses provides some indication for each of the above. If we are ever to fully understand the forest decline issues, we will need to understand the role of viruses in forest health and development.

TYPES OF VIRUSES

Viruses are infectious agents composed primarily of nucleic acid and protein. They are considered lifeless by some because they do not possess the complement of normal metabolic enzymes used by organisms for synthesis and degradation of organic compounds. Others consider them primitive organisms because they possess functional nucleic acid, which is essential for continuity of genetic information for all living organisms. One can derive additional arguments for each point of view, but the final solution to the question is dependent upon which criteria you wish to define as essential for life.

Viruses are a rather diverse group of disease agents. The simplest viruses consist of a short strand of RNA surrounded by a protein coat. The most complex incorporate nucleic acid and proteins within a membrane capsule derived from the host-cell plasma membrane.

Some viruses, for example bacteriophage, are rather complex in structure and function. Some phage have a DNA core, a protein coat, and a limited number of enzymes combined into a complex structure.

Many plant viruses are comparatively simple in structure and function. The infectious nucleic acid, usually RNA, is contained within a protein shell composed of large numbers of identical protein subunits.

Most plant viruses operate as individual infectious units. Infection of a plant cell by one virus may prohibit infection by others. But some plant viruses require other viruses to infect. For example, tobacco satellite virus (SV) cannot develop in tobacco plants in the absence of infection by tobacco necrosis virus (TNV). Apparently, SV lacks the gene for synthesis of a necessary protein and therefore relies on TNV for that function.

Another infectious agent, a viroid, is composed only of a short piece of RNA. It was reported for the first time early in the 1970s as the cause of spindle tuber of potato. The RNA strand of viroids is so short that it does not have enough nucleic acid to code for even the smallest known protein. Viroids are believed to function as regulatory RNAs, causing disease by upsetting the regulation of gene expression.

Plant viruses as observed with the electron microscope appear as rigid rods, flexuous rods, or icosahedra (Fig. 6-1). Different viruses may be of different sizes, but all the infectious particles of a given virus will generally be of the same size.

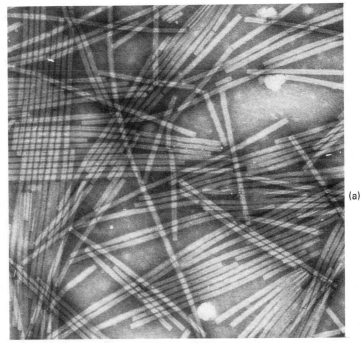

(a)

(b)

(c)

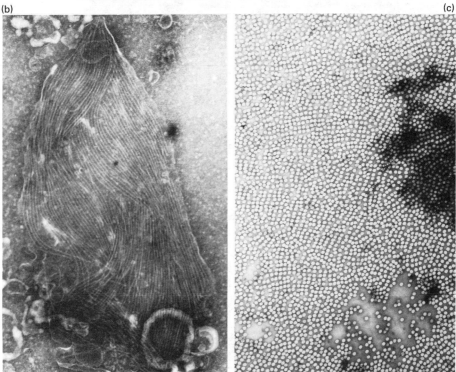

Figure 6-1 Examples of plant viruses as observed with the electron microscope: (a) rigid rods of tobacco mosaic virus; (b) flexuous rods of aphid borne cowpea mosaic virus; (c) icosahedra of tobacco ringspot virus. (Photographs compliments of Dr. Garry Lahey and Dr. John Castello.)

Some viruses are multicomponent, meaning that the infectious agent is made up of more than one particle.

The hosts for viruses appear to be limited only by our ability to recognize them. There is no reason to expect that any taxonomic group of organisms is free of viruses. The lack of viruses in certain taxonomic groups of plants, such as ferns and mosses, probably simply reflects the fact that they have not been extensively looked for in these groups.

The conifers have a limited number of viruses primarily because of the difficulty of working with viruses in these species. Observations of virus-like particles from conifer materials suggest that viruses are present. Transmission of the agents to indicator plants and then back to the trees has proven to be a very difficult task. The standard techniques used for virus work with other plants do not seem to work. It may be that these are different viruses that will require different techniques, or it may be that there are inactivators or inhibitors in the extracts from these tissues.

Virus diseases of plants have been recognized since 1892, when Iwanowski demonstrated that the infectious agent of tobacco mosaic virus (TMV) would pass through a bacteria-proof filter (Fig. 6-2). Figure 6-3 shows an example of tulip breaking, a virus disease of tulips recognized many years before viruses were known to exist. But it was not until the 1930s when the electron microscope was invented that viruses could be seen.

Figure 6-2 Tobacco mosaic virus, one of the earliest and most extensively studied plant viruses.

Figure 6-3 Tulip breaking, a virus disease of tulips that was recognized long before viruses were known to exist.

In 1935, Stanley demonstrated that viruses were composed of crystalline protein. It was three years later, when Bawden and Pirie showed that Stanley's virus protein was not a pure substance but one also containing a nucleic acid, that the true structure of viruses became known.

The naming of plant viruses is confusing. We use common names for viruses based on the symptoms they produce in the first host in which they were described. Unfortunately, different viruses can produce similar symptoms in the same host, and the same virus can produce different symptoms in different hosts. For example, tobacco ringspot virus in ash may look more like chlorotic line patterns or chlorotic spots than like chlorotic rings (Fig. 6-4).

MODE OF ACTION OF PLANT-PARASITIC VIRUSES

Plant viruses are infectious, obligate parasites that require living cells for their multiplication. Nucleic acids, amino acids, and enzymes necessary for production of virus particles are derived from the plant cell. This increases the demand on the cell for metabolic synthesis of these materials and disrupts normal plant metabolic functions.

The overall effect of plant viruses is slowly to starve the cell and the organism of nutrients and energy. Plant viruses, in contrast to animal and bacterial viruses, seldom produce mortality. The usual effect is so subtle that it is difficult to distinguish morphological differences between virus-infected and uninfected plants.

Figure 6-4 Line patterns in ash associated with tobacco ringspot virus.

Although viruses generally produce nonlethal effects on plants, it is usually possible to find some plants that are highly sensitive to virus infection. These plants produce necrotic spots, termed local lesions, where virus infection occurs. The necrosis of infected cells prevents further invasion and is therefore one type of resistance. The use of local lesion hosts in virus research will be discussed in the section on methods.

VIRUS DISEASE CYCLE

Virus particles enter cells through slight injuries or are introduced by insect, nematode, or fungus vectors. Within living cells the virus RNA molecule causes the cell to produce a complementary RNA strand, which then produces more virus RNA in the normal manner of nucleic acid synthesis. The virus RNA molecules also act as messenger RNA (mRNA). In conjunction with ribosomes, the virus RNA directs the synthesis of various enzymes required for virus replication and the virus protein coat (Fig. 6-5).

Virus RNA and coat protein assemble into virus particles. The completed virus particles may accumulate in storage vacuoles within cell organelles or in the cytoplasm. These accumulations may form crystalline structures which are sometimes observable with the light microscope.

Viruses move from cell to cell through protoplasmic strands called plasmodesmata. They can also become distributed throughout the plant in the phloem or xylem vascular system.

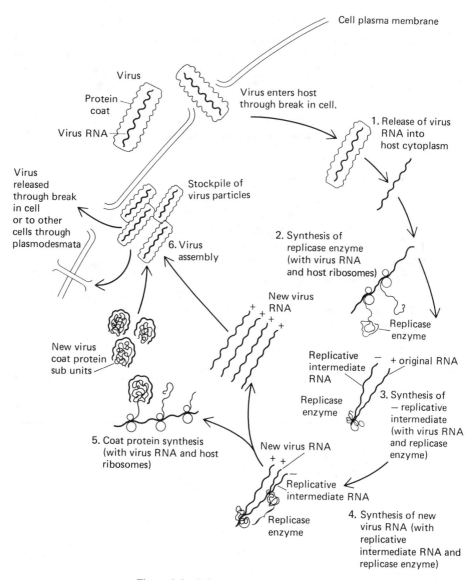

Cell plasma membrane

Virus

Protein coat

Virus RNA

Virus enters host through break in cell.

1. Release of virus RNA into host cytoplasm

Virus released through break in cell or to other cells through plasmodesmata

Stockpile of virus particles

6. Virus assembly

New virus RNA
+ +
+ +
+

2. Synthesis of replicase enzyme (with virus RNA and host ribosomes)

Replicase enzyme

New virus coat protein sub units

Replicative intermediate RNA

Replicase enzyme

− + original RNA

3. Synthesis of − replicative intermediate (with virus RNA and replicase enzyme)

5. Coat protein synthesis (with virus RNA and host ribosomes)

New virus RNA
+ +
−
Replicative intermediate RNA

Replicase enzyme

4. Synthesis of new virus RNA (with replicative intermediate RNA and replicase enzyme)

Figure 6-5 Tobacco mosaic virus "life cycle."

Viruses require outside assistance to get from one plant to another. Horticultural activities such as grafting, vegetative propagation, and even cultivation sometimes move viruses from one plant to another. Other viruses rely on specific insect vectors, such as aphids, leafhoppers, white flies, thrips, and beetles.

Mites, nematodes, and fungi can also serve as vectors. Some viruses are carried in seeds from one generation to the next, and a few are carried in pollen. Even though pollen and seed transmission are infrequent means of virus dissemination, the occurrence is sufficient to initiate virus in a population for subsequent short-range dispersal and intensification by insect and nematode vectors.

The need for vectors to move from plant to plant caused viruses to evolve varying types of associations, primarily with sucking insects such as aphids and leafhoppers. These associations with insects include mechanical transmission of viruses, which contaminate the insect mouthparts, to ingestion and multiplication within the insect prior to transmission to uninfected plants.

Aphids are ideal vectors of virus (Fig. 6-6). The aphids' perennial host provides a long-term virus reservoir. The aphids' succulent annual host is ideally suited for maximum virus increase. Seasonal shifts in aphid feeding ensure virus survival.

SYMPTOMS OF VIRUS DISEASES

The effects of viruses on their hosts result in a variety of symptoms and in some cases no visible symptoms. Typical symptoms involve flower color streaks; foliage ring spots, flecks, necrotic lesions, mosaic or mottling patterns (Figure 6-7), chlorosis, line patterns, curling, dwarfing, stem morphology changes, excessive branching (witches' brooming), and rosettes. The subtle effects of symptomless or hidden viruses may produce only yield losses. If nearly all of the plants of a given species are affected, it is almost impossible to recognize a virus by its effect on the host.

The role viruses play in ecological community relationships is totally unex-

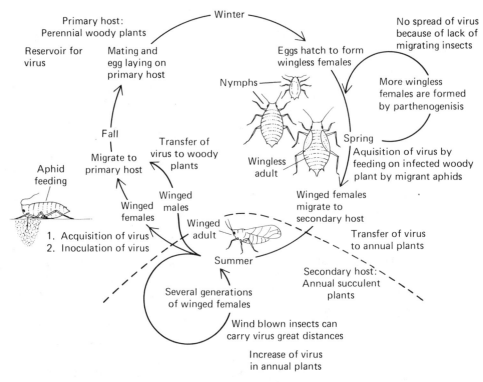

Figure 6-6 Aphid life cycle and its relationship to virus disease of plants.

Figure 6–7 Bean yellow mosaic virus is common in garden beans.

plored. Even though we have virtually no evidence as to the effects of viruses on natural communities, we can speculate from our knowledge of agricultural problems that they should have a profound effect.

METHODS FOR RECOGNITION AND VERIFICATION OF VIRUS DISEASES

We may think that a particular set of symptoms is caused by virus, based on the characteristic nature of virus symptoms, but other disease agents can cause some of the same symptoms. Therefore, transmission of the suspected causal agent to a healthy plant, followed by the development of symptoms in the formerly healthy plant, is a beginning step in the verification of virus etiology.

The demonstration of pathogenicity for viruses requires a modification of Koch's postulates. The first step of constant association of the pathogen and the disease usually requires an indirect procedure to identify the presence of the virus. The second step, involving isolation and growth in pure culture, needs to be modified. For viruses, it is necessary to separate the virus from other contaminating pathogens by transmitting the virus to another propagation host. The virus needs to multiply within the propagation host. The virus is then purified from the plant sap of this propagation host. It is then characterized as to its intrinsic properties through electron microscopy and various chemical and physical tests. When inoculated into the original host the virus must reproduce the disease symptoms to satisfy the third step. Lastly, the virus must be reisolated from the inoculated host and recharacterized to verify that it is the same agent that was used for inoculum.

Methods for Recognition and Verification of Virus Diseases 73

Transmission Studies

The suspected causal agent should be transmitted to and produce symptoms in healthy hosts, such as selected varieties of bean, cucumber, tobacco, *Datura, Gomphrena,* and a number of other indicator plants (Fig. 6–8). Based on the host range and symptoms produced on these herbaceous hosts, the suspected virus can be related to known viruses.

Transmission to herbaceous hosts usually involves extraction of plant juice from the diseased plant and mechanical inoculation of the test plant with the extract. Carborundum or other fine abrasives are used to produce small injuries for virus infection. Precautions to prevent denaturation of proteins and destruction of nucleic acid during extraction must be exercised. Buffers to control pH and additives to inhibit oxidation often are used.

Figure 6–8   Indicator plants used in virus transmission studies.

Purification of Virus

The methods used for the purification of viruses are those of the protein chemist. Buffers are used to prevent denaturation of the virus protein and to enhance denaturation of specific plant proteins in the extract. Low-speed centrifugation is used to separate cell fragments from the virus in solution. High-speed centrifugation is then used to concentrate the virus into a pellet, which is then resuspended in buffer. Low-molecular-weight compounds, such as sugars, amino acids, and salts, are separated from high-molecular-weight substances such as proteins and viruses (protein and nucleic acid) by dialyzing the small molecules through a semipermeable membrane. The addition of ammonium sulfate to a partially purified protein solution causes precipitation of specific proteins, depending upon the concentration of ammonium sulfate used, so that some plant proteins in the extract are separated from the virus by this method. After each step, a bioassay of the extract is performed on the local lesion host to verify which fraction contains the virus. Virus in solution will absorb ultraviolet light at a wavelength of 260 nm. The amount of absorption of ultraviolet light in a spectrophotometer is therefore used as a qualitative and quantitative assay tool during the purification process.

Sedimentation Properties

Purified virus particles suspended in solution will sediment in an ultracentrifuge at different rates depending on the mass size and shape of the particle. Usually, the virus preparation is layered on top of a centrifuge tube containing a series of layered increasing concentrations of sucrose solutions from the top to the bottom. Higher concentrations of sucrose produce a more dense liquid through which the virus migrates more slowly. The gravimetric force on the virus particle caused by high-speed centrifugation plus the interacting migration-retarding effect of increasing density of solution produce very good separation of particles. Identification of the virus layers is accomplished by scanning the centrifuge tube with a spectrophotometer. The virus layer is recognized by its absorbance at 260 nm and verified by infectivity on the local lesion host. The size of the virus particle can be calculated from the sedimentation properties.

Electron Microscopy

Crude extracts and partially or highly purified virus preparations can be examined with the electron microscope. Differential staining with electron-dense materials or shadow casting with heavy metals is necessary to observe the virus. The size and shape of the virus is determined.

Serological Assay

Antigen–antibody reactions in a serological assay are used to determine the similarity of the newly observed virus to other, known viruses. The test is not absolute proof of identity but rather an indication of similarity. The procedure is as follows (see Figs. 6-9 and 6-10). A purified virus preparation of a known virus is injected into the bloodstream of an animal such as a rabbit. The foreign virus protein is an antigen which induces the rabbit to produce a specific protein antibody to inactivate the foreign substance. Antibodies bind to antigen in the first step of inactivation. More antibodies are produced than are necessary. Therefore, one can purify, from the blood of the rabbit, a high concentration of the specific antibodies. The antigen-antibody interaction within animals is the basis of acquired resistance by exposure or immunization used by the medical profession. The virologist can place a portion of a purified antiserum preparation, made by him or purchased from a central supplier, in a small depression in a petri plate containing agar. A sample of the original virus is placed in another depression in the agar and a sample of the unknown virus is placed in a third depression. The proteins in the antiserum, as well as the virus, diffuse into the agar. Where they come in contact, the antibodies of the antiserum

1. Plant extract is purified to isolate virus.

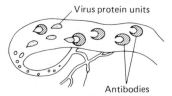

Extract plant juice

Centrifuge to remove cell fragments

Sugar solution
High   Low
conc. conc.

Density gradient centrifugation

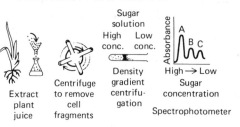

Absorbance

High → Low
Sugar concentration

Spectrophotometer

2. Inject purified virus protein into rabbit bloodstream.

3. Antibodies are formed in the bloodstream to the foreign protein.

Virus protein units

Antibodies

4. Bleed the rabbit and purify the fraction containing the antibodies.

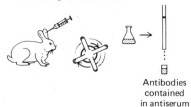

Antibodies contained in antiserum

Figure 6–9 Preparation of antibodies to a virus.

will chemically bind to the virus protein. A line of precipitated antibody–virus protein antigen will appear in the agar plate between depressions with serologically similar virus and animal antiserum.

Serological properties of viruses have been used to develop a number of extremely sensitive assays. One procedure, ELISA (enzyme-linked immunosorbent assay), has greatly simplified virus detection. Polystyrene microtiter plates with ninety-six 0.4-ml wells or test vessels are sequentially filled and washed in four steps as follows:

1. Purified virus or plant sap samples are added to the wells. The proteins of the virus stick to the polystyrene. The wells are emptied and washed.
2. Specific virus antibodies produced in a rabbit are added next. The antibodies will attach only to the virus that is stuck on the walls of the well. Washing again removes nonattached materials.
3. The next step involves adding a second antibody produced in a goat against rabbit antibody. This second antibody was previously conjugated to an enzyme. Washing again removes all goat antibody and enzyme that is not attached to the rabbit antibody.
4. Substrate for the enzyme is added and the reaction allowed to progress. The presence and amount of specific virus in the original sample will be reflected in the intensity of a color reaction product that is evaluated visually or with a colorimeter.

The procedure just described is called indirect ELISA. A direct ELISA procedure attaches the enzyme directly to the rabbit antibody. With the indirect

1. Extract sap from suspected virused plants.

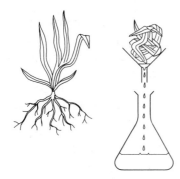

2. Place antiserum in center well and plant extracts in outer wells.

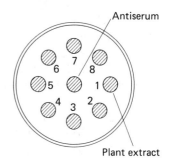

Antiserum

Plant extract

3. Examine plates for precipitation zones; extract from plants 2 and 5 contain the virus specific for the antiserum.

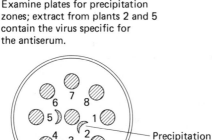

Precipitation zone

4. The precipitation is an antigen-antibody reaction.

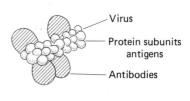

Virus

Protein subunits antigens

Antibodies

Figure 6-10 Serological test of plant extracts for a specific virus.

procedure one standard goat antibody preparation with enzyme attached can be used for a number of different viruses.

The advantage of ELISA is its ability to produce a measurable amount of reaction product from a very small amount of virus. The microtiter plates with ninety-six wells can be used to assay many plant samples as well as controls and knowns. For rapid screening of large numbers of samples ELISA has become the standard procedure.

Vectors

A final step in characterizing a virus is to identify the vectors. Once the virus is identified, the identification of the vector is somewhat easier. Known vectors of similar viruses are first checked to see if they are involved in the disease cycle in the plant under study. For example, serological and other properties of ringspot virus from ash demonstrated a relationship to tobacco ringspot virus. Dagger nematodes (*Xiphinema* spp.) are known vectors of tobacco ringspot virus. Therefore, it was logical to check out nematodes as vectors of the ash virus. The dagger nematodes were shown to be the vectors of the ash virus.

EXAMPLES OF VIRUS DISEASES OF TREES

The selection of which viruses to use as examples of virus diseases in trees is rather difficult. In Europe the poplar mosaic virus is a significant factor in poplar breeding and management. It has been reported in Canada but has never been seriously evaluated in North America. Another European advancement is with the viruses of beech. Cherry leafroll virus and brome mosaic virus are identified from beech. These and other infectious agents may play a role in the beech decline of Europe. Beech viruses in North America have never been seriously investigated. In North America, ash viruses, including tobacco ringspot virus, tobacco mosaic virus, and tomato ringspot virus, have been identified as factors in the ash dieback problem, but their specific impact is still poorly understood.

It is interesting to me to look through the list and see that some of the very wide host range viruses of other crops are also seen in trees; for example, tobacco mosaic virus (Fig. 6–2) has been demonstrated to infect oaks, ash, and chestnut. Other examples of viruses with wide host ranges include tobacco necrosis virus of oaks, poplar, birch, larch, spruce, and ash; apple mosaic virus of mountain ash, birch, and horsechestnut; cherry leafroll virus of elm, cherry, walnut, ash, beech, and birch; tobacco ringspot virus of ash, elm, and cypress; and tomato ringspot virus of ash and elm. Other viruses seem to be very specific; for example, walnut ringspot virus, robinia true mosaic virus, poplar mosaic virus, and elm mottle virus.

Many of the 108 viruses of trees are poorly described. They may be associated with particular symptoms on only a few trees. Some are identified based only on electron microscopic evidence of virus-like particles. The role of most viruses on the health and development of trees and forests has yet to be well documented.

CONTROL OF VIRUS DISEASES

The control of plant viruses is generally that of prevention. Diseased plants usually do not recover and are therefore discarded. In agriculture, considerable effort is put into producing virus-free planting stock. A grower may spray with insecticides to reduce vectors once the crop is planted. Heat treatment is sometimes used by plant breeders to eliminate virus from foundation stock. Virus-free plants can sometimes be generated from apical meristems abscised from rapidly growing plants. Tissue culture media and hormones are used to induce rooting of the abscised meristems to develop virus-free plants. In the future, antiviral chemicals may be available for eliminating viruses in breeding or foundation plant materials.

REFERENCES

BALL, E.M. 1974. Serological tests for the identification of plant viruses. The American Phytopathological Society, St. Paul, Minn. 31 pp.

BAWDEN, F.C., and N.W. PIRIE. 1938. A plant virus preparation in a fully crystalline state. Nature *141*:513–514.

BOSS, L. 1981. Hundred years of Koch's postulates and the history of etiology in plant virus research. Neth. J. Plant Pathol. *87*:91–110.

HANSEN, A.J. 1989. Antiviral chemicals for plant disease control. Crit. Rev. Plant Sci. *8*:45–88.

MATTHEWS, R.E.F. 1970. Plant virology. Academic Press, Inc., New York. 788 pp.

MATTHEWS, R.E.F. 1984. Viral taxonomy for the nonvirologist. Annu. Rev. Microbiol. *39*:451–474.

NIENHAUS, F., and J.D. CASTELLO. 1989. Viruses in forest trees. Annu. Rev. Phytopathol. *27*:165–186.

SMITH, K.M. 1972. A textbook of plant virus diseases. Academic Press, Inc., New York. 684 pp.

STANIER, R.Y., M. DOUDOROFF, and E.A. ADELBERG. 1970. The microbial world, 3rd ed. Prentice-Hall, Inc., Englewood Cliffs, N.J. 873 pp.

STANLEY, W.M. 1935. Isolation of a crystalline protein possessing the properties of tobacco mosaic virus. Science *81*:644–645.

VAN DER PLANK, J.E. 1975. Principles of plant infection. Academic Press, Inc., New York. 216 pp.

VANLOON, L.C. 1987. Disease induction by plant viruses. Adv. Virus Res. *33*:205–252.

VOLLER, A., A. BARTLETT, D.E. BIDWELL, M.F. CLARK, and A.N. ADAMS. 1976. The detection of viruses by enzyme-linked immunosorbent assay (ELISA). J. Gen. Virol. *33*:165–167.

7

BACTERIA AS AGENTS OF TREE DISEASES

- *TYPES OF BACTERIA*
- *MODE OF ACTION OF PLANT-PATHOGENIC BACTERIA*
- *DISEASE CYCLE OF PLANT-PATHOGENIC BACTERIA*
- *SYMPTOMS OF BACTERIAL DISEASES*
- *METHODS FOR RECOGNITION AND VERIFICATION OF BACTERIAL DISEASES*
- *EXAMPLES OF BACTERIAL DISEASES*
- *ROLE OF SOIL BACTERIA IN NUTRIENT CYCLING AND INDIRECTLY IN PLANT HEALTH*
- *CONTROL OF BACTERIAL DISEASES*

Bacteria are found in a wider variety of habitats than any other group of organisms. *Clostridium botulinum,* a soil-inhabiting bacterium, can survive the boiling temperatures used in food processing and develop even when sealed inside an airtight container. A toxin is produced under these conditions that is so potent that it would take only a few grams to kill every person on the planet.

Millions of bacteria occupy every cubic centimeter of topsoil. They are found inside and outside healthy as well as diseased plants and animals. Bacteria are found in association with infections caused by organisms such as fungi and nematodes and may influence the development of these disease-causing agents.

TYPES OF BACTERIA

Bacteria are unicellular procaryotic microorganisms in the Kingdom Monera (Figs. 7–1 to 7–3). They may be spiral, rod-like, or spherical in shape as seen with the light microscope. Other morphological shapes, such as spring-like, mycelial, and pleomorphic, are also possible. Special stain procedures demonstrate that some bacteria have flagella on the outer surface.

Bacteria lack mitochondria, Golgi bodies, and the endoplasmic reticulum found in eucaryotic organisms. There is also no membrane around the nuclear region which consists of one circular double-stranded DNA. Bacteria also have a number of smaller fragments of DNA called plasmids dispersed within the cytoplasm. Ribosomes and a nuclear region are the main features of bacteria as observed with the electron microscope.

Species of higher organisms are characterized by groups of interbreeding morphologically similar individuals. Vegetatively propagating bacteria cannot be grouped by interbreeding criteria, and the range of morphological variation is too limited to adequately separate dissimilar groups. Another complicating factor in classifying the bacteria is the ease with which they transfer genetic material, in the form of plasmids, between widely varying species. Biochemical, physiological, and ecological criteria are superimposed on the limited structural diversity to help classify bacteria into species.

A major separation into two large groups is made on the basis of a cell-wall Gram-stain procedure. The Gram-stain binds crystal violet with iodine to the cell walls of gram-positive bacteria. The crystal violet does not bind to the cell walls of gram-negative bacteria.

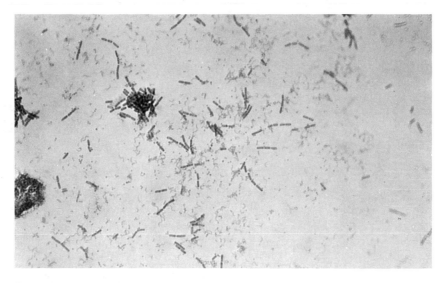

Figure 7–1 Gram positive rod and gram negative spherical and spiral bacteria as seen under high magnification with the light microscope.

Types of Bacteria

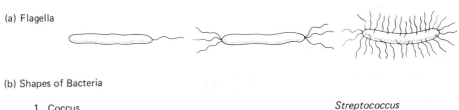

(a) Flagella

(b) Shapes of Bacteria

1. Coccus   *Streptococcus*

2. Rod   *Bacillus*   *Rickettsia*

3. Spiral

4. Mycelial   Conidia   *Actinomycetes*

                                Mycelium

5. Pleomorphic   *Mycoplasma*

6. Springlike   *Spiroplasma*

Figure 7–2 Bacterial morphology.

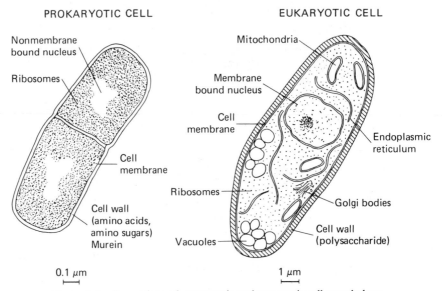

PROKARYOTIC CELL   EUKARYOTIC CELL

Nonmembrane bound nucleus

Ribosomes

Mitochondria

Membrane bound nucleus

Cell membrane

Cell membrane

Endoplasmic reticulum

Ribosomes

Golgi bodies

Cell wall (amino acids, amino sugars) Murein

Vacuoles

Cell wall (polysaccharide)

0.1 μm   1 μm

Figure 7–3 Comparison of procaryotic and eucaryotic cell morphology.

Morphological variation of bacteria is limited. Oxygen, nitrogen, and carbon requirements; fermentation products; gas evolution when grown on specific media; liquefication of gelatin medium; and many other criteria are used.

MODE OF ACTION OF PLANT-PATHOGENIC BACTERIA

Plant-pathogenic bacteria produce disease by enzymatic maceration of plant cell walls and by secretion of toxic substances. The primary walls and middle lamella of succulent plant cells are dissolved by cellulase and pectinase enzymes, while the plasma membrane, which controls the osmotic concentration of the protoplasm, is disrupted by bacterial toxins.

Another mode of action of bacteria involves interference with cell division and differentiation. This mechanism is described in more detail for crown gall later in the chapter.

DISEASE CYCLE OF PLANT-PATHOGENIC BACTERIA

Bacteria have no means of actively penetrating and infecting plants. They rely on wounds made by other organisms or natural openings such as stomata or hydathodes in leaves or openings in flower parts for ingress into plants. Once inside, a general softening and dissolving of cell walls and membranes by bacterial extracellular enzymes allows bacteria access to additional tissues.

Bacterial dispersal from one plant to another generally requires splashing rain or an insect vector, because plant-pathogenic bacteria will not survive the desiccation of wind dissemination. Some bacteria (none of the plant parasites) survive in desiccating environments by producing an extracellular polysaccharide slime outer shell or by production of endospores.

Bacteria generally reproduce by transverse binary fission. Duplication of the single circular chromosome and plasmids is followed by separation into two daughter cells.

Genetic change in bacteria has been studied extensively, but few of the studies have used plant-pathogenic bacteria. The genetic information in bacteria is carried on one circular DNA chromosome but may also be carried on one or more small strands of DNA called plasmids. Some plasmids function independently of the chromosomal DNA. Others are involved in replication of the chromosomal DNA and in mediating transfer of the genetic information from one bacterium to another.

Genetic change in bacteria occurs by one of four mechanisms.

1. *Plasmid transfer* from one bacterium to another can occur between and within species of bacteria. This extrachromosomal transfer of genetic information can be used in the laboratory to study gene functions and to produce new bacteria, which may be very useful in the production of industrial and pharmaceutical chemicals. It may also result in the formation of a new bacterium with resistance to antibiotics and other properties which may or may not generate future problems.

2. *Conjugation* involves one cell, acting as a donor cell, which injects a copy of its chromosomal DNA strand into a receptor cell. The DNA is incorporated into the receptor genetic information so that some of the characteristics of the donor cell are expressed by the fission products of the acceptor cell. Plasmids mediate both the replication and transfer of the chronosomal DNA.

3. *Transduction* involves transfer of DNA from one bacterium to another by a virus (bacteriophage). Virus multiplication in one cell incorporates some of the bacterial DNA into the virus genome. Release of the virus following lysis of the cell and subsequent virus infection of another cell introduces the virus and donor-cell DNA into a receptor cell.

4. *Mutation* occurs in all organisms. These errors in the replication of genetic material are usually lethal or of a selective disadvantage to the organism. When the genetic change is a selective advantage, the mutation or recombinants produced through any of the foregoing mechanisms may provide the bacterium with a way of adapting to a changing environment.

SYMPTOMS OF BACTERIAL DISEASES

Bacteria produce diseases of various plant parts with characteristic symptoms for each. Bacterial pathogens of leaves and shoots produce water soaking initially, followed by necrosis and maceration of tissues. Water soaking produces a shiny translucent appearance to leaves. Maceration dissolves the tissues into a dark nonstructured mass. Bacterial infection of vascular tissue produces wilting. Mycoplasma-like organism (MLO) infection of phloem causes yellowing and witches' brooming. Crown gall bacterial infection is characterized by hypertrophy of stem tissue. Bacterial infection of wood results in increased permeability due to perforation of pit membranes.

METHODS FOR RECOGNITION AND VERIFICATION OF BACTERIAL DISEASES

Bacterial infection of succulent tissues is often diagnosed based on characteristic symptoms. Microscopic examination of diseased tissue should reveal masses of bacteria streaming from infected tissue. Darkfield and phase contrast illumination enhance the visibility of bacterial streaming.

Thorough diagnosis of bacterial diseases usually involves isolation on culture media. Infected tissue is shaken in sterile water or liquid media to separate the bacteria from the tissue. Dilutions of the initial bacterial suspension are mixed in agar media held at about 60° C. The media is solidified by cooling and the bacteria are allowed to develop for a few days. Colonies of shiny cream to light-colored spots are characteristic for bacteria. Individual colonies are subcultured to agar slants.

Identification of species of bacteria requires specific staining procedures to observe morphology and selected chemical tests to determine growth requirements.

Bergey's Manual (Krieg and Holt, 1984) is the standard reference used for identification of bacteria.

If the bacterial disease is not described in the literature it is necessary to go through Koch's proof of pathogenicity, as described in Chapter 1. Specialized methods are necessary for the recognition of MLO diseases. These are described for elm yellows later in the chapter.

EXAMPLES OF BACTERIAL DISEASES

There are fewer than 200 species of plant-pathogenic bacteria, with only a few causing problems in trees. A few selected plant pathogenic bacteria are listed in Table 7-1. *Rhizobium* is included in the table to show its taxonomic relationship to plant-pathogenic bacteria. Nitrogen fixation by *Rhizobium* and other bacteria is discussed later in the chapter. The question marks after *Rickettsia* and *Mycoplasma* are there to emphasize our incomplete understanding of these two groups of plant pathogens. They have been identified primarily on the basis of electron microscopy of diseased plant tissue.

A cursory examination of the bacterial diseases of forest and shade trees listed in a USDA handbook, *Diseases of Forest and Shade Trees of the United States,* shows that there are fewer than 100 reported for the 215 species of trees covered in the book. Closer examination of the references indicates that very few present-day forest pathologists are working on bacterial diseases, since many of the references are at least 20 years old. A possible conclusion from these observations might be that

TABLE 7-1 TAXONOMIC CLASSIFICATION OF BACTERIA ASSOCIATED WITH PLANTS

Class	Order	Family	Genus	Disease
Schizomycetes	Pseudomonadales	Pseudomonadaceae	*Pseudomonas*	Bacterial canker and gummonis of stone fruits
			Xanthomonas	Leaf blight of walnut
	Eubacteriales	Rhizobiaceae	*Agrobacterium*	Crown gall and hairy root
			Rhizobium	Nitrogen fixation
		Enterobacteriaceae	*Erwinia*	Fire blight, wet wood, and soft rot of vegetables
		Rickettsiaceae	*Rickettsia?*	Pierce's disease of grape, phony peach, scorch
		Corynebacteriaceae	*Corynebacterium*	Ring rot of potato
	Actinomycetales	Streptomycetaceae	*Streptomyces*	Potato scab
Mollicutes	Mycoplasmatales	Mycoplasmataceae	*Mycoplasma?*	Phloem necrosis
			Spiroplasma	Citrus stubborn

bacterial diseases do not represent a serious threat to forest and shade trees. But this does not reflect the true significance of bacterial pathogens. Most present-day forest pathologists are very inadequately trained in the modern techniques of bacteriology. They generally avoid bacterial problems and rely on the poorly founded concepts of a former generation, which recognized bacterial diseases and studied them as well as was possible. Bacteria have evolved to fit into almost every niche available, so there is no reason to assume that they have not developed pathogenic capacity for trees.

The tree will be divided into functional parts, and examples of bacterial disease will be presented to characterize the role that bacteria play in each of the parts.

Shoot Blights

The succulent tissues of newly forming shoots are ideally suited for bacterial invasion. Pectolytic enzymes of the pathogen readily macerate the succulent nonlignified cell walls.

A good example of a shoot blight disease is fire blight, caused by *Erwinia amylovora* (Figs. 7-4 and 7-5). This was the first bacterial disease to be studied by plant pathologists and represents an example of a disease agent that we exported to Europe. Fire blight is one of the most serious diseases of apple and pear trees. It is largely responsible for the elimination of major pear production in the northeast and the establishment of pear growing as a major crop in the Pacific northwest. In addition to apple and pear, quince, hawthorn, mountain ash, sweet cherry, and serviceberry are affected by the bacterium.

Work done about the turn of the century established the disease cycle (Fig. 7-6): survival of the bacterium over winter in stem cankers (see Fig. 7-5); oozing of bacteria from the cankers in the spring; dissemination of the bacterial inoculum from cankers to young shoots and flower parts by splashing rain and insects;

Figure 7-4 Fire blight on apple causes a shoot blight recognized by the blackened wilted leaves and the curled shoot tip.

Figure 7-5 Canker caused by *Erwinia amylovora* is the site for overwintering of the bacterium.

secondary rapid spread of the bacterium to additional flowers by bees and other insects; infection through hydathodes of the flower and stomates of the leaf; and invasion and maceration of parenchyma and cambium tissues of the leaf, petiole, shoot, and branch.

There are still some rather puzzling features to fire blight. In some years, the disease can be very destructive in the northeast. These serious years sometimes catch growers unprepared, because there may have been a long period without much impact. The interaction of inoculum levels, weather, injuries, vectors, and host development are extremely complex and therefore make prediction and application of control procedures rather difficult.

In California, the control of fire blight involves application of copper or antibiotic materials at 5-day intervals during the flowering period. The weather conditions may draw out the flowering period for up to 3 months. Therefore, the growers regularly use many sprays. Recent research has demonstrated that by monitoring mean temperature in the orchard, one can predict when conditions are ready for rapid spread of the disease. Using this approach they can identify when to start applying chemicals in the spring and have saved as many as three sprays.

Control of fire blight is accomplished by pruning out the cankers to remove the source of inoculum to prevent future spread and spraying the trees with copper or antibiotic materials to prevent infection in the spring. Pruning should be done in late summer or winter to avoid spreading bacteria by means of pruning tools. Another recommended precaution is to dip tools in Clorox after each cut.

Species other than *E. amylovora* are probably involved with shoot blights in

Cause: *Erwinia amylovora*
Hosts: Apple, hawthorn, European
mountain ash, and others

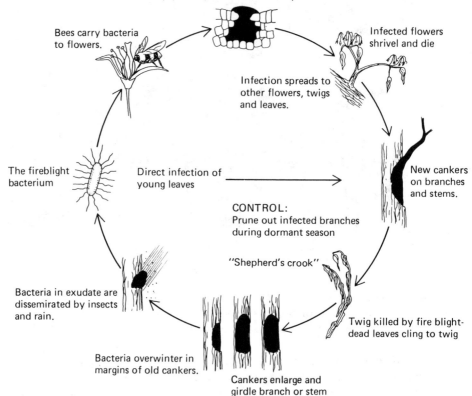

Bacteria penetrate flowers through hydathodes
and leaves through wounds or stomates. They
multiply and spread intercellularly.

Bees carry bacteria
to flowers.

Infected flowers
shrivel and die

Infection spreads to
other flowers, twigs
and leaves.

The fireblight
bacterium

Direct infection of
young leaves

New cankers
on branches
and stems.

CONTROL:
Prune out infected branches
during dormant season

"Shepherd's crook"

Bacteria in exudate are
disseminated by insects
and rain.

Twig killed by fire blight-
dead leaves cling to twig

Bacteria overwinter in
margins of old cankers.

Cankers enlarge and
girdle branch or stem

Figure 7–6 Fire blight disease cycle.

other plants. *Pseudomonas syringae,* a bacterium that causes a serious foliage disease of beans, is reported as causing shoot blight of cherry.

Foliage Blights

There are very few foliage diseases of trees caused by bacteria. Bacteria such as *E. amylovora,* which parasitize shoots, also cause foliage blotches, leaf spots, and death. *Pseudomonas syringae,* also mentioned above, is a foliage pathogen of citrus. Other Pseudomonads cause leaf spots of bigleaf maple, tung, California laurel, and red mulberry.

It is curious that so few bacterial foliage pathogens are reported for trees because serious bacterial foliage diseases of agricultural crops, ornamental shrubs, and vines are rather general in occurrence.

Stem Vascular Disorders

There are two groups of bacterial diseases of the vascular system. One group, found in the xylem vessels, was first identified as rickettsia-like bacteria because of their small size and rough outer wall. This group of bacteria is difficult to characterize properly because many of them cannot be grown in culture. At the present time, the bacteria from the xylem of plants are called fastidious xylem-limited bacteria (FXLB).

FXLB may cause scorch-like symptoms to the leaf margins of affected plants. When the outer margins of the leaves die and turn brown this is called scorch. Scorch occurs on Norway maple and other broadleaf species following hot, dry spells or following root disturbance. FXLB-induced scorch has a yellowed transition margin between the browned outer portion of the leaf and the green inner portion of the leaf, while mechanically caused scorch, as with Norway maple, has an abrupt transition between the browned and green areas. Another symptom of FXLB is stunting or reduced growth.

Some of the more intensively studied FXLB diseases are Pierces' disease of grape, phony peach, plum leaf scald, elm leaf scorch, oak leaf scorch, and sycamore leaf scorch. An interesting degeneration of larch in Germany has been shown to be associated with FXLB. The degenerating larch produce extensive witches' brooming. A common feature of FXLB diseases is that they are found in areas with warm climates but are absent in colder areas, even though the same plants are grown in both places.

The vectors for many of the FXLB have not been identified, but leafhoppers and spittlebugs are vectors for some. The bacteria propagate in the foregut of the insects. The vectors remain infectious only until they molt and shed the foregut.

The management strategies of the FXLB diseases are still being developed. In commercial hosts it is important to recognize the possible role of wild symptomless plants as reservoirs for the agent and vectors. In some instances, tolerant cultivars are available.

The second group of bacterial diseases causing vascular disorders are restricted to the phloem of the plant. The bacteria, in this instance, are called mycoplasma-like organisms (MLOs), because they are again very difficult to characterize properly in relation to other known bacteria. The first MLO was described in 1967 as the agent responsible for aster yellows, a disease assumed, until that time, to be caused by a virus. We now recognize many diseases of agricultural crops and trees with mycoplasma etiology, including elm yellows (elm phloem necrosis), ash yellows, black locust witches' broom, and pecan bunch.

Mycoplasmas have been recognized to be disease-causing agents of animals since the early part of this century. Both parasitic and nonparasitic mycoplasmas are part of the normal microflora of the human mouth.

Mycoplasmas are procaryotic microorganisms that have no cell wall. They usually require sterol in the growth medium and are sensitive to tetracycline antibiotics.

L-form bacteria also do not have a cell wall, but they can revert back to walled normal bacteria. A mycoplasma is always without a cell wall. Two other bacteria lacking cell walls, bedsonia and the pitacycosis agent, are separated from mycoplasmas by their specific growth requirements.

Mycoplasmas observed with the electron microscope are generally variable in shape (pleomorphic) because the outer membrane is not rigid (Fig 7-7). The three-layered outer unit membrane surrounds a protoplasm consisting of granular ribosomes and a nuclear region.

One group of mycoplasmas are spiral in shape. The spiroplasmas are the only group of plant mycoplasmas that have been grown in culture. Their shape and movement can be observed with a phase contrast microscope.

Elm yellows is one example of an MLO-induced disease of trees. The agent is presumably vectored by phloem feeding insects such as the leafhopper, *Scaphoideus luteolus* (Fig. 7-8), but in central New York, where there is a serious epidemic, this insect is rather uncommon. The identities of the most important vector or vectors are therefore still unknown, even after extensive sampling.

Detailed surveys of elm populations in central New York have shown that the disease moves as a wave front through an area. Within four years of first observation of symptoms, one population of elms was virtually eliminated by this extremely rapid-moving epidemic. Unlike the better recognized Dutch elm disease, elm yellows kills essentially every elm in the area, from seedlings to large trees. Five years after the wave moved through one area, with a large population of wild elm developing in hedgerows and in abandoned agricultural fields, we still cannot find a single elm.

The characteristic symptoms of elm yellows and many other MLO-induced diseases is a chlorosis or yellowing of the foliage (Fig. 7-9). Another symptom of MLO in many hosts is the development of branch clusters called witches' brooms

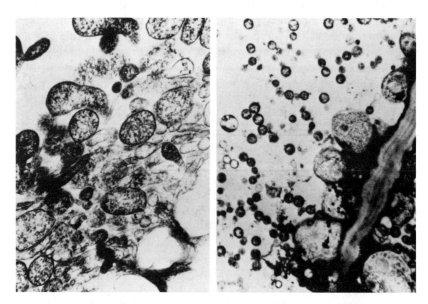

Figure 7-7 Mycoplasma-like organisms seen in the phloem with an electron microscope. (Photograph from C. R. Hibben and B. Wolanski, Phytopathology *61*:153.)

Figure 7–8 *Scaphoideus luteolus* leafhopper, a vector of elm yellows. (Photograph courtesy of the U.S. Forest Service.)

(Fig. 7–10) on the main stem or throughout the crown. Another diagnostic symptom of elm yellows is a wintergreen odor to the phloem of infected trees. The odor is detected in mid- to late summer from freshly cut phloem from the base of an infected tree. It is helpful to warm the tissue by enclosing it in a container or the fist of your hand before attempting to detect the odor. Rapid browning of phloem tissue is also a feature of elm yellows.

The specialist, attempting to diagnose MLO as the suspected causal agent of a problem, would use a specific fluorescent staining procedure for DNA and then

Figure 7–9 American elm dying of elm yellows in New York is characterized by a chlorotic thin crown.

Examples of Bacterial Diseases

Figure 7-10 Witches' broom associated with mycoplasma-infected white ash. (Photograph compliments of Dr. Craig Hibben.)

under the microscope differentiate the foreign bacterial DNA from the normal DNA of the phloem. Additionally, electron microscopy of phloem of the original plant or of plants that have been grafted with the original plant should reveal the characteristic membrane-bound agent (Fig. 7-7).

Elm yellows and presumably other mycoplasma diseases result in degeneration of feeder roots. Disruption of downward transport of photosynthetic products can be expected to cause degeneration of roots. The death of a tree is like a slow decline, because of the effects of the mycoplasma agent on roots.

The relationship of yellows to decline has become particularly confusing with the ash yellows/ash dieback situation. The witches' brooming symptom is common to both diseases. Current research is attempting to define the roles of the yellows agent and viruses in the problem that was originally called ash dieback and is today sometimes called ash decline. It is tempting to suggest that ash yellows is the specific cause of dieback and death of ash (see Chapter 18).

Remission of symptoms of MLO diseases can usually be accomplished by the introduction of tetracycline antibiotics into the plant. The antibiotic does not kill the pathogen but just slows it down or affects the host such that the witches' brooming and yellowing of the foliage is reduced. After one growing season the effects of the antibiotic wear off and the symptoms reappear. Very little is known about the environmental and host factors affecting the development of these diseases, so that additional approaches to control are not known.

Stem Cankers

Cankers or localized necrosis of vascular cambium of stems and branches can be caused by bacteria. The fire blight bacterium, already discussed as a shoot pathogen, also has a canker phase. Other bacteria, such as *Xanthomonas pruni* and *Pseudomonas syringae,* cause cankers of cherries but also cause shoot and foliage blights. Bacterial canker of poplars in Europe, caused by *Xanthomonas populi,* is probably the most important bacterial canker of trees. Infection is mainly through stipule scars produced by shedding of stipules during the growing season. Some infection occurs through bud-scale scars and leaf scars. This disease causes rough callusing cankers on the stem and branches of some of the important poplar clones. No one has seriously looked for the disease in North American poplars, but we can predict that importation of the pathogen, if it is not already here, may produce a future problem.

Stem Galls

The bacterium *Agrobacterium tumefaciens* induces a cancer-like uncontrolled proliferation of cells of most species of dicotyledon plants, which includes the broadleaf trees (Fig. 7–11). It is a major problem in nurseries producing ornamental plants. The grasses (monocots) and conifers (gymnosperms) are not affected by the pathogen. The bacterium survives in the soil and gains access to living tissues by means of wounds. The bacterium infects damaged plant cells and then from this position

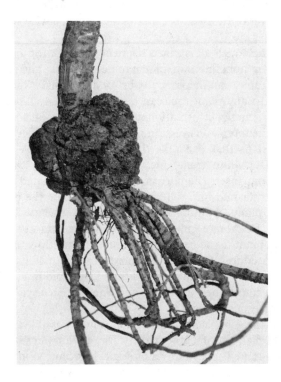

Figure 7–11 Crown gall on *Prunus.* (Photograph compliments of Dr. Wayne Sinclair.)

releases a Ti plasmid which is picked up by adjacent living cells. A section of the plasmid called T DNA incorporates its genetic information into the genome of the plant. Part of the genetic information on the plasmid induces the host responses leading to gall formation. Other parts of the genome code for the synthesis of nopaline and/or octapine, two unusual amino acids that are used by *A. tumefaciens.*

Agrobacterium causes disruption of control of DNA replication within the plant host. Host DNA replication is not synchronized with cell division, so that multiple sets of chromosomes accumulate within some nuclei. Hyperplasia (abnormal increase of cells) and hypertrophy (abnormal increase in cell size) result in gall formation.

The bacterium sloughs off with a few outer cells and returns to the soil. The uncontrolled cell multiplication will continue on in the absence of the bacterium and may even occur at other locations in the plant remote from the original infection center. At one time it was thought that the secondary galls were caused by translocation of the bacterium within the plant, but it is now recognized that the plasmid genome is translocated.

The incorporation of the T DNA segment of the Ti plasmid into the plant cell genome, on the one hand, produces a serious disease called crown gall and, on the other hand, is the foundation of most present-day genetic engineering of plants. To convert a disease situation to a potentially useful improved plant involves a number of rather interesting steps. One of the first steps is to delete the genes for gall formation from the T DNA segment of the Ti plasmid. Mutant forms of *A. tumefaciens,* which do not induce gall formation, have been identified. These lines have still been shown to transfer the T DNA because plants infected with such a line produce nopaline and/or octapine.

These *Agrobacterium* lines, with a disarmed segment of DNA for gall formation but an intact segment for nopaline and/or octapine synthesis and ability to incorporate into the plant genome, can theoretically be used to transfer any desired piece of genetic information to any plant that can be infected by *A. tumefaciens.* The trick is to identify the genetic information that you desire in some other organism by some specific phenotypic expression. The rest of the procedures are rather standard mechanical steps that just take time: Isolate the DNA from the desired organism, split the DNA into fragments with endonucleases, catalog the fragments by size and function, insert the correct segment into isolated disarmed plasmid, and incorporate the plasmid into the bacterium, inoculate the plant with the bacterium, and screen for plants containing the desired gene. The use of specific endonucleases that split the DNA molecule in specific places, marker genes to simplify screening in culture, plant tissue culture procedures to multiply the desired plants, and a few other procedures facilitate the process. Fortunately, bacteria readily exchange plasmids among various species and therefore rapid multiplication of desired plasmids can be accomplished using selected lines of *Escherichia coli,* a common laboratory workhorse bacterium.

The potential output of the genetic engineering that is now possible because of our understanding of the mechanism of pathogenicity for *A. tumefaciens* boggles the mind. At the present time there is essentially a biological revolution in progress.

A very interesting approach to the control of crown gall has recently been discovered and is already in widespread use. Another bacterium, *Agrobacterium radiobacter* strain 84, produces a bacteriocin molecule (agrocin 84) which presumably inhibits the *A. tumefaciens* tumer-inducing plasmid from transferring to the host plant. This means of control has been shown to be effective and economically acceptable to the nursery industry.

Root Problems Associated with Bacteria

Root diseases caused by bacteria are very limited in number. The bacteria that cause vascular wilts and the crown gall bacteria infect trees through wounds in the roots and lower stem. Root rot fungi and nematode diseases always have been associated with bacteria.

The only root disease of trees in which the bacterium is the primary agent is hairy root, caused by *Agrobacterium rhizogenes.* It is not known in the northeast but is a problem of ash in the Great Plains states. Excessive proliferation of roots is characteristic of this disease.

Decomposition of Wood Tissue by Bacteria

Bacteria colonize and affect woody tissues of both living trees and wood after trees are cut. A common condition known as wetwood in elm, poplar, fir, and other species is associated with populations of bacteria. Although early reports suggested that single species of bacteria was involved, more recent work has identified as many as 14 species of bacteria in wetwood tissues. Large populations of anaerobic bacteria, including nitrogen-fixing and methanogenic species, are involved. Some of the species are members of the following genera: *Clostridium, Bacteroides, Erwinia, Edwardsiella, Klebsiella, Lactobacillus,* and *Enterobacter.*

Activities of the bacteria in the inner sapwood/outer heartwood region result in high concentrations of bacterial fermentation products, including methane. Vessel-to-ray pit membranes are destroyed. Water and pressure increase to the point that liquid and gas are forced out of any wound that intersects the wetwood region (Fig. 7–12). A stream of liquid and gas will sometimes spew forth from an increment borer wound when the core is pulled from the borer. Tree trimmers are sometimes doused by the foul-smelling liquid as they cut branches from the top of wetwood-infected elms or poplars. The characteristic odor of elm and poplar lumber as it is being cut is due mainly to the bacterial products.

Wetwood is reported to cause death in elms, but surveys of elm populations have, on at least two occasions, demonstrated that every tree in the population was infected with wetwood. Therefore, when a few trees die, it is almost impossible to demonstrate that death is caused by the wetwood bacteria. In white fir it is suggested that wetwood is the result of an internal nonpathological osmotic process in which the bacteria are nonspecific casual participants.

Despite inadequate proof of pathogenicity the belief among arborists that this is a potential disease causes them to attempt to prevent infection by painting wounds in trees with wound dressings. Another popular activity is to drill a hole in the elm in the zone of wetwood and insert a pipe to release the liquid and pressure and allow

Figure 7-12 Wetwood in slippery elm causes a black odoriferous liquid to flow from a pruned branch.

the cambium to heal over a festering wound. The value of either of these two activities is questionable.

A serious defect resulting from wetwood occurs as wood from wetwood-infected trees is dried prior to its use as lumber. In the absence of carefully controlled drying conditions, the wood will irregularly collapse and warp, to form almost useless lumber. Warp and collapse associated with elm, poplar, and fir lumber is very common, and therefore these species and others were not commonly used for lumber. Today, controlled drying schedules prevent warp and collapse, so that fir and poplar are now acceptable building materials.

An interesting positive side to wetwood in white fir suggests that wetwood may restrict or prevent decay in the tree by root decay fungus, *Heterobasidion annosum*. This phenomenon may also occur in other trees.

Degradation of Wood Products

Bacteria invade logs submerged in holding ponds for long periods of time and logs that sink during the rafting of logs to mills. The bacteria cause degeneration of pit membranes between cells. No measurable reduction in the strength of wood cut from such invaded logs is evident, but the destruction of pit membranes allows liquids to move in and out irregularly and rapidly. This will become a problem if the wood is exposed to moisture in its final use, because decay fungi rapidly decay wood that is easily wetted.

Increased permeability of wood by bacterial deterioration of pits is a possible asset in wood that is to be treated with a wood preservative. Maximum uptake of preservative and deep penetration by the preservative are characteristics of ideal preservative treatments. But the bacteria do not uniformly invade the wood, and therefore the preservative treatment is not uniformly distributed throughout the wood.

Soft Rots of Fleshy Plant Products

A last role of bacteria that is very important in agricultural crops but is not a problem in trees is soft rot bacterial decomposition of fruits and vegetables. *Erwinia carotovora* can decompose a bushel of carrots or potatoes to a pile of viscous, foul-smelling mush in a short period of time.

ROLE OF SOIL BACTERIA IN NUTRIENT CYCLING AND INDIRECTLY IN PLANT HEALTH

Biosynthetic processes of plants and animals tie up carbon, nitrogen, and sulfur. Bacteria in the soil are essential degraders of organic material, thereby recycling the bound-up elements. It has been estimated that the supply of carbon dioxide on this planet would be exhausted within 20 years if it were not for the bacteria and other microorganisms that release carbon dioxide by degrading organic matter. Most students have been exposed to the carbon, nitrogen, and sulfur cycles in basic biology courses. I will attempt to relate the subject to plant health.

Carbon Cycle

Although we can speculate on what would happen if carbon were not recycled, there is no documented case where a shortage of carbon dioxide is a major plant problem. The addition of more carbon dioxide to the atmosphere of an enclosed chamber such as a greenhouse enhances the growth of plants — so we can say that plants could use more carbon dioxide but get along well enough at the present levels.

Nitrogen Cycle

Nitrogen recycling by bacteria, on the other hand, is variable from one soil to another, and plant health is delicately balanced by the variable supply of nitrogen. The nitrogen cycle (Fig. 7–13) basically involves oxidation of ammonia in two steps to nitrate. Plants, in turn, reduce the nitrate immediately upon intake to the amino form, attach the amino form of nitrogen to transport amino acids, and then further synthesize additional amino acids and other nitrogen-containing compounds by chemically transferring the amino group.

The cycle thus far presented results in no increase or decrease of nitrogen in the

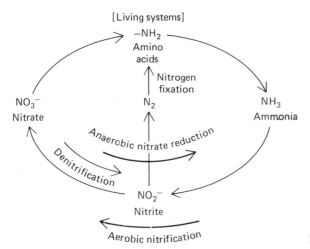

Figure 7-13 Nitrogen cycle.

biological system. In actual practice, the biologically recycled nitrogen is also interchanging with atmospheric nitrogen, the major element of our atmosphere.

Many aerobic bacteria can reverse the cycle by reducing nitrate to nitrite, but nitrite is toxic, so that the amount reduced is limited. A few normally aerobic bacteria such as *Pseudomonas* and *Bacillus,* can utilize nitrate under anaerobic conditions and reduce it beyond nitrite to molecular nitrogen. This type of reaction is one of the effects of anaerobic soil conditions which were discussed in Chapter 2.

The littleleaf disease of shortleaf pine is discussed at length in Chapters 16 and 18. Littleleaf sites are characterized by shallow topsoil on top of hardpan. It is important to recognize the possible role of anaerobic bacteria. High soil-moisture conditions during portions of the year reduce oxygen and probably cause the facultative anaerobic bacteria to shift to utilization of nitrate, thereby removing nitrogen from the soil in the form of nitrogen gas. Addition of nitrogen fertilizers reverses the yellowing decline symptoms of littleleaf disease, but fertilization has only temporary effects which are counteracted during the next period of high moisture.

A well-recognized contribution of nitrogen to the soil occurs as a result of bacteria in symbiotic relationships with legumes such as peas, beans, alfalfa, and clover. Nitrogen fixation by symbiotic association of bacteria and legume can fix as much as 400 lb of nitrogen per acre per year (448 kg/ha). Symbiotic nitrogen fixation results from invasion of tetraploid (four sets of chromosomes) cortical root cells of the legumes by species of *Rhizobium.*

The *Rhizobium* bacterium invades by penetration of a root hair. Bacteria within cortical cells stimulate the cells to divide, thereby forming a nodule. In the nodule is found leghemoglobin, which neither the plant nor the bacteria can synthesize when grown separately. The leghemoglobin is thought to play a role in reversible oxygen binding to maintain the appropriate reduction potential for fixation of nitrogen. After a period of time the bacteria disappear, presumably because they are absorbed by the host plant.

Nitrogen fixation by symbiotic associations of the bacterium *Frankia* and nonlegume plants such as *Alnus, Myrica, Casuarina, Comptonia,* and others has recently been demonstrated. These plants are pioneer shrubs and trees that are important in land reclamation.

Other bacteria, such as *Klebsiella, Azotobacter,* and *Clostridium,* are estimated to contribute about 6 lb of nitrogen per acre per year (6.7 kg/ha). Some reports suggest that *Clostridium* and other anaerobic nitrogen-fixing bacteria may play a role in supplying wood decay fungi with sufficient nitrogen to survive in the nitrogen-deficient environment of the xylem of trees.

Sulfur Cycle

Sulfur is another required element for plant growth that is cycled by bacteria in the soil. The normal bacterial cycling of sulfur from decomposing organic matter involves oxidation of hydrogen sulfide to elemental sulfur and then further oxidation of sulfur to sulfate (Fig. 7–14). The plant takes up sulfate and reduces it for incorporation into the amino acid cysteine. Hydrogen sulfide oxidation results in the production of hydrogen ions. Therefore, sulfur or organic amendments commonly added to alkaline soils are used to increase acidity. Anaerobic bacteria utilize sulfate as an electron acceptor. The product hydrogen sulfide can be smelled in mud at the bottom of ponds and bogs. The black color of mud is the accumulation of ferrous sulfide.

The hydrogen sulfide odor is very distinctive and can be utilized to diagnose anaerobic conditions of recently planted trees. If trees are planted too deep or are watered excessively, anaerobic conditions prevent proper development of roots. The crowns of the trees will not fully leaf-out and twigs will die. Excavation to the roots of such trees will clearly identify the cause. The rotten egg smell of hydrogen sulfide will induce you to fill in the hole quickly.

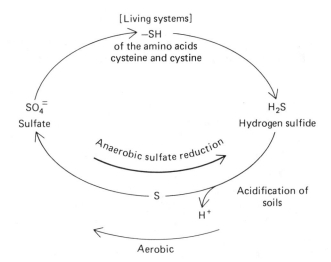

Figure 7–14 Sulfur cycle.

CONTROL OF BACTERIAL DISEASES

Plant-pathogenic bacteria are spread by insects, splashing water, and human beings. Insect control will prevent some bacterial diseases. Elimination of diseased individuals or diseased parts of individuals in a population avoids the short-range dispersal by splashing rain. Transport and storage of infected plants and plant parts and infested soil by human beings perpetuates and distributes the bacteria of agricultural crops. Quarantines on the movement of plant materials and soil do much to prevent this problem.

Direct control of bacterial diseases of plants by application of antibiotics is much less successful than is control of bacterial diseases of human beings and other animals, because plants do not readily absorb and efficiently distribute the antibiotics.

Manipulation of bacteria involved in nutrient cycling usually requires alteration of soil aeration. Obviously, one cannot change soil aeration over large forest areas. But it is possible to affect aeration around individual or groups of ornamental trees. Drainage tiles and gravel can be used to advantage in these situations.

Anaerobic bacterial production of plant toxic ammonia and hydrogen sulfide is a particularly serious problem when attempting to grow trees and other vegetation on reclaimed refuse dumps. Experimental tests using drainage tiles and planting on mounds of soil are being conducted because direct plantings into topsoil covering the refuse have failed. In the long run, it may be appropriate to screen various plant species for tolerance to ammonia and hydrogen sulfide. Development of tolerant plant materials for reclamation of refuse sites could become a major priority as cities and parks continue to expand onto these sites.

REFERENCES

AHO, P.E., R. J. SEIDLER, H.J. EVANS, and P.N. RAJU. 1974. Distribution, enumeration and identification of nitrogen-fixing bacteria associated with decay in living white fir trees. Phytopathology *64*:1413–1420.

BARTON, K.A., and M.D. CHILTON. 1983. *Agrobacterium* Ti plasmids as vectors for plant genetic engineering. Methods Enzymol. *101*:527–539.

BINNS, A.N., and M.F. TOMASHOW. 1988. Cell biology of *Agrobacterium tumefaciens* infection and transformation of plants. Annu. Rev. Microbiol. *42*:575–606.

BOVE, J.M. 1984. Wall-less prokaryotes of plants. Annu. Rev. Phytopathol. *22*:361–396.

CHILTON, M.D. 1983. A vector for introducing new genes into plants. Sci. Am. *248*(6):50–59.

DEKAM, M., and S.H. HEISTERKAMP. 1987. Comparison of two methods to measure susceptibility of poplar clones to *Xanthomonas populi*. Eur. J. For. Pathol. *17*:33–46.

GILMAN, E.F., I.A. LEONE, and F.B. FLOWER. 1977. Vegetating the completed sanitary landfill. Proc. Am. Phytopathol. Soc. *4*:188.

GREMMEN, J., and R. KOSTER 1972. Research on poplar canker (*Aplanobacter populi*) in The Netherlands. Eur. J. For. Pathol. *2*:116–124.

HEPTING, G.H. 1971. Diseases of forest and shade trees of the United States. USDA For. Serv. Agric. Handb. 386. 658 pp.

HIBBEN, C.R., and B. WOLANSKI. 1971. Dodder transmission of a mycoplasma from ash witches' broom. Phytopathology *61*:151–156.

KERR, A. 1987. The impact of molecular genetics on plant pathology. Annu. Rev. Phytopathol. *25*:87–110.

KNUTH, D.T., and E. MCCOY. 1962. Bacterial deterioration of pine logs in pond storage. For. Prod. J. *12*:437–442.

KRIEG, N.R., and J.G. HOLT. 1984. Bergey's manual of systemic bacteriology. Williams & Wilkins, Baltimore.

LACY, G.H., and J.V. LEARY. 1979. Genetic systems in phytopathogenic bacteria. Annu. Rev. Phytopathol. *17*:181–202.

LANIER, G.N., D.C. SCHUBERT, and P.D. MANION. 1988. Dutch elm disease and elm yellows in central New York. Plant Dis. *72*:189–194.

LIGON, J.M., and J.P. NAKAS. Isolation and characterization of *Frankia* sp. strain FaC1 genes involved in nitrogen fixation. Appl. Environ. Microbiol. *53*:2321–2327.

MATTEONI, J.A., and W.A. SINCLAIR. 1985. Role of the mycoplasmal disease, ash yellows, in the decline of white ash in New York state. Phytopathology *75*:355–360.

MOORE, L.W., and G. WARREN. 1979. *Agrobacterium radiobacter* strain 84 and biological control of crown gall. Annu. Rev. Phytopathol. *17*:163–180.

MURDOCH, C.W., and R.J. CAMPANA. 1983. Bacterial species associated with wetwood of elm. Phytopathology *73*:1270–1273.

NIENHAUS, F. 1985. Infectious diseases in forest trees caused by viruses, mycoplasma-like organisms and primitive bacteria. Experientia *41*:597–603.

NIENHAUS, F., and R.A. SIKORA. 1979. Mycoplasmas, spiroplasmas, and rickettsia-like organisms as plant pathogens. Annu. Rev. Phytopathol. *17*:37–58.

RAJU, B.C., and J.M. WELLS. 1986. Diseases caused by fastidious xylem-limited bacteria and strategies for management. Plant Dis. *70*:182–186.

SCHINK, B., J.C. WARD, and J.G. ZEIKUS. 1981. Microbiology of wetwood: role of anaerobic bacterial populations in living trees. J. Gen. Microbiol. *123*:313–322.

SCROTH, M.N., S.V. THOMSON, D.C. HILDEBRAND, and W.J. MOLLER. 1974. Epidemiology and control of fire blight. Annu. Rev. Phytopathol. *12*:389–412.

SHERALD, J.L., S.S. HEARON, S.J. KOSTKA, and D.J. MORAN. 1983. Sycamore leaf scorch: culture and pathogenicity of fastidious xylem-limited bacteria from scorch-affected trees. Plant Dis. *67*:849–852.

SMITH, R.S. 1975. Economic aspects of bacteria in wood. *In* Biological transformation of wood by microorganisms, ed. W. Liese. Springer-Verlag, Berlin, pp. 89–102.

THOMSON, S.V., M.N. SCHROTH, W.J. MOLLER, and W.O. REIL. 1982. A forecasting model for fire blight of pear. Plant Dis. *66*:576–579.

TORREY, J. 1987. Nitrogen fixation by actinomycete-nodulated angiosperms. BioScience *28*:586–592.

WHITBREAD, R. 1967. Bacterial canker of poplars. Ann. Appl. Biol. *59*:123–131.

WILSON, C.L., C.E. SELISKAR, and C.R. KRAUSE. 1972. Mycoplasma-like bodies associated with elm phloem necrosis. Phytopathology *62*:141–143.

WORRALL, J.J., and J.R. PARMETER, JR. 1983. Inhibition of wood decay fungi by wetwood of white fir. Phytopathology *73*:1140–1145.

8

INTRODUCTION
TO FUNGI

- *FUNGI IN THE FOREST COMMUNITY*
- *CHARACTERISTICS OF FUNGI*
- *TAXONOMIC CLASSIFICATION AND LIFE
  CYCLES OF FUNGI*
- *DISEASE CYCLE OF PATHOGENIC FUNGI*

Some estimates indicate that there are close to 100,000 species of fungi. Each has special environmental requirements and occupies a certain niche. Almost all niches and organic substrates have been experimented with and found suitable by some fungus. It has been said that if these meek organisms have not yet exactly inherited the earth, they do eventually inherit most of the living things on the planet as well as the goods and products made from them.

Only in the twentieth century have we been able to produce organic substrates that seem to defy the fungal ingenuity for variability. Some of plastics and styrofoams seem to be immune to the digestive enzymes of fungi and bacteria. We can all hope that the fungi and bacteria have not given up in this race with us and that they eventually come forth with a champion to degrade this portion of human residue.

FUNGI IN THE FOREST COMMUNITY

Fungi have many roles in the forest community (Fig. 8–1). For the most part, fungus activity is highly beneficial to both human beings and the forest community.

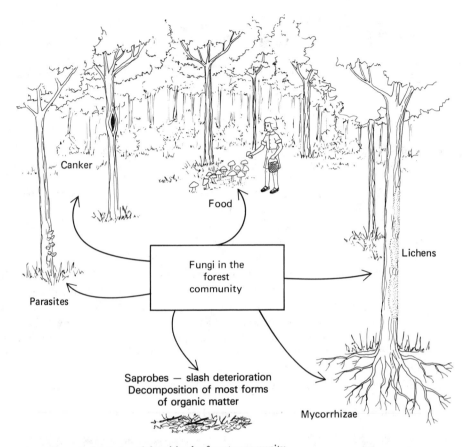

Figure 8-1 Roles of fungi in the forest community.

Saprobic Fungi

The most important role of fungi is in the saprobic decomposition of cellulose and lignin residues. In Chapter 7 it was noted that in the absence of carbon recycling, the supply of carbon dioxide would be used up in 20 years. How much of the carbon is recycled by fungi and how much is recycled by bacteria is not known. The fungi are well adapted for very rapid enzymatic breakdown of massive cellulose–lignin accumulations in woody stems. Fungal hyphae actively penetrate cell walls and permeate the whole substrate. Bacteria, on the other hand, passively move into a substrate and enzymatically macerate the tissues in localized areas.

Parasitic Fungi

A minor role of fungi in the forest community is the parasitic role. The basic difference between the parasite and the saprobe is that the parasite encounters and deals primarily with the defenses of the living host, whereas the saprobe has to compete with other saprobes for dead organic matter.

Fungi in the Forest Community

A fraction of the parasites of the forest community are detrimental to their hosts; these are the pathogens. This small group of the total fungi is the group the forest pathologists have traditionally emphasized in their studies on wood decay, rust diseases, canker diseases, foliage diseases, wilt diseases, and root rot diseases. Much of the remaining chapters will deal with these problems.

Fungi as Food for Other Organisms

In the forest community it is necessary to recognize that fungi serve as sources of food for both human beings and many other forms of life. Groups of invertebrates have specialized in the utilization of fungi for food. Instead of knocking off the next wood decay fungus fruit body you encounter, stop and observe the array in insect and other life that centers around the fruit body environmental niche.

Vertebrates, including man, recognize the culinary delicacy of fungi. The Vikings found another use for fungi. They ate a species of *Amanita* mushroom before going into battle. The mushroom produced hallucinogenic effects, which heightened their savagery in dealing with the enemy. Mushrooms in the genus *Amanita* have other noteworthy properties. Poisoning from eating mushrooms commonly occurs as a result of eating certain species of *Amanita*. A general rule to follow in selecting edible mushrooms is to choose those enclosed within a can or jar. If you are going to eat wild mushrooms, you had better have absolute confidence in the person who collected the mushrooms. Everyone should read the account of a family that was poisoned by mushrooms as described by Pilát and Ušák in the book *Mushrooms*. Mushroom poisoning is a very painful way to go.

Symbiotic Fungi in Lichens and Mycorrhizae

Fungi in the forest community are symbiotically associated with algae in lichens. Most biologists recognize the large foliose and fruiticose lichens that provide the array of colors and shapes to bark and limbs of trees, but few look closely at the bark of trees and recognize the subtle evidences of small crustose lichens. The fungus–algal association, lichens, are so common that we take them for granted. Human pollution of the environment results in changes in numbers and species of lichens. Lichens can be used as biological indicators of cumulative effects of air pollution.

To complete the roles of fungi, it is important to mention mycorrhizae. Mycorrhizae will be more fully discussed in Chapter 9, but for now it is important to recognize that fungi associated with plant feeder roots cause morphological changes in the roots. The morphologically modified roots called mycorrhizae are better nutrient- and water-absorbing organs. Most of the higher plants have developed this type of relationship with fungi.

CHARACTERISTICS OF FUNGI

Heterotrophic growth requirements (lack of chlorophyll), mycelial growth form, and reproduction by means of spores are the specific characteristics of most fungi.

Yeasts, water molds, and fungi symbiotic with *arthropods* may have a simplified budding growth form rather than mycelium.

Heterotrophic Growth Requirements

Inability to synthesize their own food as chlorophyllous plants do forces fungi to live as heterotrophs on products synthesized by other organisms (Fig. 8-2). As already mentioned, most fungi consume the remains of dead organisms, while a few parasitize living organisms. Others straddle the fence, occasionally lending a hand to the demise of the host and then readily consuming the remains. Fungi that utilize dead substrates are called saprobes. Those that parasitize only living hosts are called obligate parasites, and the fence straddlers are called faculative parasites or faculative saprobes, depending upon their affinity for dead or living hosts, respectively.

Each fungus has its specific nutrient requirements and optimum environmental conditions for development. Cellulose, starch, and many other organic compounds are utilized by fungi for energy and for raw material for synthesis. Organic

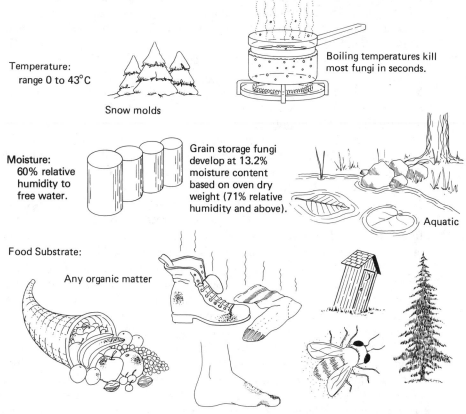

Temperature: range 0 to 43°C

Snow molds

Boiling temperatures kill most fungi in seconds.

Moisture: 60% relative humidity to free water.

Grain storage fungi develop at 13.2% moisture content based on oven dry weight (71% relative humidity and above).

Aquatic

Food Substrate:

Any organic matter

Note: Each fungus has its own specific requirements for temperature, moisture, and substrate.
Oxygen: Most require some oxygen (O_2), though some tolerate high concentrations of CO_2.

Figure 8-2 Requirements for growth of fungi.

nitrogen in the form of amino acids or ammonia is the usual source of nitrogen. Other mineral requirements are similar to higher plants. Many fungi are deficient in the vitamins thiamine and/or biotin.

Most fungi require oxygen, but some rumen-inhabiting fungi are obligate anaerobes. Some water molds are indifferent to oxygen. The yeasts used in the brewing industry generate alcohol in the absence of oxygen, but growth and development of the population capable of fermentation to 12% alcohol first took place in an environment with oxygen. The lack of oxygen during fermentation eventually results in the death of the yeasts. It is interesting to note that almost every civilization has exploited fungi for fermentation of local fruits, grains, potatoes, and many other substrates.

Most fungi require some free water to be able to grow in or on a substrate. A wooden table in a classroom will not be decayed by fungi because of lack of water. The same table placed outside will readily decay, because moisture, from rain and from the soil, absorbed by the wood is sufficient for fungal growth.

A few fungi, the most notable examples being the grain storage fungi, can grow in an environment where the substrate is in equilibrium with a relative humidity of 71%. At this relative humidity there is no free water. The grain moisture content is 13.2% based on an oven-dry weight. A few other fungi can grow at 60% relative humidity.

You can readily recognize how difficult it is to prevent grain storage fungi from developing if the minimum requirements are a humidity of at least 71%. Moisture not dried out of the grain before it is placed in storage is a source of some moisture. A drop in temperature within a closed storage elevator or ship also will increase the relative humidity.

In large piles of grain, metabolism of the initial invading fungi adds moisture and heat. Additional fungi can become established as well as thermophilic bacteria. Eventually, if the grain is not dried and stored properly, spontaneous combustion takes over when the temperature exceeds that acceptable to the thermophiles. Today, we seldom hear of grain elevators burning because of spontaneous combustion, because the high value of grains has forced us to a better understanding of fungus activity. Every year, though, a few barns burn because high-moisture hay is invaded by storage fungi, which start the process toward spontaneous combustion.

Today, we recognize another problem associated with fungus invasion in stored grain. Toxic chemical products produced by grain storage fungi induce accumulative deleterious effects when ingested by livestock or human beings. These products, called aflatoxins and other compounds have become serious concerns in animal feeds and in products processed for human consumption.

The other extreme in moisture requirements is represented by the aquatic fungi. Some fungi are totally aquatic in their existence. Leaves and other organic material that finds its way into well-aerated streams is decomposed by an interesting group of fungi called aquatic hyphomycetes. The curious characteristic of these fungi is their star-shaped spores, spores with long appendages presumably for easy dispersal in rapidly moving water. Lack of oxygen in stagnant water prevents development of these fungi, thereby allowing organic debris to accumulate.

Another requirement for fungus growth is an environment with a temperature range between freezing and about 43°C. Most fungi have an optimum near 24°C, but some operate at the extremes of the range.

Snow molds (Fig. 8–3) can develop below the snow in a temperature very near the freezing point and avoid much competition by other microorganisms. These fungi will grow at higher temperatures, but at higher temperatures, they have to compete with a wide array of other microorganisms. Therefore, an ecologically competitive advantage is exploited by the snow mold fungi growing at or near the freezing point.

Another group of fungi find the ecological niche of higher temperatures a less competitive environment. A maximum temperature of about 55 to 60°C is the upper limit for a fungus, *Chaetomium thermophile*, which can be found in the waters of the cooler geysers. Temperatures at the boiling point will kill all fungi in a very few minutes. Thermophilic bacteria, on the other hand, can survive at temperatures in excess of 60°C, and some produce resting spores that can survive boiling temperatures.

Mycelial Growth Form

Fungi have a branching filamentous thread-like growth form. Individual threads averaging 1 to 5 μm in diameter are called hyphae (plural) or hypha (singular); see Fig. 8–4. A collection of hyphae produces a thallus or a colony of mycelium (singular) or mycelia (plural).

The thread-like hyphae grow at the tips and penetrate between and within cells of the substrate. Enzymes released from the fungus hyphae cause specific digestive

Figure 8–3 Mycelial felt of a snow mold looking like a thick spider web covering the forest floor. The snow melting back at the right exposes the fungus, which had developed under the protective snow covering.

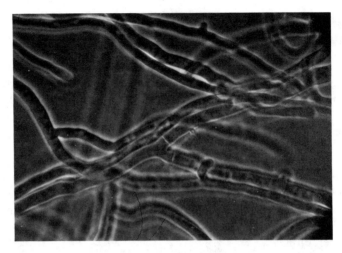

Figure 8–4 Fungus hyphae as seen under high magnification with a light microscope.

reactions to take place which solubilize the substrate. The soluble form of the substrate is diffused through the cell wall of the fungus and is then further degraded within the fungus using the standard metabolic enzyme systems of living organisms. Microscopic hyphae of some fungi aggregate and produce the more conspicuous fungal structures we are more familiar with, such as mushrooms.

There is not much variation in hyphae from one fungus to another. Most fungi have hyphae made up primarily of chitin, glucans, and mannans. Chitin is a long-chain polymer made up of *N*-acetylglucosamine building blocks. Chitin is also a major component of the insect exoskeleton. Glucans are 1,3- and 1,6-linked glucose molecules. Mannans are polymers made up of mannose and glucose.

Oomycetes and other fungi in the subdivision Mastigomycotina differ from most fungi in that the cell wall is made up primarily of cellulose rather than chitin. This major difference and other features of the group suggest to some that these organisms are not really fungi. In some classification systems these are placed in the Kingdom Protista rather than the Kingdom Fungi.

Hyphae may be pigmented or without pigment (hyaline); they may have one continuous cytoplasm with many nuclei and organelles and no cross walls (coenocytic), or they may split up into numerous cells by cross walls (septate). Cytoplasmic movement between cells is not eliminated in septate hyphae because the septa have a central pore which allows some movement of organelles and materials from one cell to another. The pore may become plugged if a cell is injured, thereby preventing the death of the entire hypha.

A septum with a central pore with swollen margins is called a dolipore septum. The dolipore septum is thought to be characteristic of the *Basidiomycotina* fungi and is, therefore, one hyphal character with potential application in identifying fungi. The dolipore septum is best seen with the electron microscope.

Another hyphal characteristic unique to some of the Basidiomycotina is the clamp connection (Fig. 8–5). A hyphal branch from the terminal cell turns back and fuses (anastomoses) with the cell just behind the septum. One nucleus is transferred through the hyphal branch to the cell just behind the terminal cell. In this way, each cell maintains two nuclei. The reason for its existence is still not known.

Reproduction by Means of Spores

Spores are the reproductive structures of fungi (Fig. 8–6). They differ from seeds of higher plants in that a spore cannot be subdivided into various tissues, such as embryo and endosperm. Spores are microscopic, one- to many-celled, and may be of diverse shapes and colors.

The macroscopic fungal fruiting structures that we often use to recognize fungi are evolutionary developments for more efficient spore production. Examples of the types of spore-producing structures of the major groups of fungi will be presented.

Fungi generally produce extremely large numbers of spores because of the random or only partially directed dispersal of the spores to other sites. I have sampled and counted spores from the fruiting body of the wood decay fungus *Phellinus tremulae (Fomes igniarius* var. *populinus)*. More than 100,000 spores per square millimeter of undersurface per day was calculated from the sample. You can estimate the seasonal spore production of such a fungus using a 4 by 10-cm lower surface for the period between April 15 and October 15. Very few of these spores actually infect another tree. They are buffeted about by localized turbulence and wind until rain or settling in still air brings them to the forest floor.

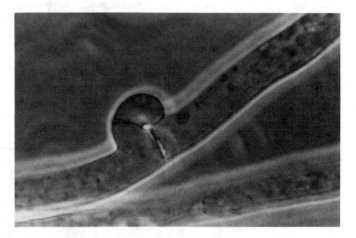

Figure 8–5 Clamp connection of a basidiomycete fungus hypha as seen under high magnification with a light microscope. (Photograph compliments of Dr. Edson Setliff.)

(a)

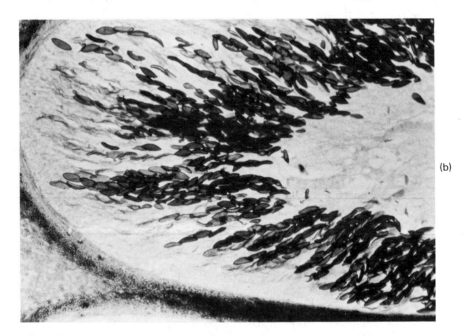

(b)

Figure 8–6 Examples of fungus spores as seen with the light microscope: (a) rought-walled light-colored single-celled spores; (b) dark-colored single-celled spores; (c) various fungus spore shapes.

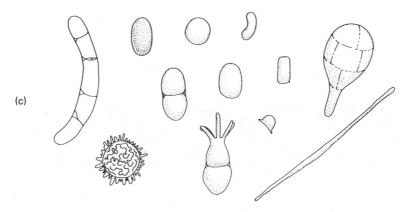

(c)

Figure 8-6 (continued)

TAXONOMIC CLASSIFICATION AND LIFE CYCLES OF FUNGI

The plant pathologist needs to be familiar with the details of fungus structures, life cycles, and taxonomy. The diagnosis of disease is often based upon identification of a specific fungus observed fruiting on the diseased tissue. Control procedures are more effective when applied to certain stages in the life cycle of a fungus, so an understanding of the life cycle is required.

Fungal populations are genetically variable and changeable, just like other populations of organisms. Therefore, we must understand the sexual activities of fungi to comprehend why disease-resistant varieties of plants do not always remain resistant.

Mycology is a subject that is not easily presented in a single chapter. You will become more proficient in the details presented here as we apply them to specific diseases in subsequent chapters. As these chapters are presented, it may be worthwhile to refer back to this chapter to see where the specific details fit into the scheme of things.

The taxonomic classification of any group of organisms provides a compilation of standardized names for organisms. This facilitates communication among people working with similar organisms. The rules for naming and grouping organisms into a taxonomic classification were devised by man. Ideally, they should represent natural biological groupings, but seldom is the ideal realized. Therefore, we continue to try and improve our system of organization.

To someone not intimately involved in taxonomic considerations, changes aimed at improving the system may seem more like changes to confuse than to improve. The biologist, continually confronted with changes, can resist the changes or adopt the new names, but in any case must be aware of both the old and the new systems.

The taxonomic treatment of fungi in this book will generally follow *The Dictionary of the Fungi*, Hawksworth et al. (1983). Anyone greatly interested in fungi will quickly discover why the older system is inadequate.

The reader interested in current concepts in taxonomy of fungi is directed toward the taxonomic review of *The Fungi*, Vols. IVA and IVB, edited by Ainsworth et al. (1973), and *The Dictionary of the Fungi*.

Basidiomycotina

The fungi of the Basidiomycotina subdivision of the Kingdom Fungi (simply basidiomycetes) are characterized by basidiospores and basidia (Fig. 8-7). The basidium is a specialized terminal hyphal cell originating from a fertile layer called a hymenium or from a specialized fertile cell called a teliospore.

Within the basidium, two nuclei ($n + n$) fuse to form a diploid nucleus ($2n$) which immediately undergoes meiosis to form four haploid nuclei ($1n$). Each of the nuclei migrates to a small projection, sterigma, from the basidium, and a spore forms by what appears to be a blowing out of the tip of the sterigma (Fig. 8-8). When mature, the basidiospore is forcibly ejected from the sterigma to be picked up by wind currents and disseminated to another host or substrate.

Some Basidiomycotina, the class Hymenomycetes, produce their basidia and basidiospores in a hymenium lining the flat, pored, gilled, or tooth-like outer surface of the fruit bodies (basidiomata). A few Basidiomycotina (Gasteromycetes) produce their basidia enclosed within a closed chamber. In this case, the basidia disintegrate and the basidiospores are disseminated upon breakage of the chamber. The puffballs are good examples of the latter type and mushrooms of the former type. Two other classes of the Basidiomycotina, Urediniomycetes and Ustilaginomycetes, produce septate basidia (sometimes called promycelia) from teliospores. The Urediniomycetes, rust fungi, are discussed in detail in Chapter 11.

Figure 8-7 Dark brown basidiospores formed on nonseptate basidia in a hymenium lining both sides of a gill of a mushroom as seen under high magnification with a light microscope.

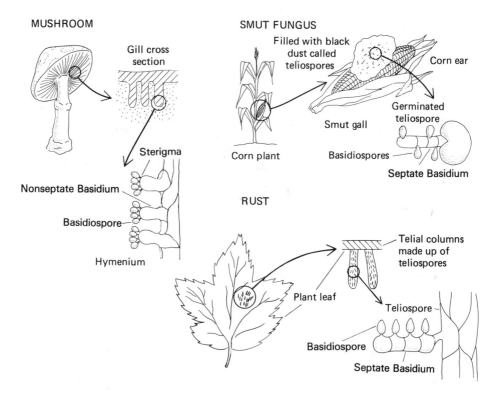

Figure 8-8 Basidiomycotina reproductive structures.

The Ustilaginomycetes, smut fungi, are important pathogens of grain crops, but since these fungi do not affect trees, they are not discussed in this book.

Earlier, when describing the clamp connection, it was pointed out that each cell contains two nuclei. The standard basidiomycete life cycle begins when one of the basidiospores containing a single haploid nucleus ($1n$) (the product of meiosis) germinates to produce hyphae. Nuclear division, mitosis, results in multiplication of the single genome within the developing septate thallus. If the hyphae come in contact with another thallus developing from another basidiospore, a fusion of the hyphae may occur to introduce a second genome to the thallus. If the nuclei in this heterocaryon are sexually compatible, fruiting may eventually occur. The introduced nucleus in each thallus multiplies rapidly and migrates throughout the thallus out to the growing tips. The hyphae, which up until fertilization produced only simple septa, may then begin to produce clamps each time a new cell wall is laid down. The nuclei mitotically divide at the same time (conjugate nuclear division). Therefore, from the time when fertilization takes place until some time in the future when the fungus produces a fruiting body and sporulates, the two genomes ($n + n$) are maintained within each cell. The completion of the life cycle to production of spores following meiosis has already been discussed.

Some Basidiomycotina produce asexual spores following mitotic division. These vegetatively produced spores called oidia or conidia are very much like the conidia of the Ascomycotina, which will be discussed next.

Ascomycotina

Fungi of the subdivision Ascomycotina (simply ascomycetes) are characterized by the production of ascospores within asci. Asci are sac-like structures usually containing eight ascospores.

The life cycle of Ascomycotina usually starts with germination of haploid (ln) ascospores (Fig. 8–9) to form branching septate hyphae. Fertilization, the bringing together of two ln genomes, is somewhat different from that in Basidiomycotina. Ascomycotina may produce two morphologically distinct fertilization structures, ascogonia and antheridia. The nuclei of the antheridia are transferred to the ascogonium and pair with nuclei of the ascogonium. Pairs of nuclei migrate into developing ascogenous hyphae. A crozier or hook forms on the tips of the ascogenous hyphae. The crozier is very similar to the clamp connection of the Basidiomycotina. Other forms of fertilization may occur.

The two nuclei of the subterminal cell of the crozier fuse. This terminal cell will enlarge and become the ascus. Meiosis produces four haploid nuclei in this developing ascus. Each nucleus then undergoes a mitotic division to produce eight

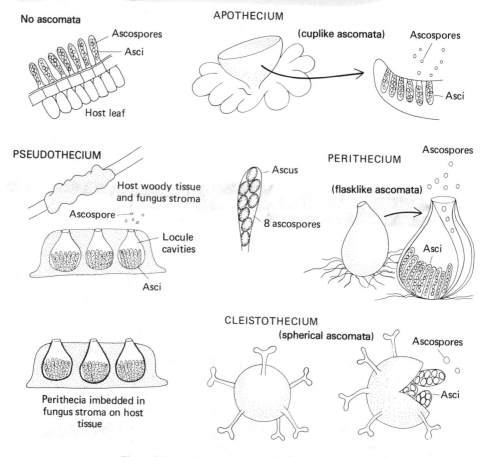

Figure 8–9 Ascomycotina reproductive structures.

nuclei. The ascus elongates during the meiotic and mitotic divisions. The contents of the ascus become incorporated into the ascospores that form around each nucleus.

The asci of the Ascomycotina are aggregated into a number of different morphological structures (fruit bodies or ascomata; see Figs. 8–10 to 8–13). In some classification systems the different types of ascomata were the primary basis of classification. The classification system today is a bit more confusing because the primary separation of the Ascomycotina is based on the number of wall layers of the ascus (unitunicate ascus of bitunicate ascus). The types of ascomata are still very useful in identifying these fungi and therefore they will be described without reference to any specific classification system.

Some Ascomycotina form asci directly on the surface of the substrate. Others form asci in spherical ascomata called cleistothecia that release ascospores only when broken. A third type forms asci in a flask-shaped structure called a perithecium that has an opening at the top to release the spores. A fourth type produces asci in an open cup-like fungus structure called an apothecium. A fifth type forms asci in cavities (locules) in fungus stroma. These are called pseudothecia since they look like perithecia but do not have a perithecial wall. Perithecia can also be formed in fungus stroma, but the wall of the perithecium is distinct enough to be recognized (see Figs. 8–6 and 8–9).

The ascospores of the unenclosed asci, perithecia, apothecia, and pseudothecia are often forceably ejected a few millimeters from the mature ascus through a rupture in the tip of the ascus. These small light spores are dispersed through localized turbulence and wind much like smoke. Some ascospores are oozed from the ascomata to be water drop or insect dispersed.

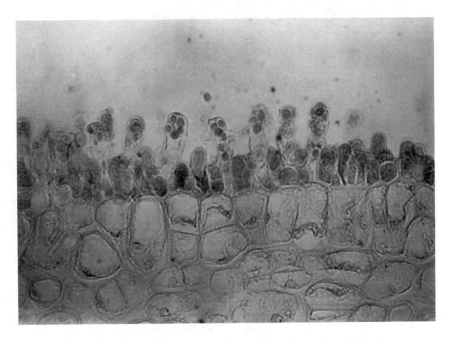

Figure 8-10   Asci with ascospores borne free on the surface of a leaf.

Figure 8–11 Cleistothecia of *Microsphaera penicillata (M. alni)* with asci and ascospores.

(a)   (b)

Figure 8–12 (a) Perithecia of *Nectria galligena* as seen under the dissecting microscope; (b) cross section through a perithecia as seen with a compound microscope. Ascospores occur in groups of eight. The ascus wall is not readily visible in this section. The exit pore to the perithecium is also not shown in this cross section.

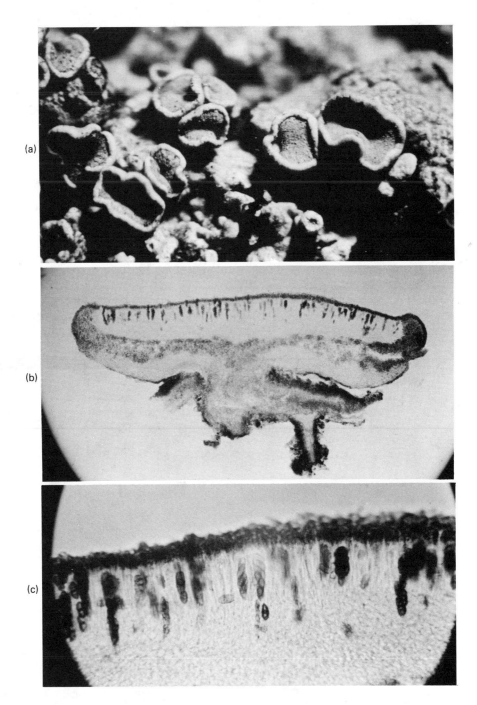

(a)

(b)

(c)

Figure 8-13 Apothecia of a lichen as seen with the dissecting microscope; (b) cross section through an apothecium as seen under low magnification with a compound microscope; (c) higher magnification of the apothecium cross section showing clusters of ascospores (the asci are not readily visible in this section).

Taxonomic Classification and Life Cycles of Fungi

117

Deuteromycotina or Fungi Imperfecti

A difficult concept for most beginning biologists to understand about the fungi is that many fungi have more than one spore stage. The ascospores of the Ascomycotina, called sexual or perfect spores, are produced following meiotic nuclear division. Ascomycotina and other fungi also can reproduce asexually. The asexual or imperfect spores of Ascomycotina, called conidia, are produced following mitotic division.

To make things even more confusing to the beginner, we have developed a second classification system for fungi that produce conidia. A given fungus may therefore have more than one name. One name is for the sexual stage and the other is for the asexual stage. (You will note two names for many canker and foliage fungi discussed in later chapters.)

Some fungi produce only asexual spores, or at least we have never found a sexual spore stage. Those fungi that do not produce sexual spores and the asexual spore stages of Ascomycotina and Basidiomycotina are grouped into the Deuteromycotina or Fungi Imperfecti.

The major characteristic of the Fungi Imperfecti is the production of conidia on conidiophores (Figs. 8-14 and 8-15). The conidiophore is a modified hypha which gives rise to spores either by segmentation or by a blowing out of spores from the tip or sides. Conidiophores bearing conidia can be aggregated inside flask-like structures called pycnidia. They may be aggregated in cavities made up of both fungus and host tissue (acervuli). The fungi with pycnidia or acervuli enclosing conidia and conidiophores are in the class Coelomycetes.

A second class of the Fungi Imperfecti is the Hyphomycetes, in which the conidia and conidiophores are born unenclosed. They may arise individually or

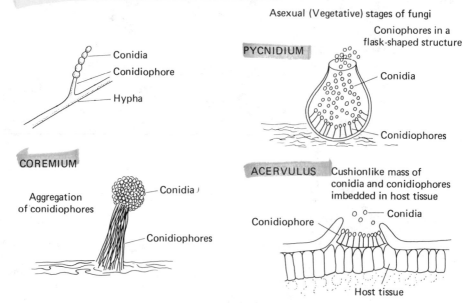

Figure 8-14 Fungi Imperfecti reproductive structures.

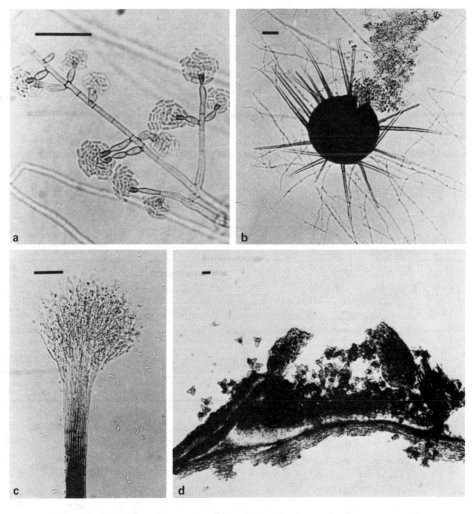

Figure 8-15 Photomicrographs of Fungi Imperfecti reproductive structures (the bar in each photograph is 25μ m); (a) conidiophores and conidia of *Phialophora americana;* (b) pycnidium of *Chaetomella* sp.; (c) coremium of the Imperfect stage of *Ceratocystis ulmi;* (d) acervulus of *Asterosporium asterospermum.* (Photographs compliments of Dr. Chun-Juan Wang.)

aggregate into tight clusters of conidiophores called synnemata or coremia. Another group of Hyphomycetes produce conidiophores on a cushion of fungus hyphae called a sporodochium.

 ## Oomycetes and Zygomycetes

The Oomycetes, Zygomycetes, and a few other fungi were, at one time, grouped under the class Phycomycetes. This group was distinguished from the other fungi by the infrequent septations in the hyphae. There are very distinct differences in the spores and chemistry of the hyphae of these fungi, and therefore they are better recognized as individual classes within different subdivisions of the Kingdom Fungi.

The Oomycetes are a class in the subdivision Mastigomycotina. Fungi in the Oomycetes are very different from other fungi in that the cell wall is primarily cellulose rather than chitin. They are found in water, in soil, or on upper portions of plants under very wet conditions. The asexual spores of the fungus have flagella that give the fungus motility in water. The asexual spores called zoospores are produced in a saclike structure called a sporangium. The spores are able to sense specific chemical stimuli of plant roots or openings in the upper portions of plants and move in a film of water to the infection site. The spores germinate and the infrequently septate hyphae invade the host. An interesting feature of Oomycetes is that the sporangium can be disseminated, and under slightly warmer temperatures, the sporangia may initiate infection by germination to form hyphae rather than motile spores.

The sexual reproduction of the Oomycetes begins with the fertilization of a specialized cell called an oogonium by nuclei from another, smaller, different-shaped specialized cell called an antheridium. Oogonia and antheridia are sites of meiosis in this diploid fungus. Fusion of these dissimilar gametangia brings the nuclei of two compatible thalli into the oogonium, which then rounds up into a resting spore called an oospore. Within the oospore nuclear fusion takes place. Under the proper environmental conditions the oospore will germinate to form diploid hyphae.

There are a number of very important Oomycete fungi. *Phytophthora infestans* causes late blight, a very serious disease of potatoes that caused the death or migration of one half of the population of Ireland in the mid-1800s. Most Americans with Irish backgrounds can probably trace their roots to this devastating epidemic. Other Oomycetes are involved with damping-off and root rot problems.

Another group of fungi with infrequently septate hyphae is the class Zygomycetes of the subdivision Zygomycotina. There are no serious tree diseases caused by this group, but they are very common mold fungi found on all forms of organic matter. If one attempts to culture fungi from soil or diseased plant parts, it is very common to recover Zygomycete fungi. Some Zygomycetes are associated with vesicular arbuscular (VA) mycorrhizae of most crop plants and many broadleaf trees and shrubs. Mycorrhizae are discussed more thoroughly in Chapter 9.

The Zygomycetes produce asexual sporangiospores in a saclike cell called a sporangium. Like most fungi, these spores have no flagella and are disseminated passively by wind, water, and anything that moves. If the spores land on the correct substrate, they will germinate to produce infrequently septate hyphae which invade and contribute to the decomposition of the material. The sexual stage of these fungi involves the fusion of two similar gametangia to form a resting zygospore. Meiosis within the zygospore followed by germination and production of sporangia and sporangiospores packages and disseminates the 1n genome products of meiosis.

A summary of the classes of fungi with typical examples is presented in Table 8-1.

TABLE 8-1 SIMPLIFIED SUMMARY OF THE KINGDOM FUNGI

I. Sexual reproduction by basidiospores formed on clublike structures called basidia (Figs. 8-7 and 8-8). Asexual reproduction sometimes through conidia (*see* Deuteromycotina).

subdivision — BASIDIOMYCOTINA

A. Basidia formed in a basidiomata (fruit body) from an external fertile layer called a hymenium (Figs. 8-7 and 8-8). Hyphae sometimes with clamp connections at septations (Fig. 8-5).

class — Hymenomycetes

1. Basidiomata (fruit bodies, Figs. 14-28, 14-29, etc.) with the hymenium lining the walls of pores in the lower surface.

order — Aphyllophorales (wood decay fungi, Chapters 14 and 16)

2. Basidiomata with hymenium lining the walls of gills (Fig. 8-8).

order — Agaricales (mushrooms, saprobic, parasitic, or mycorrhizal fungi)

3. Jelly-like basidiomata. Basidiospores produced on long sterigmata from septate basidia.

order — Tremellales (jelly fungi, mostly saprobic)

B. Basidiomata enclosing an indistinct hymenium.

class — Gasteromycetes (puffballs, saprobic or mycorrhizal fungi) (stinkhorns and bird's nest fungi, saprobic)

C. Basidiomata lacking. Septate basidia arising from a teliospore. Basidiospores produced on sterigmata and forcibly discharged.

class — Urediniomycetes (rust fungi, obligate parasites; see Chapter 11)

D. Basidiomata lacking. Septate basidia arising from a teliospore. Basidiospores not on sterigmata and not forcibly discharged.

class — Ustilaginomycetes (smut fungi, important in grain crops)

II. Sexual reproduction by ascospores produced in sac-like structures called asci (Figs. 8-6, 8-9 to 8-13). Asexual reproduction through conidia or hyphal fragments (*see* Deuteromycotina).

subdivision — ASCOMYCOTINA

1. Cavity-like ascomata called pseudothecia within fungus stroma (Fig. 8-9). bitunicate ascus

Foliage and shoot pathogens

2. Ascomata (fruiting bodies) absent (Figs. 8-9, 8-10, 10-9; unitunicate ascus)

Taphrina (leaf blister)

3. Ascomata present. unitunicate ascus.

a. Spherical ascomata called cleistothecia that release spores when broken (Figs. 8-9, 8-11, 10-6, 10-7)

Powdery mildews

b. Flask-shaped ascomata called perithecia that release spores through a pore (Figs. 8-9, 8-12, 12-6, 12-9, 12-11, 12-18)

Canker and foliage pathogens and saprobic fungi

c. Cup-like ascomata called apothecia (Figs. 8-9, 8-13)

Many lichens, canker fungi, and saprobic cup fungi

III. Asexual reproduction only or asexual stages of Ascomycotina or Basidiomycotina fungi.

subdivision — DEUTEROMYCOTINA or FUNGI IMPERFECTI

A. Conidiophores and conidia formed within pycnidia or acervuli (Figs. 8-14, 8-15)

class — Coelomycetes

Taxonomic Classification and Life Cycles of Fungi

TABLE 8-1 CONTINUED

B. Conidiophores single or sometimes aggregated into synnemata (coremia) or on sporodochia (Figs. 8-14, 8-15).

    class—Hyphomycetes

IV. Sexual reproduction by oospores produced from the fertilization of dissimilar gametes, a large oogonium, and a small antheridium (Fig. 8-16). Asexual reproduction by motile, flagillate zoospores produced in a sporangium (Fig. 8-16).

    subdivision—Mastigomycotina
    class—Oomycetes
    (root rots and damping-off)

V. Sexual reproduction by zygospores produced from the fusion of two similar gametes (Fig. 8-16). Asexual reproduction by nonflagellate sporangiospores produced in sporangia (Fig. 8-16).

    subdivision—Zygomycotina
    class—ZYGOMYCETES
    [saprobic fungi and vesicular arbuscular (VA) mycorrhizae]

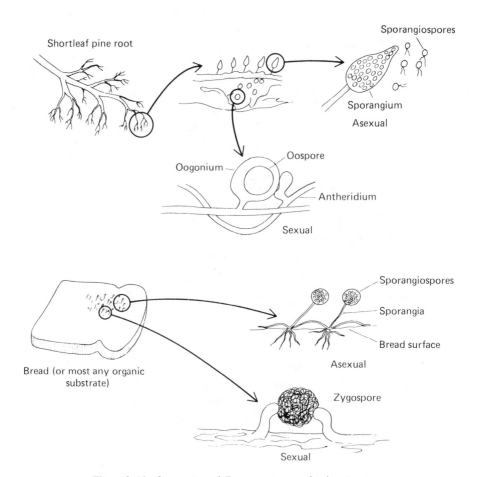

Figure 8-16 Oomycete and Zygomycete reproductive structures.

DISEASE CYCLE OF PATHOGENIC FUNGI

The disease cycle of a pathogenic fungus is an interaction of the life cycles of both the plant and the fungus. Placement of spores or other vegetative reproduction propagules in an infection court on the host is called *inoculation*. *Incubation* of the spore involves uptake of water, possibly nutrients, and initiation of metabolic processes leading toward *germination* or the transition from the resting spore to dynamically growing mycelial phase of the fungus. *Infection* occurs when the fungus germ tube makes contact with the host cells and begins deriving nutrients from the enzymatic digestion of host material rather than utilizing stored reserves within the spore. *Invasion* is the ramification of fungus hyphae within the host tissues. *Disease symptoms* result from fungus invasion and host response. Further invasion may occur to intensify the disease symptoms. The fungus must go through a metabolic transition from vegetative expansion to growth termination and *sporulation* somewhere after invasion. In the case of foliage fungi, this transition may occur in just a matter of days, whereas in heart rot decay fungi it may take decades. Fungi that cause diseases of crops in temperate regions must accommodate their life cycles to survive the freezing conditions of winter. Fungi of other climates must survive high temperatures or drought. Therefore, some type of *dormant stage* is necessary. Some fungi combine sporulation and dormancy; others just stop mycelial growth and then begin mycelial growth again when temperature conditions are more favorable. Other fungi produce sclerotia. Sclerotia are tight aggregates of vegetative fungal cells surrounded by an impervious outer layer. Sclerotia may remain dormant in the soil until stimulated to grow by the presence of a susceptible host plant. Each fungus must have some method of *dissemination* of propagules from the diseased plant to another healthy plant. Some fungi have adopted insects as specific vectors for dissemination of the spores; others use wind or splashing water. The random or specifically directed placement of the fungus spore in the infection court completes and begins the disease cycle again.

Each type of disease discussed in future chapters will include a discussion of a typical disease cycle. You will soon recognize how little we really know about the disease cycles of fungi that cause diseases of trees. Hopefully, some of you may be stimulated to investigate some of the unknown aspects of diseases of trees. The rest of you will at least be reasonably capable of evaluating what is supposedly known.

REFERENCES

AINSWORTH, G.C., F.K. SPARROW, and A.S. SUSSMAN, eds. 1973. The fungi: an advanced treatise. Vol. 4A, 621 pp.; Vol 4B, 504 pp. Academic Press, Inc., New York.

ALEXOPOULOS, C.J. 1962. Introductory mycology. John Wiley & Sons, Inc., New York. 613 pp.

CHRISTENSEN, C.M. 1961. The molds and man: an introduction to the fungi, 2nd ed. University of Minnesota Press, Minneapolis. 238 pp.

DEVERALL, B.J. 1969. Fungal parasitism. Studies in biology 17. St. Martin's Press, Inc., New York. 57 pp.

FARR, D.F., G.F. BILLS, G.P. CHAMURIS, and A.Y. ROSSMANN. Fungi on plants and plant products in the United States. APS Press, St. Paul, Minn. 1252 pp.

HAWKSWORTH, D.L., B.C. SUTTON, and G.C. AINSWORTH. 1983. Ainsworth & Bisby's dictionary of the fungi. Commonwealth Mycological Institute, Kew, Surrey, England. 445 pp.

HUDSON, H.J. 1972. Fungal saprophytism. Studies in biology 32. Edward Arnold (Publishers) Ltd., London. 67 pp.

MOORE-LANDECKER, E. 1982. Fundamentals of the fungi. Prentice-Hall, Inc., Englewood Cliffs, N.J. 578 pp.

PILÁT A., and O. UŠÁK. 1951. Mushrooms. Spring Books, London, pp. 59–61.

SKYE, E. 1979. Lichens as biological indicators of air pollution. Annu. Rev. Phytopathol. *17*: 325–341.

TALBOT, P.H.B. 1971. Principles of fungal taxonomy. St. Martin's Press, Inc., New York. 274 pp.

9

FUNGI AS SYMBIONTS OF TREE ROOTS: MYCORRHIZAE

- *TYPES OF MYCORRHIZAE*
- *MODE OF ACTION OF MYCORRHIZAE*
- *MYCORRHIZAL ASSOCIATION CYCLE*
- *IMPORTANCE OF MYCORRHIZAE*
- *RECOGNITION OF MYCORRHIZAE*
- *EXAMPLES OF MYCORRHIZAE*
- *ESTABLISHMENT OF MYCORRHIZAE ON SEEDLINGS*

Mycorrhizae (fungus roots) are the normal means by which almost all higher plants take up water and minerals from the soil. Only a few higher plant genera are known to lack mycorrhizal associations. One might question the insertion of this topic in a book on disease concepts, because mycorrhizae represent supposedly mutualistic symbiosis. The mycorrhizal fungus parasitizes the cortical cells of the root, but the presence or absence of mycorrhizae will dramatically affect seedling form and size and tree development.

Mycorrhizae are usually emphasized for their beneficial effect on root function. They are also important as barriers to infection by other destructive root pathogens in the nursery and in the field.

TYPES OF MYCORRHIZAE

The mycorrhizal associations of fungi and plants can be cataloged into three groups: Endomycorrhizae (Fig. 9–1), ectomycorrhizae (Figs. 9–2 and 9–3), and ectendomycorrhizae (Fig. 9–4). Vesicular-arbuscular (VA) mycorrhizae are a major sub-

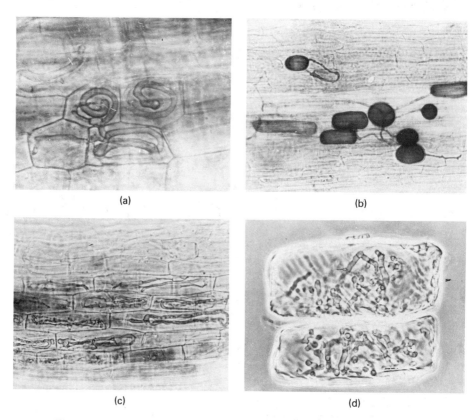

(a)

(b)

(c)

(d)

Figure 9-1 Endomycorrhizal root showing (a) coiled hyphae, (b) vesicles, (c) and (d) intracellular hyphae. (Photographs compliments of Dr. Hugh Wilcox.)

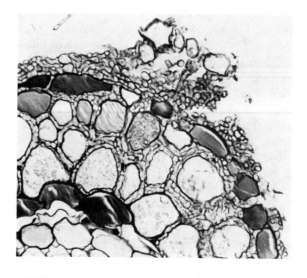

Figure 9-2 Ectomycorrhizal root cross section, showing a Hartig net and mantle. Photograph compliments of Dr. Hugh Wilcox.)

Figure 9-3 Examples of ectomycorrhizal roots of red pine. (Photographs compliments of Dr. Hugh Wilcox.)

Types of Mycorrhizae

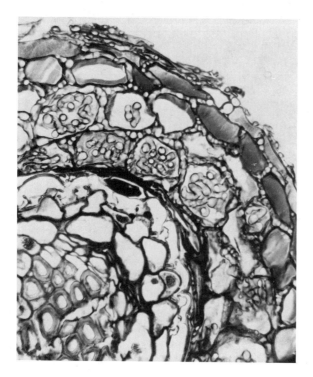

Figure 9-4 Ectendomycorrhizal root cross section, showing both intra- and intercellular invasion of cortical cells and a thin mantle. (Photograph compliments of Dr. Hugh Wilcox.)

group of the endomycorrhizae. The VA mycorrhizal association occurs with a wide array of plants, including grasses, legumes, shrubs, and hardwood trees. The fungus invades the cortical cells. The terminology "vesicular arbuscular" describes the fungus–host relationship. Vesicles are formed within and exterior to cortical cells (Fig. 9-1b). Arbuscules describe the fine branching pattern to the intracellular hyphae. Coiling and course branching hyphae also occur in the VA mycorrhizal association (Fig. 9-1).

Fungi in the family Endoginaceae of the subdivision Zygomycotina produce VA mycorrhizae. These are grouped into four genera, *Acaulospora, Gigaspora, Glomus,* and *Sclerocystis.* These fungi may produce sporocarps 1 to 25 mm in diameter in the soil or litter. Within the sporocarps are formed either sexual zygospores, vegetative resting chlamydospores, or asexual sporangia which give rise to sporangiospores. The vesicles described above may be immature chlamydospores or sporangiospores, but the exact connection has not been established.

VA mycorrhizae do not form a fungal mantle on the surface of the root. There is no outward difference between a VA mycorrhizal root and a nonmycorrhizal root, and therefore they are only identified through microscopic observations of the structures described above.

Ectomycorrhizae have a fungus–host relationship where the fungus surrounds the roots, forming a mantle, and invades the roots but does not penetrate the cortical cells. A Hartig net (named for Robert Hartig, who first described the association) of hyphae forms between the cortical cells (Fig. 9-2). Conifers, in the family Pinaceae and some hardwood mycorrhizae are of this type (Fig. 9-3).

The fungi found in ectomycorrhizal associations are commonly members of the Basidiomycotina in the order Agaricales (mushrooms) or in the class Gasteromycetes (puffballs). A few members of the Ascomycotina also form ectomycorrhizae.

The third type, called ectendomycorrhizae, combines a Hartig net and a thin mantle with invasion of cortical cells (Fig. 9–4). One group of ectendomycorrhizae is found in well-fertilized nurseries. Both conifers and broadleaf plants have this type of mycorrhizae. This type of ectendomycorrhizae is often referred to as E-strain mycorrhizae, based on the trivial name given to dark fungi involved. Other groups of ectendomycorrhizae are specifically specialized to certain groups of plants. The E-strain fungi have been difficult to classify properly, but they have been recently placed in a new cup fungus genus, *Wilcoxina*, of the Ascomycotina.

The characteristics of the three types of mycorrhizae are summarized in Table 9–1.

TABLE 9-1 CHARACTERISTICS OF MYCORRHIZAL TYPES

	Mantle	Hartig net	Intracellular fungus invasion	Intercellular fungus invasion
Ectomycorrhizae	Yes	Yes	No	Yes
Endomycorrhizae (VA)	No	No	Yes	No
Ectendomycorrhizae	Yes	Yes	Yes	Yes

MODE OF ACTION OF MYCORRHIZAE

The mycorrhizal association of fungi and plant roots is a two-way mode of action. Fungus invasion causes a major proliferation of short roots. In turn, the increase in roots increases the moisture- and mineral-element-absorbing surface and provides the plant with a better capacity to survive and grow.

In addition, the fungus–root association is a more efficient mineral-element-absorbing organ. More nitrogen and phosphorus are absorbed from the soil and accumulated in plants with mycorrhizae, as compared to plants without mycorrhizae. The fungus capacity for extraction of elements from the soil organic matter is assumed to be part of the increased efficiency.

The fungus parasitizes the plant for most of its carbohydrate and vitamin needs. According to some mycorrhizal specialists, the fungus may actually be obligately parasitic on plant roots.

Why are some fungi mycorrhizal, some pathogenic, and some saprobic? These differences are probably based on the mode of action of the organism. The saprobic fungi cannot penetrate the physical and chemical defense barriers of living roots. The parasitic fungi penetrate and cause a series of disruptive reactions leading toward the death of the host. The mycorrhizal fungi, like other obligate parasites, induce a limited host chemical defense reaction which is not totally disruptive to the host and is tolerated by the fungi.

If we could identify the important chemical reactions in mycorrhizal fungus root interactions, we might be able to utilize these natural chemical defense mechanisms for the prevention of plant diseases.

MYCORRHIZAL ASSOCIATION CYCLE

Most of the known ectomycorrhizal fungi are basidiomycetes, so initial infection of roots is probably by basidiospores. Reasonable mycorrhizal colonization of fumigated nursery beds usually occurs during the first growing season. The source of inoculum appears to be basidiospores from mushroom fruit bodies associated with tree roots around the nursery.

Mycorrhizal roots develop and generally function for a single growing season. Presumably, the fungus survives in the root association and reinfects new roots produced along the expanding root system of the following year.

Fruit bodies of Agarics (gilled mushrooms), Boletes (pored mushrooms), and Gasteromycetes (puffballs) are formed above ground. Hyphal strands connect the fruit body with the mycorrhizal root nutrient source.

The VA is initiated, in some instances, by germination of a vesicle or chlamydospore. The chlamydospore is a relatively large, thick-walled, resting fungus spore produced inside the cortical cells or on the surface of endomycorrhizal roots. These vesicles and chlamydospores can be collected from soil through special washing and sieving techniques. They appear to be well distributed in the upper layers of soil.

It is not known how sporocarps are disseminated to infect additional plants. One can infest a soil by adding spores to soil or introduction of soil from an area where plants were recently grown. Planting cover crops such as grasses or legumes will also initiate a population of VA mycorrhizal fungi.

With ectendomycorrhizae initial invasion of root tips is like that of ectomycorrhizae, but eventually endomycorrhizal invasion of cortical cells occurs. There seems to be a progression involving initially ecto, then ectendo, and then primarily endolike invasion of lateral roots. Higher levels of soil fertility appear to favor ectendomycorrhizal associations, because this type of association is reported most commonly from relatively fertile conifer nurseries.

IMPORTANCE OF MYCORRHIZAE

Physiological and silvicultural aspects of mycorrhizae are important to foresters and others working with plants. Experiments in which seedlings are grown with and without the mycorrhizal fungi demonstrate the effect of mycorrhizae on plant growth and vigor. Plants with mycorrhizae are much larger and much more vigorous than nonmycorrhizal plants.

Mycorrhizal benefits to plants include (1) increased mineral and water availability resulting from nutrient exchange with the fungus, and increased root surface area; (2) resistance to root pathogens through mantle formation and reduced carbohydrate levels in the plant root; and (3) increased populations of nonpathogenic microorganisms in the rhizosphere of the plant, resulting from mycorrhizal activities.

It is apparent that mycorrhizal associations on plant roots are beneficial to the plant's well-being. Thus foresters, wishing to maximize timber and/or fiber yield in the shortest possible time, and other land managers, interested in the reclamation of highly disturbed sites, would be wise to investigate the effects of mycorrhizae on seedling and mature plant development as a natural means of increasing productivity and establishing new forests. Arborists should also consider mycorrhizae in the establishment and maintenance of ornamental plantings.

RECOGNITION OF MYCORRHIZAE

Ectomycorrhizae are rather variable in color and form. Some mycorrhizae are just single short stubs, others are the more characteristic fork-branched short roots, and others are coralloid clusters of branching roots.

Ectendomycorrhizae have similar gross morphological characteristics, but VA have no specific gross morphological characteristics. Detailed observations of mycorrhizae may require microscopic examination of cleared, stained, and sectioned material.

Proof of mycorrhizal association for specific fungi is accomplished in the laboratory using aseptically grown seedlings. A pure culture of the suspected mycorrhizal fungus is introduced into the rooting medium of the aseptically grown seedling, and the development or lack of development of a mycorrhizal association is evaluated after a period of time.

The usual method of obtaining pure cultures of fungi for such tests of mycorrhizal capacity is from basidiospores or from pieces of fungus tissue of fruit bodies observed in association with tree roots. If one attempts to culture from actual mycorrhizal roots, a different collection of nonsporulating mycorrhizal fungi and many common saprobic soil fungi are recovered.

Identifiable basidiomycete fungi are seldom cultured from mycorrhizal roots, but those found fruiting in the forest often will form mycorrhizae in laboratory cultures. A vast array of unidentified fungi cultured from roots form equally good mycorrhizae. What are the actual roles and importance of these groups of fungi? It can only be left as a question at this time.

EXAMPLES OF MYCORRHIZAE

Specific mycorrhizal associations are being exploited extensively now to enhance survival of conifers planted in highly deficient sites. This is just the beginning of a promising future for mycorrhizae in tailormaking seedlings for specific sites.

One fungus that has been studied in detail throughout the United States and in other countries is *Pisolithus tinctorius*. This puffball is commonly observed in association with and isolated from trees growing on very deficient sites, such as mine spoils. Other fungi including a number of species of *Rhizopogon* are also being evaluated with bare root and container grown planting stock.

Examples of Mycorrhizae   **131**

The fungi can be grown in quantity on a sterilized substrate and introduced into well-fumigated nursery beds at planting time. Seedlings in such infested beds develop the desired mycorrhizae rather than *Thelephore terrestris* or other mycorrhizae because the massive inoculum outcompetes the normal *T. terrestris* recolonization of the bed. Spore inoculation procedures are also being tested.

Survival and growth comparison of seedlings with *P. tinctorius* show striking superiority of the *P. tinctorius* seedlings in deficient sites but not on highly fertile sites. *Rhizopogon* spp. have been associated with improved drought tolerance.

ESTABLISHMENT OF MYCORRHIZAE ON SEEDLINGS

Nature usually does an acceptable job of bringing together compatible fungi and roots in mycorrhizal associations. Lack of mycorrhizae accounts for a limited number of examples of nonsurvival of conifers. One such example was the attempt to establish conifer nurseries in the central plains. Only after introduction of some soil from forested areas were the prairie nurseries able to produce acceptable seedlings.

Natural infestation of fumigated nurseries from spores produced on fruit bodies around the nursery has been taken for granted. But there are potential advantages in controlling the types of fungi producing mycorrhizae. Introduction of mycorrhizal fungi inoculum appears to work if the fumigation is properly done and the moisture and temperature conditions are favorable for the fungus . Some work is going into developing less bulky inoculum by using spores rather than mycelium infested substrate.

Because the topic of mycorrhizae is just starting to get the practical attention it deserves, we will probably see additional methods for controlling various types of mycorrhizal fungi. Maybe we will learn to recognize good mycorrhizal sites for establishment of nurseries or maybe we will learn to modify present nursery sites to exploit appropriate natural inoculum.

Establishment of mycorrhizae on ornamental plant materials has not been a concern of the producers of such materials. One might speculate that concern and research on the topic of mycorrhizae could improve the survival and development of some of the ornamental materials presently being utilized.

Mycorrhizae and their role in agricultural production is presently a fruitful topic of research. With increasing costs of fertilizers, the use of mycorrhizal inoculation to improve plant vigor has been considered.

REFERENCES

BRUNDRETT, M.C., Y. PICHE, and R.L. PETERSON. 1985. A developmental study of early stages in vesicular-arbuscular mycorrhizae formation. Can. J. Bot. *63*:184–194.

CASTELLANO, M.A., and J.M. TRAPPE. 1985. Ectomycorrhizal formation and plantation performance of Douglas-fir nursery stock inoculated with *Rhizopogon* spores. Can. J. For. Res. *15*:613–617.

CASTELLANO, M.A., J.M. TRAPPE, and R. MOLINA. 1985. Inoculation of container-grown Douglas-fir seedlings with basidiospores of *Rhizopogon vinicolor* and *R. colossus*: effects of fertility and spore application rate. Can. J. For. Res. *15*:10–13.

GERDEMANN, J.W., and J.M. TRAPPE. 1974. The Endogonaceae in the Pacific northwest. Mycol. Mem. 5. 76 pp.

HACSKAYLO, E. 1971. Mycorrhizae. USDA For. Serv. Misc. Publ. 1189. 255 pp.

HACSKAYLO, E. 1972. Mycorrhizae: the ultimate in reciprocal parasitism. BioScience *22*:577–582.

HARLEY, J.L., AND S.E. SMITH. 1983. Mycorrhizal symbiosis. Academic Press, Inc. (London) Ltd., London. 483 pp.

HATCH, A.B. 1936. The role of mycorrhizae in afforestation. J. For. *34*:22–29.

HEPTING, G.H. 1971. Diseases of forest and shade trees of the United States. USDA For. Serv. Agric. Handb. 386. 658 pp.

MARKS, G.C., and T.T. KOZLOWSKI, eds. 1973. Ectomycorrhizae: their ecology and physiology. Academic Press, Inc., New York.

MARX, D.H. 1972. Ectomycorrhizae as biological deterrents to pathogenic root infections. Annu. Rev. Phytopathol. *10*:429–454.

MARX, D.H., and W.C. BRYAN. 1975. Growth and ectomycorrhizal development of loblolly pine seedlings in fumigated soil infested with the fungal symbiont *Pisolithus tinctorius*. For. Sci. *21*:245–254.

MARX, D.H., W.C. BRYAN, and C.E. CORDELL. 1977. Survival and growth of pine seedlings after two years on reforestation sites in North Carolina and Florida. For. Sci. *23*:363–373.

MILLER, O.K., JR. 1982. Taxonomy of ecto- and ectendomycorrhizal fungi. *In* Methods and principles of mycorrhizal research, ed. N.C. Schenck. The American Phytopathological Society, St. Paul, Minn., pp. 91–101.

MOSSE, B. 1973. Advances in the study of vesicular-arbuscular mycorrhizae. Annu. Rev. Phytopathol. *11*:171–196.

PARKE, J.L., R.G. LINDERMAN, and C.H. BLACK. 1983. The role of ectomycorrhizae in drought tolerance of Douglas-fir seedlings. New Phytol. *95*:83–95.

SINCLAIR, W.A. 1974. Development of ectomycorrhizae in a Douglas-fir nursery. II. Influence of soil fumigation fertilization and cropping history. For. Sci. *20*:57–63.

SLANKIS, V. 1974. Soil factors influencing formation of mycorrhizae. Annu. Rev. Phytopathol. *12*:437–457.

TRAPPE, J.M., and N.C. SCHENCK. 1982. Taxonomy of the fungi forming endomycorrhizae. *In* Methods and principles of mycorrhizal research, ed. N.C. Schenck. The American Phytopathological Society, St. Paul, Minn., pp. 1–9.

WILCOX, H.E. 1983. Fungal parasitism of woody plant roots from mycorrhizal relationships to plant disease. Annu. Rev. Phytopathol. *21*:221–242.

WILCOX, H.E., and R. GANMORE-NEUMANN. 1974. Ectendomycorrhizae in *Pinus resinosa* seedlings. I. Characteristics of mycorrhizae produced by a black imperfect fungus. Can. J. Bot. *52*:2145–2155.

10

FUNGI AS AGENTS OF TREE DISEASES: FOLIAGE DISEASES

- *TYPES OF FOLIAGE DISEASES*
- *MODE OF ACTION OF FUNGUS-CAUSED FOLIAGE DISEASES*
- *DISEASE CYCLE OF A FUNGUS FOLIAGE PATHOGEN*
- *SYMPTOMS OF FOLIAGE DISEASES CAUSED BY FUNGI*
- *METHODS FOR RECOGNITION OF SPECIFIC FOLIAGE DISEASES*
- *EXAMPLES OF FOLIAGE DISEASES CAUSED BY FUNGI*
- *CONTROL OF FOLIAGE DISEASES CAUSED BY FUNGI*

Foliage diseases are of concern because of their effects on photosynthetic activity and, therefore, on the growth of the plant. But the effects of reduced photosynthetic activity on plant vigor may influence the susceptibility of the plant to invasion by other fungi and insects, thereby compounding the potential losses. Another effect of foliage diseases that is difficult to quantify is the psychological trauma caused by such an obvious disorder on the homeowner with a limited number of well-cared-for plants.

Foliage diseases are caused by a wide array of biotic and abiotic agents, such as fungi, insects, mites, air pollutants, viruses, mycoplasmas, bacteria, nematodes, and other factors (Fig. 10-1).

Foliage diseases caused by fungi are sometimes separated into groups based on the taxonomic position of the causal fungus. Basidiomycete rust fungi represent one group that will be discussed in Chapter 11. This chapter will emphasize foliage diseases caused by Ascomycotina fungi or imperfect stages of Ascomycotina.

A foliage disease and tuber rot of potato, late blight, caused by the Oomycete fungus *Phytophthora infestans*, produced one of the most devastating plant disease epidemics in recorded history. But foliage diseases of trees caused by Oomycete fungi are almost unknown in the temperate forests.

Sooty mold is one common foliage problem that does not fit properly within the typical categories of this chapter or the rust chapter. This situation occurs whenever sucking insects such as aphids feed on plant foliage. The sucking insects produce sticky exudate secretions. A number of black fungi in the Ascomycotina and Fungi Imperfecti groups develop on the secretions. The black color of these fungi may affect the marketability of Christmas trees or may be a nuisance to the homeowner when the sooty mold develops on the lawn furniture or automobile that is parked under insect-infested broadleaf trees. General distribution of a black sooty appearance over foliage, branches, and anything in the vicinity is a diagnostic indicator of sooty mold. The obvious way to handle the problem is to control the insects.

Figure 10-1 Nipple gall mite on sugar maple.

Types of Foliage Diseases

MODE OF ACTION OF FUNGUS-CAUSED FOLIAGE DISEASES

Some fungi that parasitize leaves produce intercellular or surface mycelium which extracts nutrients from living cells through intracellular haustoria (Fig. 10-2) that form within the cell wall but do not rupture the plasma membrane. Nutrients are absorbed through the plasma membrane of the plant cells into the haustorium. Parasitism of this type usually does not kill the cells immediately but rather causes the infected cells to act as a pumping system for nutrients from the plant. If one traces movement of elements by introduction of radioactive phosphorus or carbon to a plant, one finds that elements are preferentially translocated to the infection site.

Other fungi parasitize leaves by direct penetration of parenchyma and mesophyll cells and extraction of available nutrients. Rapid death of cells occurs.

The leaves being parasitized react with specific chemical responses to both types of infections. Studies on host response to foliage infection and invasion show specific and genetically controlled reactions.

DISEASE CYCLE OF A FUNGUS FOLIAGE PATHOGEN

The disease cycle of a typical Ascomycotina foliage pathogen of trees involves the production of ascomata either during the fall or early spring on infected needles or leaves. Ascospores are shot from these ascomata about the time that new foliage is coming out in the spring. Those few infections that take place produce large numbers of conidia which intensify the infection. The fungus generally overwinters as mycelium or ascomata in diseased leaves. Some foliage pathogens invade small branches and therefore overwinter in the infected shoots. Inoculum produced from

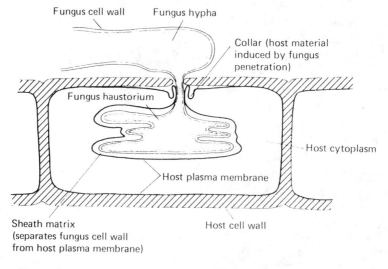

Figure 10-2 Relationship between the fungus haustorium and the host cell.

these shoots effectively distributes the pathogen to newly emerging foliage in the spring.

Ascomycotina fungi are well suited as foliage pathogens because they generally have a sexual ascospore stage and an asexual conidial stage. The ascospores are formed in asci inside some type of ascomata. Mycelium in infected tissue or the ascomata may provide the fungus with a stage for survival during adverse environmental conditions such as winter. The ascospore or sexual stage also allows for genetic recombination to take place. Genetic recombination is essential for diversity in the parasite, and diversity is important for parasitic virulence on a host with wide genetic variability and for the ability to survive under varying environmental conditions.

The asexual spore stage provides the fungus with the capacity to rapidly increase its numbers once a virulent ascospore infects a susceptible host. In the immediate area of infection, thousands of vegetatively produced conidia are formed. These need disperse only a short distance to find an infection point on susceptible tissue—often the same host. The selection and increase of the more virulent lines provide foliage diseases with explosive potential for causing disease epidemics.

Ascomycotina fungi, except for mildews, can survive both as parasites and as saprobes. This is necessary for survival on fallen or dead leaves of the host during the dormant season. The obligate parasitic mildews survive as cleistothecia rather than hyphae in dead host material.

Some Ascomycotina simplify their life style by infrequently utilizing the sexual stage. Therefore, many foliage pathogens are better known by their Fungi Imperfecti names.

For foliage pathogen disease buildup, wet conditions are necessary. Moisture is required for the triggering of ascospore or conidia release as well as for germination and infection. Foliage diseases are most serious during years when cool, wet conditions occur.

The infection process of foliage fungi is by direct penetration of the leaf surface or through stomates (Fig. 10-3). The diversity of structure and chemical composition of the cuticle and leaf cell wall prevents direct penetration by all but a few fungi. The foliage fungus must penetrate these layers either by force or enzymatic action or by both to get inside the leaf, where it derives nutrients by causing cells to leak materials into the intercellular spaces or by intracellular invasion and nutrient absorption from the cytoplasm of living cells.

SYMPTOMS OF FOLIAGE DISEASES CAUSED BY FUNGI

Infection by foliage fungi can result in total necrosis and shriveling of leaves or discrete localized necrotic areas. Massive infection or invasion at yearly intervals may cause the tree to shed its leaves and produce a second crop. Reduced terminal growth and tip dieback sometimes occur. Adventitious buds may produce new shoots and foliage, forming a clustering of branches into witches' brooms. Associated with necrotic symptoms, one usually finds signs of fungus fruiting structures.

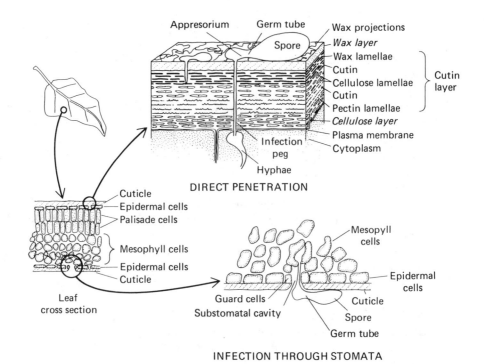

Figure 10-3 Infection by foliage fungi is by direct penetration or through stomates.

METHODS FOR RECOGNITION OF SPECIFIC FOLIAGE DISEASES

Foliage problems caused by fungi are not always easily distinguished from those caused by insects, mites, air pollutants, virus, mycoplasmas, bacteria, nematodes, and other causes. Insect problems of foliage are usually characterized by some type of distinctive feeding symptom, the presence of larva or other insect stage, or the fecal pellets of former insect activity. Mites sometimes cause distinctive color and morphological changes, particularly in hardwoods. In conifers the symptoms of mites are subtle and difficult to recognize with certainty. Shortening and curling of some of the needles in a fascile and chlorosis is sometimes associated with mites. Air pollutants produce flecking that resembles the early symptoms of some fungal diseases. The hypersensitive reaction of a virus infection results in flecking. Other viruses cause irregular ring spots. Mycoplasma-induced chlorosis and witches' brooming of twigs may occasionally resemble fungal diseases. Bacterial leaf spots are distinguished from fungal diseases based on the presence of bacteria as seen with dark-field illumination of the microscope or by isolation on agar media. There are very few nematode diseases of foliage, and these are readily diagnosed by the presence of the nematode. Other natural phenomena, such as normal fall coloration and drop of conifer needles, or man-made problems resulting from phytotoxicity of chemicals applied to plants, are sometimes confused with foliage diseases.

Foliage diseases may be confused with late spring frosts. Frosts occur on leaves throughout the crown or in the upper crown of trees, while foliage diseases are usually more serious in the lower crown. The microclimatic conditions, particularly high relative humidity of the lower crown, favor infection by foliage pathogens.

Injury caused by rapid temperature changes during the winter may also be confused with foliage disease (see Chapter 3). If the air temperature stays well above freezing for a number of days in January or February, the trees start to come out of dormancy. If a rapid and dramatic drop in temperature follows this warm period, freezing injury occurs. Dead foliage is seen associated with injured portions of the tree in early spring. By midsummer, these dead needles will be invaded by saprobic fungi. These fungi may sporulate, thereby confusing the issue as to the cause of the problem.

Fungus induced foliage diseases usually produce distinctive symptoms and signs for easy identification. It is often necessary to make a freehand section and observe the fungus microscopically to make a positive identification. Using a fungus-host index, it is usually possible to identify the disease.

A good way to identify a biologically induced foliage problem from a non-biologically induced problem is to look at the transition zone between diseased and healthy tissue. An invading biological agent will usually induce a gradual change or at least a different-colored zone line between healthy and diseased tissue. With nonbiotic agents the transition is more often very abrupt.

If one cannot find reference to a previous association of the observed fungus and host in the literature, it is advisable to prove pathogenicity using Koch's postulates. Some foliage fungi are obligate parasites, so that growth on culture media may not be easily accomplished. In this case it is necessary to inoculate with spores obtained directly from a host plant.

EXAMPLES OF FOLIAGE DISEASES CAUSED BY FUNGI

Hardwood Foliage Diseases

Foliage diseases have traditionally been assumed to be of little importance to hardwoods except where they are used as ornamentals. As an ornamental, any defect that reduces the appearance quality of a tree is of importance, not so much to the tree but to the owner or observer of the tree.

The reason foliage diseases of hardwoods are seldom important is that the tree appears to produce much more leaf surface area than it can efficiently fully utilize for photosynthetic production.

Foliage diseases can be expected to increase in importance with intensive management of hardwoods. In Europe, *Melampsora* rusts and *Marsonina* leaf spot are some of the most serious problems for their extensively used poplar varieties. If approximately 60% of the leaf area is destroyed by either fungus or insects, growth reduction will occur and sometimes another crop of leaves will be formed.

If defoliation takes place early in the season, there may be a slight overall reduction in growth that year, but the tree will refoliate and generally suffer no consequences from the defoliation. But if defoliation occurs during the middle part of the summer, the tree will attempt to refoliate and may not become fully hardened-off for cold weather when fall comes. Under these conditions, there may be injury to the next year's buds by frost and an overall reduction in food reserves stored in the plant for initiation of growth the next year. Late summer or fall defoliation in the absence of refoliation does not seriously affect food reserves, so that refoliation rather than defoliation is the serious aspect of the problem. Reduction in food reserves seems to predispose a tree to attack by *Armillaria*, root-infecting fungi discussed in Chapters 16 and 18.

The fungi that cause foliage diseases will often parasitize young shoots and twigs. The twig and shoot blight foliage fungi are very similar to canker fungi in their interaction with the host cambium. Twig cankers and dead shoots sometimes provide a method of overwintering for the fungus.

Hardwood foliage diseases are loosely organized into various groups, such as leaf spot, scab, anthracnose, blotch, powdery mildew, leaf blister, tar spot, rust and sooty mold. The first four grade one into another, with the leaf spot and scab used for distinctive circular spots and the anthracnose and blotch used for irregular-shaped necrotic areas. The next four leaf disease types are individually quite distinct, as is noted below. Sooty mold, as has already been discussed, is not really a foliage disease. Sooty mold is caused by black fungi colonizing secretions from insects on the surface of the plants.

The foliage diseases presented below are representative examples of various foliage disease types. For more detailed information on specific foliage diseases in any particular area, one should consult local sources such as bulletins from federal and state agencies.

Septoria leaf spot and canker. *Septoria musiva* causes a foliage and canker disease of eastern cottonwood. The pathogen is of minor concern except when cottonwoods and hybrid poplars are grown in intensive culture.

The future for intensive culture for poplars in the United States looks rather promising. The trees grow very rapidly on good sites, producing up to 2.5 cm of radial increment per year. They can be readily propagated as clones by cuttings.

Septoria foliage and canker problems have been recognized on clones of both cottonwood and hybrid poplars. Growth loss is related to disease intensity. Variation in susceptibility suggests possible genetic control of resistance.

The easiest thing to do is to select a resistant plant for production of cuttings, but this type of practice will lead to disaster. It is important to recognize how the resistance is inherited in the host as well as the pathogen before moving into any improvement program. We have often seen resistant varieties of agricultural crops eventually destroyed by devastating diseases. Selection for resistance generally narrows the genetic variability. As discussed in Chapter 20, it is important to maintain genetic variability to avoid rapid disease development.

Another way to minimize loss due to *Septoria* and other similar diseases is to evaluate the effects of growing site and stand density on disease development. Environmental conditions influencing both the plant and the pathogen may be utilized to advantage in reducing disease losses. Unfortunately, initial impressions that *Septoria* was less of a problem in wider-spaced plantings may not provide an answer for the management of susceptible poplar clones. The stem canker phase of this pathogen causes stem breakage and wider spacings do not seem to reduce stem cankering to acceptable levels.

Phyllosticta leaf spot.   Phyllosticta leaf spots are characterized by circular yellowish-brown necrotic areas surrounded by a darker sometimes purple boundary (Fig.10-4). The disease is sometimes called shot hole because the necrotic centers of the spots sometimes fall away leaving holes in the leaves. Black pycnidia of the causal agent form in the necrotic area. The Ascomycotina stage of *Phylosticta* is generally not known, so infection requires conditions conducive for dispersal of rain splashed conidia.

Superficial examination of leaf spots on red maple and other trees may confuse this disorder with one caused by the insect *Cecidomyia ocellaris.* The larval stage of the midge can sometimes be found inside the leaf spot but later in the season only a cavity remains. This situation points out the need to examine foliage problems carefully for signs of insects as well as fungi.

Anthracnose.   Anthracnose has been a convenient but artificial pool of a number of different foliage diseases. One group of anthracnose diseases are caused by Ascomycotina fungi in the genus *Apignomonia (Gnomonia).* These fungi have an

Figure 10-4 Shot hole disease of red maple.

Examples of Foliage Diseases Caused by Fungi

imperfect stage in the genus *Discula (Gloeosporium)*. In this group are *A. veneta* of sycamore, *A. guercina* of oaks, and *A. errabunda* of ash.

An anthracnose of flowering dogwood caused by a *Discula* sp. is currently a problem of major concern for the survival of dogwood. The disease produces flower spots, leaf spots, twig dieback, and mortality of forest and ornamental dogwoods.

Another group of anthracnose problems are caused by *Kabetiella* spp. of the Fungi Imperfecti. The perfect stage of these is unknown or rare. *Kabetiella apocrypta* is the anthracnose pathogen of maples. Both the *Discula* and *Kabetiella* fungi produce small single-celled conidia in acervuli. Typical symptoms are irregular dead blotches that merge together to cause death of most of the leaf and often result in defoliation (Fig. 10-5 and 10-6). If defoliation does not occur, the leaves remain distorted. On sycamore and oaks, shoot infection causes the death of twigs and the emergence of lateral buds to form witches' brooms on the heavily infected trees. Probably, very little damage occurs in natural stands of hardwoods, but with future intensification in the planting and management of hardwoods, we can expect foliage diseases to develop to the point where more effort will be necessary to reduce growth losses. The best approach with intensively managed hardwoods may be to select for genetically controlled resistance.

Figure 10-5 Antracnose disease of bur oak results in almost complete defoliation early in the season.

Figure 10–6 Close-up of anthracnose-affected bur oak foliage.

Powdery mildew.   Powdery mildew is another common foliage disease of many trees, shrubs, agricultural crops, and grasses. In trees and in shrubs the disease often develops late in summer or fall and does not cause significant damage to the plants. A characteristic white cast is seen on the leaves due to the mycelium of the fungus growing entirely on the outer surface (Figs. 10–7 and 10–8). Black cleistothecia often appear as distinct dots on the white surface.

The powdery mildew fungi produce distinctive appendages on the cleistothecia. This makes specific identification of the fungus very easy, as can be seen from the following key to the genera of fungi causing powdery mildews.

1. Cleistothecial appendages simple hyphae
      2. One ascus per cleistothecia   *Sphaerotheca* spp.
      2. Several asci per cleistothecia   *Erysiphe* spp.

1. Cleistothecial appendages branched or bushlike at tips
      3. One ascus   *Podosphaera* spp.
      3. Several asci   *Microsphaera* spp.

Examples of Foliage Diseases Caused by Fungi   **143**

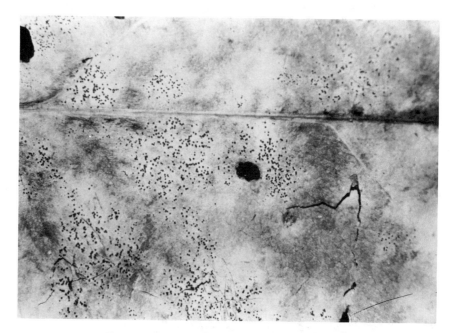

Figure 10-7 Mildew cleistothecia on a leaf surface.

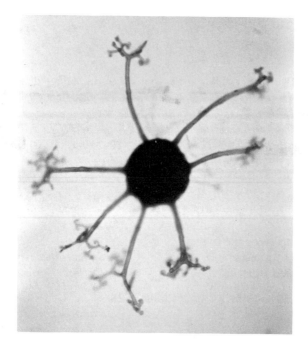

Figure 10-8 Close-up of single *Microsphaera* sp. cleistothecium.

1. Cleistothecial appendages curled at tip, several asci

Uncinula spp.

1. Cleistothecial appendages with swollen bases, several asci

Phyllactinia spp.

The powdery mildew fungi are unique in that the mycelium does not penetrate the host. Specialized nutrient-absorbing structures called haustoria are inserted into the leaf cells but do not break through the plasma membrane, so that the cell remains functional. The haustoria absorb nutrients from the living cell through the plasma membrane to supply the mycelium and sporulation on the surface of the leaf.

Another unique feature of many of the mildews is their ability to germinate and infect leaves in the absence of free water. Mildews can therefore be problems on crops in arid regions.

Because mildew generally develops late in the year, control is not warranted. If a problem occurs earlier in the season, fungicide sprays can be effectively utilized on high-value trees. The mycelium on the plant surface is readily killed with fungicides.

Leaf blister. Leaf blister, common on oaks (Fig. 10–9), is caused by *Taphrina caerulescens*. *Taphrina* is a primitive Ascomycotina that produces its asci on the surface of the host in no specialized ascomata structure. The eight ascospores

Figure 10–9 Leaf blister disease of red oak.

often undergo additional mitotic divisions, so that the ascus may have many ascospores (Fig. 10-10). This fungus differs from the typical foliage pathogen in that the ascospores are shot in the fall. The spores overwinter on the bark and bud scales of the tree and are washed onto the newly emerging foliage in the spring. A single infection period occurs each year. Therefore, a strong fungicide applied during the dormant season will eliminate much of the potential disease inoculum for the next season.

The *Taphrina* fungus produces disruption of cell division and enlargement,so that a distorted blistered leaf is produced. Some of the *Taphrina* spp. infect flower parts and petioles of alder and other trees and produce enlargement and distortion of these parts.

Tar spot of maple. The tar spot disease (Fig. 10-11) is one of the easiest diseases to identify and remember. The name is very appropriate. The tar-like spot is fungus stromata, which develops during the summer on infected leaves. In the spring apothecia are formed in the stromata on fallen leaves. Cracks expose the apothecia and allow the spores to be ejected.

Two species of *Rhytisma* cause tar spots on maples. *R. acerinum* produces individual large tar spots. *R. punctatum* produces clusters of small tar spots.

The incidence of the disease varies considerably from one year to another probably in response to climatic conditions. Black tar spots on maples can be rather disturbing to people but probably have little importance for the development of the trees.

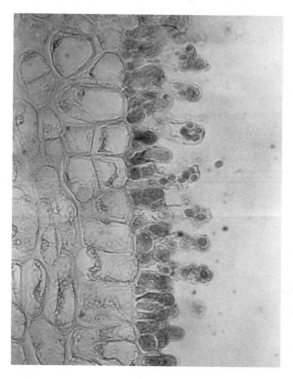

Figure 10-10 Asci and ascospores of *Taphrina* sp. found on the surface of the blistered leaf as seen in the microscopic section.

Figure 10-11 Tar spot disease of red maple. (Photograph compliments of Dr. Wayne Sinclare.)

Conifer Foliage Diseases

Foliage diseases of conifers are important because, when severe, they reduce growth and even cause death. Conifers do not have the ability to refoliate, and therefore reduced photosynthetic surface causes the tree to become less and less vigorous.

Some of the most serious diseases of exotic conifers planted in various locations are foliage diseases. These diseases are often caused by fungi that cause minor damage in the natural range of the species but, when introduced into a new area, develop very rapidly to limit the potential of the exotic tree species.

Dothistroma blight. *Dothistroma septospora (D. pini)* is an example of a damaging foliage disease to exotic Monterey pine (*Pinus radiata*) grown in the southern hemisphere. The fungus can infect over 30 pine species as well as Douglas-fir and European larch. In North America, the disease occurs on a number of pines in the west but is confined to young plantations and ornamental plantings of Austrian and ponderosa pines in the east. *Dothistroma* is the asexual stage of an Ascomycotina fungus *Mycosphaerella pini (Scirrhia pini)* which is reported only from the west. The story behind the *Dothistroma* disease is most interesting because it demonstrates the type of problem encountered in attempting to maximize forest production.

Plantings of Monterey pine were first initiated near the end of the nineteenth century in Chile, Africa, New Zealand, and Australia. Excellent growth and form in these locations encouraged additional plantings and the development of lumber and paper industries, particularly in Chile and New Zealand.

Examples of Foliage Diseases Caused by Fungi

No major diseases occurred on these plantings until 1957, when the *Dothistroma* blight became an important threat. *Dothistroma* was known from native Monterey pine in California but only as a minor pest. Examination of historical material documents the presence of the fungus as early as 1940 in Chile, parts of Africa, and New Zealand, but the extent of the problem at that time apparently did not warrant serious concern.

No one knows what triggered the explosive epidemic of *Dothistroma* in Chile, New Zealand, and East Africa, but not in Australia and South Africa. It could be associated with a new virulent race of the fungus or particular weather conditions. In any case, the extensive plantings were ideal for the rapid development of a major problem.

An example of the explosive potential for this disease is shown in a study in California. Foliage infection of Monterey pine increased from 2.7% to 94% in 2 years, and mortality increased from 0.2% to 67.6%.

Disease increase is directly related to climatic conditions. Cool, moist conditions favor dispersal and infection by the pathogen. Seedlings and saplings are most susceptible to the disease and are often killed. Larger trees experience growth reduction in relation to the degree of foliage infected.

Excellent control is obtained with the use of copper fungicides. In New Zealand, fungicides are applied to large areas with aircraft on young plantations in three or four operations over a period of 15 years. After 15 years the trees have sufficient natural resistance to maintain themselves.

One might use this disease epidemic as an example of the difficulties encountered in maximizing forest production. Problems of this type might exemplify the need for managing natural forest systems. Actually, there is more to be learned from the *Dothistroma* problem than "don't fool with Mother Nature."

It is generally accepted that natural forest management systems are less likely to be upset by epidemic pest problems. The real question is whether the natural forest can supply the wood needed by our industrial civilization. Just as with agricultural production, it will be necessary to intensify wood production to keep up with demand. Just as with agriculture, the intensification generates new sets of problems. If the potential for problems is recognized and anticipated, disaster can be avoided or minimized.

In New Zealand, pathologists were conscious of the potential for this type of disease. Survey and research efforts by pathologists anticipated the problem. They have made spray recommendations to accommodate this disease in the short run. Over the long rung, they are also evaluating resistance and other management practices.

The *Dothistroma* problem is not a disaster; it is an example of the need to integrate pathology into intensive forest management.

Elytroderma needle blight.   Elytroderma needle blight of Ponderosa pine is caused by *Elytroderma deformans*. This disease produces a conspicuous reddened foliage in the spring. On the needles are seen elongated black fruiting structures of the fungus. Repeated or systemic infection of trees causes a clustering of branches called witches' brooming. The tips of branches turn upward.

This native pathogen appears to intensify due to specific environmental conditions. During wet climatic cycles, the disease intensifies. In areas where high moisture occurs because of local topography, localized outbreaks are common. Disease centers persist during dry periods because the fungus is able to parasitize and survive in twigs and buds. Therefore, the fungus can persist and expand its influence whenever environmental conditions are favorable.

Brown spot needle blight.   Another foliage disease of a native conifer caused by a native pathogen is brown spot needle blight of longleaf pine. *Mycosphaerella dearnessii* (*Scirrhia acicola*), the pathogen, infects needles of longleaf pine seedlings, reduces the growth of the seedlings, and prolongs the grass stage of the tree for 4 to 10 years longer than normal (Figs. 10–12 and 10–13). Once the tree starts height growth, which will not occur until the seedling has developed at least 1 inch of diameter growth at the ground line, the disease has very little effect on the tree.

This disease is typical of a disruption of natural balances caused by human attempts to utilize a plant species. Longleaf pine is very tolerant of fire even in the seedling stage, so natural fires and mixed ages of trees prevented serious buildup of the pathogen. Trees grown in nurseries are highly susceptible to the rapid spread of infection via the asexual stage. Even though chemical controls were applied in the nursery, some of the stock planted out was infected. This infected stock provided the initial source of inoculum for most plantations, which spread rapidly through the remainder of the plantation of uniform susceptible-sized trees.

Controlled burns of seedlings in the grass stage have provided some measure of control of the disease, but it is difficult to control the intensity of the fire necessary to reduce inoculum; it must not be so hot as to kill the terminal bud. One

Figure 10–12 Grass stage of longleaf pine infected by *Mycosphaerella dearnessii*. (Photograph compliments of Dr. Savel Silverborg.)

Examples of Foliage Diseases Caused by Fungi   **149**

Figure 10-13 Close-up of the brown spot disease on longleaf pine needles. (Photograph compliments of Dr. Savel Silverborg.)

often gets some reduction in stocking because of fire kill, particularly of the resistant trees that have started height growth. The amount lost to fire may or may not be about as great as would be lost due to the disease.

Another form of control for this disease may involve silvicultural stand manipulations. A shelterwood overstory of medium-density longleaf pine prevents seedling infection for up to 8 years. The disease is more serious on poor sites and on sites with minimum ground-cover density.

Therefore, it is possible to regenerate longleaf pine on good sites through a shelterwood cut. Removal of overstory within 8 years and a subsequent burn during the winter of the second season following the cut should regenerate a reasonable stand. The burn is merely to reduce infection buildup. It occurs at a time when needle litter and logging debris have deteriorated sufficiently to produce a moderate intensity burn.

Continued research on this problem has provided some major advances in the management of brown spot needle blight. The advances are in three areas.

1. The systemic fungicide, benomyl, when applied as a root dip treatment to seedlings as they come from the nursery, provides effective disease reduction for 3 years on the outplantings.
2. Stimulation of tree growth in outplantings, through weed control with shallow scalping of the planting site and the use of herbicides and supplemental

fertilization, reduces the length of time that the seedlings are in the disease vulnerable grass stage.

3. Improved seedling quality, through pollinations among superior longleaf pine lines, through hybridization with other southern pines, and by nursery management, provides top-quality nursery stock for optimum survival and growth in the field. Effective use of mycorrhizae also stimulates tree growth. Seedlings inoculated in the nursery with *Pisolithus tinctorius* (see Chapter 8) develop better root systems and are larger in stem diameter.

Integration of these three approaches provides a foundation for successful management of longleaf pine. This tree species is highly desired for lumber and poles, is resistant to fusiform rust, and is very effective at growing on dry coastal plain sites.

Christmas tree foliage diseases. Foliage diseases are a serious problem in eastern Christmas tree plantations. The Christmas tree plantation diseases sometimes begin in the nursery stock, and quickly spread throughout the uniform plantation.

Foliage diseases of current concern on Scots pine are Cyclaneusma needle cast [*Cyclaneusma minus (Naemacyclus niveus)*], Lophodermium needle cast (*Lophodermium seditiosum*), and brown spot needle blight [*Mycosphaerella dearnessii (Scirrhia acicola)*]. Cyclaneusma causes yellowish spots and eventually yellowing to browning of entire needles. Lophodermium symptoms start with a brownish band but quickly expand to total needle necrosis. Brown spot is characterized by distinctive brown bands with yellow margins on green needles. Eventually, the banding kills the needles. Severely affected trees by any of these three may be yellowed to brown in appearance. New growth each year may produce trees with green outer needles, but the thin green outer shell of needles does not produce very acceptable Christmas trees.

The fruiting bodies of each of these are quite distinctive. *Lophodermium* produces black elongate ascomata called hysterothecia that are somewhat embedded and flattened to the needle (Fig. 10–14). *Mycosphaerella* produces black conidial masses that stick out from the needle. *Cyclaneusma* produces off-white elongate apothecia embedded and flattened to the needle. The ascomata of *Lophodermium* or *Cyclaneusma* open up to release ascospores during wet periods. The conidia of *Mycosphaerella* are also splash dispersed during wet periods.

On Douglas-fir, another very popular Christmas tree, Rhabdocline needle cast (*Rhabdocline pseudotsugae* and *R. weirii*) causes red-brown bands and splotches on current-year needles and Swiss needle cast (*Phaeocryptopus gaeumannii*) causes yellowing and browning of older needles. Round-to-elongate apothecia of *Rhabdocline* fill the lesion area in the spring (Fig. 10–15). Two rows of black pseudothecia can be observed on the underside of needles affected by *Phaeocryptopus* (Fig. 10–16).

There are many other trees that are used for Christmas trees but the two noted above can serve as examples. With each there is more than one foliage problem. The

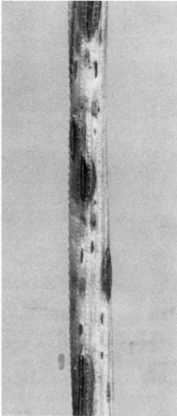

Figure 10-14 *Lophodermium seditiosum* hysterothecia on pine needles. (Photograph compliments of Dr. Wayne Sinclair.)

organisms that cause disease may be closely related to very similar saprobic fungi that do not cause disease, so very careful identification of the fungus is necessary. Control recommendations using chemical sprays are continually being updated to reflect the most recent information, so it is appropriate to check with the local forestry college or extension service to get the most current information. One general recommendation is to start with quality disease-free stock and reduce weeds and competing vegetation by mowing or herbicides. This last recommendation is aimed at reducing humidity on the lower branches of the trees. Foliage diseases in this type of crop are highly influenced by moisture conditions and therefore anything that can improve air movement in the stand can be expected to reduce the impact of diseases.

Snow molds and brown felt blight. These problems are uniquely associated with areas of heavy snow. They are particularly serious on conifer reproduction in mountainous areas but can also become a problem in northern conifer nurseries. The fungi are adapted to grow at temperatures just above the freezing point. Snow molds are recognized as a grey veil-like webbing seen on vegetation in the spring of

Figure 10-15 *Rhabdocline* apothecia on lower surface of Douglas-fir needles. (Photograph compliments of Dr. Savel Silverborg.)

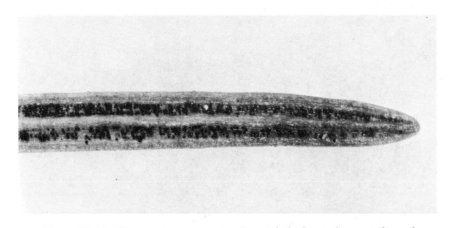

Figure 10-16 *Phaeocryptopus gaeumannii* pseudothecia on lower surface of Douglas-fir needles. (Photograph compliments of Dr. Savel Silverborg.)

Examples of Foliage Diseases Caused by Fungi

153

the year. The mycelium disappears as the season progresses. Brown felts produce a thick brown felt like hyphal growth that persists throughout the season.

Sphaeropsis tip blight. Although shoot blights could represent another class of disease, they have many features common to foliage diseases. The *Sphaeropsis sapinea(Diplodia pinea)* fungus, for example, infects through elongating shoots or needles by direct penetration, very much like a foliage pathogen.

The Sphaeropsis disease is reported for at least 11 species of hard and soft pine throughout the eastern and midwestern states. It is most serious on Austrian, Scots, mugo, ponderosa, and red pines, particularly where the species are used for ornamental plantings.

Killing back of the current year's growth detracts from the ornamental value of the trees (Fig. 10–17). Growth stunting and ragged appearance of the tree eventually reduce the ornamental value. Death of the trees is uncommon because they are usually removed before death occurs. Fungicidal sprays during the time of shoot emergence can be used to protect trees, but it is important to protect the elongating tip by more than one spray.

The evidence and impact of Sphaeropsis blight varies from one year to another, probably in response to environmental conditions. Drought conditions seem

Figure 10–17 Sphaeropsis-killed branch of Austrian pine.

to extenuate the symptoms, while wet conditions seem to favor spread and infection of the pathogen. The timing and interrelationship of these contrasting conditions limit the effectiveness of control activities for this disease.

CONTROL OF FOLIAGE DISEASES CAUSED BY FUNGI

A limited array of control procedures are known for foliage fungi of trees. In agriculture, some are controlled by fungicides and others by breeding disease-resistant varieties. Foliage diseases in Christmas trees and ornamentals can be reduced by the application of fungicides. Most treatments prevent new infections rather than eradicate older ones. Therefore, it is important to anticipate the problem and spray before infection has a chance to develop. With the present concern over pesticide pollution of the environment, it is appropriate to evaluate the benefits of application more closely rather than just spray. An example of evaluation procedures is discussed in Chapter 19.

Prevention by making proper species selection for specific sites is a much more appropriate control. Foresters and nurserymen are often more impressed by super growth or super form of an exotic than they are concerned regarding potential future problems. Based on past experiences on many continents, one can predict with reasonable accuracy that large-scale plantings of exotic trees will be seriously plagued, particularly by foliage diseases. Decisions must be made at the beginning of such an introduction program as to whether the desirability of the introduction, based on growth or form, is worth the increased risk and future costs associated with inevitable pest problems. (I say that decisions must be made, but I recognize that they are usually ignored even today by federal, state and private agencies developing and supplying new materials and methods.)

Control of foliage diseases by selection and breeding is possible but is not without problems. The subject of disease-resistance breeding is expanded in Chapter 20.

Control through site selection and manipulation will be possible choices once we learn how to quantify the qualitative elements of the environment that affect disease intensification. Quantification of qualitative elements affecting disease intensification in the late blight of potato and scab of apples shows that one can predict disease-intensification periods and therefore advise growers when to spray. The best example of disease control in trees through site selection is based on the studies of Van Arsdel on white pine blister rust. This topic is expanded upon in Chapter 11.

REFERENCES

BENYUS, J.M. 1983. Christmas tree pest manual. USDA For. Serv. North Cent. For. Exp. St. 108 pp.

BERRY, F.H., and W. LAUTZ. 1972. Anthracnose of eastern hardwoods. USDA For. Serv. For. Pest Leafl. 133. 6 pp.

BOWERSOX, T.W., and W. MERRILL. 1976. Stand density and height increment affect incidence of Septoria canker in hybrid poplar. Plant Dis. Rep. *60*:835–837.

BOYER, W.D. 1975. Brown spot infection on released and unreleased longleaf pine seedlings. USDA For. Serv. Res. Pap. 50-108. 9 pp.

CHILDS, T.W. 1968. Elytroderma disease of ponderosa pine in the Pacific Northwest. USDA For. Serv. Res. Pap. PNW-69. 45 pp.

CHILDS, T.W., K.R. SHEA, and J.L. STEWART. 1971. Elytroderma disease of ponderosa pine. USDA For. Serv. For. Pest Leafl. 42. 6 pp.

COBB, F.W., JR., B. UHRENHOLDT, and R.F. KROHN. 1969. Epidemiology of *Dothistroma pini* needle blight on *Pinus radiata*. Phytopathology *59*:1021–1022.

GIBSON, I.A.S. 1972. Dothistroma blight of *Pinas radiata*. Ann. Rev. Phytopathol. *10*:51–72.

HIBBEN, C.R., and M.L. DAUGHTREY. 1988. Dogwood anthracnose in northeastern United States. Plant Dis. *72*:199–203.

HOCKING, D., and D.E. ETHERIDGE. 1967. Dothistroma needle blight of pines. I. Effect and etiology. Ann. Appl. Biol. *59*:133–141.

KAIS, A.G. 1985. Recent advances in control of brown spot in longleaf pine. 34th Annu. For. Symp. Insects Dis. South. For. Louisiana State University, School of Forestry, Wildlife and Fisheries, Baton Rouge, La., pp. 83–90.

LIGHTLE, P.C. 1954. The pathology of *Elytroderma deformans* on ponderosa pine. Phytopathology *44:557–569.*

LIGHTLE, P.C. 1960. Brown-spot needle blight of longleaf pine. USDA For. Serv. For. Pest Leafl. 44. 7 pp.

MORTON, H.L., and R.E. MILLER. 1977. Rhabdocline needle cast in the Lake States. Plant Dis. Rep. *61*:801–802.

MORTON, H.L., and R.F. PATTON. 1970. Swiss needlecast of Douglas-fir in the Lake States. Plant Dis. Rep. *54*:612–616.

NICHOLLS, T.H., and D.D. SKILLING. 1974. Control of lophodermium needlecast diseases in nurseries and Christmas tree plantations. USDA For. Serv. Res. Pap. NC-110. 11 pp.

PETERSON, G.W. 1977. Infection, epidemiology, and control of Diplodia blight of Austrian, ponderosa, and Scots pine. Phytopathology *67*:511–514.

PETERSON, G.W., and D.A. GRAHAM. 1974. Dothistroma needle blight of pines. USDA For. Serv. For. Pest Leafl. 143. 5 pp.

SCHWEITZER, D.J., and W.A. SINCLAIR. 1976. Diplodia tip blight on Austrian pine controlled by benomyl. Plant Dis. Rep. *60*:269–270.

SINCLAIR, W.A., W.T. JOHNSON, and G.W. HUDLER. 1977. Anthracnose diseases of trees and shrubs. N.Y. State Coll. Agric. Cornell Tree Pest Leafl. A-2. 7 pp.

SKILLING, D.D., and T.H. NICHOLLS. 1974. Brown-spot needle disease biology and control in Scots pine plantations. USDA For. Serv. Res. Pap. NC-109. 19 pp.

WAKELEY, P.C. 1970. Thirty-year effects of uncontrolled brown-spot on planted longleaf pine. For. Sci. *16*:197–202.

11

FUNGI AS AGENTS OF TREE DISEASES: RUST DISEASES

- *TYPES OF RUSTS*
- *MODE OF ACTION OF RUSTS*
- *RUST DISEASE CYCLE*
- *SYMPTOMS OF RUST DISEASES*
- *DIAGNOSIS OF RUST DISEASES*
- *EXAMPLES OF RUSTS AND THEIR CONTROL*

Rust fungi produce major diseases of leaves and stems of agricultural as well as forest crops. Some aspects of the foliage infection by rust fungi are very comparable to other foliage diseases. Other aspects of the stem canker phase of rusts are very comparable to canker diseases. Why, then, do we discuss rusts as a separate group of diseases rather than include them in the foliage and canker chapters? Rust diseases are caused by a group of fungi all of which are obligately parasitic. Obligate parasites require living hosts for normal development. The fungi of the class Urediniomycetes of the subdivision Basidiomycotina have evolved a complex life cycle not found in any other group of fungi. The obligately parasitic complex life cycle, usually involving two very different plant hosts, results in a unique disease cycle worthy of separate discussion from other fungus diseases.

It is difficult to understand why a fungus should evolve such a complex life cycle as the rusts. Such complexity would require long-time mutual association of the two hosts with the pathogen, during which time both the hosts and the pathogen would experiment with numerous mutations and gene combinations for betterment of resistance or virulence positions. Increased resistance in the host would put a negative selection pressure on the pathogen, leading eventually toward extinction.

On the other hand, the pathogen with genes for increased virulence (ability to cause disease) would be at a selective advantage and, therefore, would increase in the pathogen population, but this would eventually result in a reduced host population. This evolution, involving the interrelationship of hosts and pathogens, eventually results in a standoff. Neither can gain a strong advantage over the other for any length of time. The result is a mutual tolerance of the host for a certain amount of parasitism and of the pathogen for a certain amount of inhibition in its invasion of the host. The trend is toward mutualistic symbiosis.

If rusts and their hosts have evolved a tolerance of each other, why are rusts a problem? There are three types of conditions under which rusts and other balanced disease relationships become a problem. One results from the *introduction of the fungus into a new area* where host plants, similar to those in the natural range, have not evolved genetic tolerance because they have never been exposed to the fungus. A plant that does not have the genetically evolved capacity to tolerate the fungus generally is not able to produce the correct defense mechanisms to prevent the fungus from rapidly invading. This would be similar to a small child finding a large jar of candy for the first time. The candy (the host) will become consumed and the small child (the pathogen) very sick in a short period of time. Another possibility resulting from this first introduction is extreme shock on the part of the plant to this new invader. The defense reaction mechanism may be so great that the plant kills invaded tissue and sometimes itself. In either case, we have a rapid destruction of the host plant as a result of a serious rust disease and also an eventual decrease in the pathogen. In the short run, the pathogen population initially increases very rapidly, but in the long term, the system should eventually come back to equilibrium.

Many of our most serious disease epidemics are associated with introduced pathogens. White pine blister rust, Dutch elm disease, and chestnut blight are classic examples.

The second way a balanced relationship can be upset is through our agricultural and forest practices (our eagerness to produce the maximum food or fiber per acre). Overenthusiasm with maximum production or selection of better-looking disease-resistant individuals for future crops eventually *reduces the genetic variability within the crop population*. The narrowing of genetic variability in relation to disease resistance places a selection pressure on the pathogen population for a mutant or gene combination with the capacity to overcome the resistance. It is just evolution speeded up. By the time the fungus comes forth with its champion there is often extensive acreage of the "resistant" variety. The variant strain, called a race, wastes no time in destruction of the crop, and we have a serious rust disease. In wheat, where races of *Puccinia graminis* have been looked for, researchers have found more than 200 different races of the fungus. Disease-resistance breeding is more complex than just selecting the resistant individuals. We discuss this topic in more detail in Chapter 20.

Extensive plantings of limited genetic variability are not as common in forest practice as they are in agriculture. But, present activities with hybrid poplar, sycamore, and other species could lead to reduced genetic variability and susceptibility to serious disease. The use of grafted or vegetatively propagated varieties in urban trees is increasing and can be expected to cause problems in the future.

The third way a rust can become a serious disease problem results from our utilization of new plants for the products we desire. We generally put the kiss-of-death on ecological balances as soon as we decide to utilize a particular plant. Changing timber and fiber species will bring to light a whole new array of problems, some of which will be rust diseases.

Fusiform rust is a good example of a disease problem caused by *changing ecological balances through forest management practices*. Fusiform rust is caused by a native pathogen which was a minimal problem in natural forests but is now a major problem of plantation forests.

TYPES OF RUSTS

The common name "rusts" was derived from the rusty appearance of the uredinial stage in the cereal rusts. The common name in current usage is applied to any fungus in the Urediniomycetes class of the subdivision Basidiomycotina.

Many rust fungi of trees initially infect and cause diseases of the foliage or succulent shoots. Some eventually invade the cambium and phloem regions of stems and branches. The latter group are referred to as stem rusts. Other rusts occur on cones.

Diseases of trees caused by rusts can result from infection by basidiospores, as in the case of the *Cronartium* stem rusts of pines, or from infection by aeciospores and urediniospores, as in the case of *Melampsora* leaf rust of poplars.

Alternate hosts of tree rust diseases can be herbaceous plants, woody shrubs, or other trees.

MODE OF ACTION OF RUSTS

Rust fungus hyphae penetrate between the cells of the host and derive nutrients through haustoria, as described in Chapter 10. The rust fungus is an obligate parasite dependent upon the living cells of the host for growth and development.

It was assumed for many years that obligate parasites such as rusts would not grow on culture media in the laboratory because they needed some unknown vital requirement which they got from living hosts. But plant pathologists have learned to grow rusts on general culture media enriched with peptone or yeast extract.

If rust fungi can be grown on media like other fungi, the necessity for an essential specialized nutrient does not account for host specificity or the need for living cells. The rust fungi, like the mycorrhizal fungi, probably have adapted themselves to elicit a minimal host chemical defense response which they can tolerate. The rust fungi avoid interference and competition for nutrients from other microorganisms that are held in check by the host chemical defense response.

Figure 11–1 depicts the stages in a typical heteroecious (two-host) long-cycled rust. Although rust fungi have a unique complex life cycle, the functions of the various stages are basically the same as those found in most fungi. The urediniospores stage (Fig. 11–2) of rust fungi is comparable to conidia produced by the Imperfect Fungi. Each spore has two nuclei ($n + n$). These asexual spores can reinfect the same plant or other plants of the same species to intensify the infection. Urediniospores are generally capable of withstanding the variable temperature and moisture conditions of long-range dispersal. The rust fungus can survive indefinitely in some instances, with no other spore stage. If we were to eliminate white pine from North America, we would still have a rust disease of gooseberries and currants because of urediniospores.

In the teliospore stage (Fig. 11–3), the two compatible nuclei fuse to form a diploid ($2n$) nucleus. The teliospores in the *Puccinia* genus also provide a dormant stage capable of overwintering in dry or cold climates.

At the time of nuclear fusion, the teliospore germinates to form a basidium. The diploid nucleus undergoes meiosis to produce four haploid ($1n$) nuclei. Each nucleus becomes separated from the others by the production of septa which segment the basidium into four cells. Each cell produces a basidiospore that contains a single haploid nucleus.

The basidiospores of rusts are very temperature- and moisture-sensitive. As

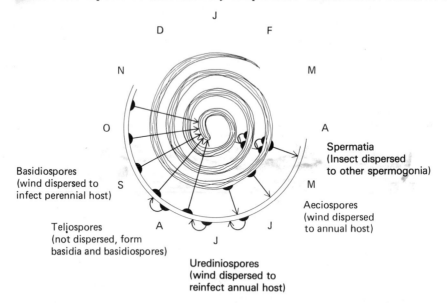

Figure 11–1 Life-cycle diagram of a *Cronartium* rust on a monthly calendar. Two hosts (═══ uredial host, ▧▧▧ aecial host) and the periods and direction of spore transfer ⬆ are depicted. Note the 3 years of development in the perennial host before production of inoculum to infect the annual host. (Modification of a life cycle diagram by Ziller, 1974.)

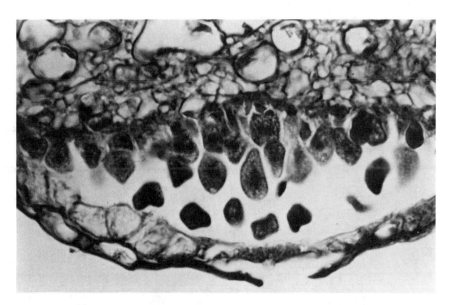

Figure 11-2 Photomicrograph of a section through a uredinium of *Cronartium ribicola* on a *Ribes* leaf.

such, they are produced only under specific environmental conditions and are disseminated short distances. They generally serve as a host-transfer stage. Basidiospores produced from germinating telia on one host will usually only infect the alternate host.

The haploid (l*n*) hyphae, produced upon germination of basidiospores, invade between the cells of the new host. Nutrients are obtained from living cells of the host without invading the cytoplasm. The metabolically active host cells allow a certain amount of nutrient to diffuse out to specialized feeder hyphae called haustoria, which penetrate the cell wall but not the plasma membrane.

Spermogonia (= pycnia) containing spermatia (= pycniospores) are produced from haploid hyphae. Spermatia do not germinate to infect plant tissue. Spermatia function as sperm to transfer genetic material from one fungus thallus to another. Spermatia are oozed from the spermogonium in a sweet exudate. Spermatia are transferred from one spermogonium to another by insects attracted to the exudate. Compatible spermatia fuse with specialized hyphae of the spermogonium and transfer a nucleus. The thallus then contains two genomes ($n + n$), which are maintained in two nuclei. This condition is called dikaryotic.

Aeciospores (Fig. 11-4) are produced from the dikaryotic hyphae and are specialized for host transfer. They are very similar to conidia but are unique in that they will not infect the host they are formed on. The dikaryotic hyphae, produced upon germination, invade the host as did the haploid hyphae. Urediniospores are formed from this hyphae and the cycle is completed.

To make things simple, urediniospores are formed in uredinia, teliospores are formed in telia or in telial columns, basidiospores are formed on basidia, spermatia are formed in spermogonia, and aeciospores are formed in aecia.

In summary, the typical rust produces two kinds of dikaryotic ($n + n$) conidia-

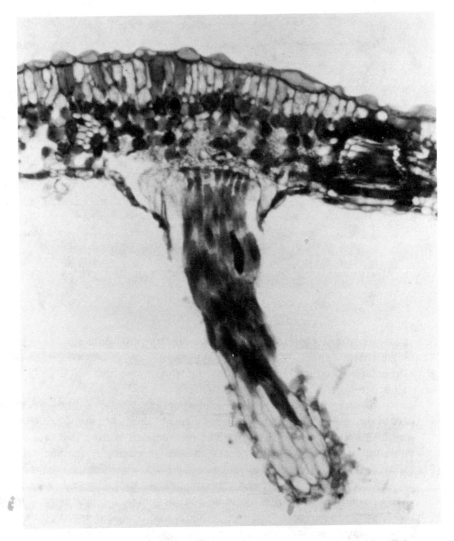

Figure 11-3 Photomicrograph of a section through a telium of *Cronartium ribicola* on a *Ribes* leaf.

like spores, aeciospores for host transfer and urediniospores for intensification of infection on one of the hosts. The rust produces two spores that do not function like spores, teliospores and spermatia. It also produces a sexual spore type, basidiospores. Teliospores function like the hymenium of the Hymenomycetes. In dikaryotic $(n + n)$ teliospores, nuclei fuse to form diploid $(2n)$ nuclei. Germination of the teliospores leads to the formation of basidia in which meiosis takes place. Four haploid (ln) basidiospores are formed on the septate basidium. Basidiospores produced from telia on one host generally infect the alternate host. The haploid (ln) mycelium resulting from infection by basidiospores produces haploid spermatia. Spermatia transfer nuclei to hyphae of compatible spermogonia. Nuclear division

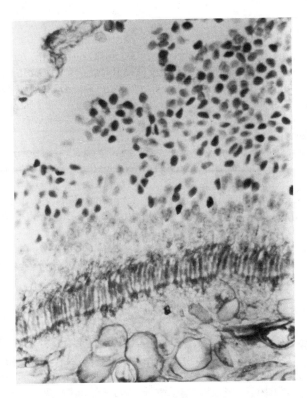

Figure 11-4 Photomicrograph of a section through an aecium of *Cronartium ribicola* on a white pine branch.

and migration of nuclei results in a dikaryotic ($n + n$) mycelium from which aeciospores are formed.

Some rust fungi shorten the five-stage life cycle by not producing one spore stage, as in the case of *Gymnosporangium* spp. (cedar apple rust). Others shorten and change the life cycle by using the host-transfer aeciospore stage for reinfection of the pine host, as in the case of *Peridermium* stem rust of pines.

SYMPTOMS OF RUST DISEASES

Rust diseases produce very distinctive aecia, uredinia, or telia signs on the infection lesion. In the absence of signs, rusts may look like many other foliage or stem disorders.

Rust specialists can recognize infection prior to sporulation. Color, shape, and size of the infection lesion when inoculated on specific host varieties are used as fingerprints for specific races of rusts.

Symptoms of stem rusts are more characteristic. The shapes of galls or cankers are somewhat characteristic. Remnants of spermogonial and aecial pustules around the margin of the canker are also specific signs for stem cankers.

Rusts reduce growth, yield, and sometimes cause death of parts or whole trees. Death of branches and tops from white pine blister rust can be easily recognized from a distance. The distinctive dead branches are called flags.

Symptoms of Rust Diseases

DIAGNOSIS OF RUST DISEASES

Diagnosis of rust diseases is based on identification of the causal organism. Arthur's *Manual of the Rusts in the United States and Canada* (1934) is the most comprehensive treatment of rusts for diagnostic purposes. *The Illustrated Genera of Rust Fungi* (Cummins and Hiratsuka, 1983) is an excellent recent treatment of the rust fungi. The book *The Tree Rusts of Western Canada* (Ziller, 1974) is another very useful reference for most tree rusts.

Rusts are obligate parasites, so one generally assumes that a rust sporulating on a diseased host is the cause of the problem. Proof of pathogenicity for rusts requires a modification of Koch's postulates. It may be necessary to find the alternate host to get basidiospore inoculum for testing pathogenicity. Although methods for growing rusts in culture are known, inoculation with a pure culture of a rust is not commonly done.

EXAMPLES OF RUSTS AND THEIR CONTROL

White Pine Blister Rust

White pine blister rust of eastern white (*Pinus strobus*), western white (*P. monticola*), sugar (*P. lambertiana*) and other five-needle pines, is caused by the fungus *Cronartium ribicola* (Figs. 11-5 and 11-6). It has been assumed that the pathogen came from Asia, that it was introduced into Europe by early plant collectors, and that it was introduced into both eastern and western North America from Europe on infected nursery stock of eastern and western white pines.

The introduction into North America on nursery stock is reasonably well accepted, but there may be some question on the origin of the white pine blister rust pathogen. In the European Alps *C. ribicola* appears to be a native pathogen on *Pinus cembra*. It also appears to be native on various pine species in the Himalayas, Yugoslavia, Korea, China, and Japan. In addition, recent work has identified other alternate hosts for the pathogen. Therefore, some of our assumptions need to be reconsidered.

It may seem peculiar that nursery stock for planting in North America with species native to North America should come from Europe. At the turn of the century, Europeans supplied both the skills and the materials for forest practice to the North American continent, where people were just beginning to recognize that exploitation could not continue indefinitely without some management.

The disease cycle of blister rust is shown in Fig. 11-7. Infection of pine needles occurs via basidiospores produced on *Ribes* bushes during late summer or fall (Figs. 11-8 and 11-9). A small yellow or orange fleck develops at the infection site on the needles within a few weeks. By the following spring the fungus hyphae have grown down the needle to the twig. It will persist in the cambium region until the branch or the tree dies. A year later spermogonia are produced on the infected twig. Aeciospores are formed on the same tissue during the spring of the next year (Fig. 11-10). Aeciospores infect the *Ribes* leaves. Urediniospores produced on the *Ribes*

Figure 11-5 Death of upper crown branches (flagging) of eastern white pine resulting from *Cronartium ribicola* infection of upper crown branches.

Figure 11-6 Blister rust stem canker on eastern white pine.

Examples of Rusts and Their Control

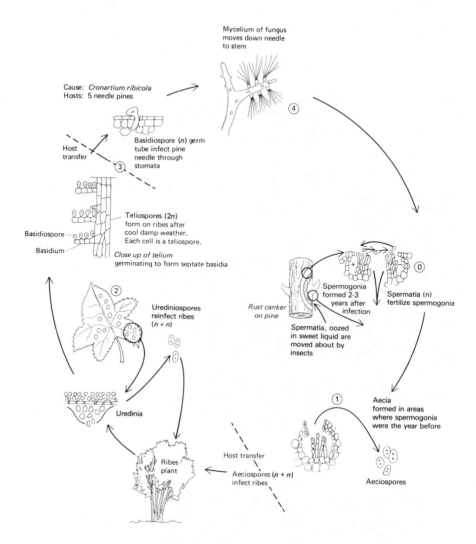

Figure 11-7 Disease-cycle diagram of white pine blister rust.

leaves increase the number of infections on *Ribes*. During the late summer and fall telial columns are found rather than uridinia. The teliospores of *Cronartium* fungi are cemented together in a column. Germination of the teliospore results in haploid basidiospores for infection of pines.

As with any introduced pathogen, it takes a number of years to recognize the new pathogen and assess the potential problem. By the middle of this century, white pine blister rust had spread throughout the ranges of the eastern and western white pine and sugar pine.

Major efforts were carried out by the U.S. Forest Service, state, and private concerns to control the disease by eradication of *Ribes*. The Forest Service began its control program to save the western white pine of the northern Rocky Mountains in

Figure 11-8 *Ribes* bush.

Figure 11-9 Close-up of undersurface of *Ribes* leaf, showing uredinial and telial spore stages of *Cronartium ribicola*.

Examples of Rusts and Their Control

Figure 11–10 Young blister rust stem canker showing oval-shaped remnants of spermogonial scars.

1930. By 1934, 11,000 workers were employed in the blister rust control program. Phasing out of the program began 30 years after it began and was ended completely by 1965.

In the east, large numbers of people were employed by the various states in blister rust control programs up until 1970. Federal monies supported this effort. By 1970, rust control funds were restricted to control efforts on only the highest-quality white pine stands.

Hindsight shows us that the control efforts were bound to be disappointing and causes us to ask why so much effort was spent so poorly. But before we criticize our predecessors for their lack of vision, we should recognize that human beings do not have perfect foresight. Rather than criticize, it would be appropriate for us to review the programs, to see both the successes and the failures, so as to enhance our own chances of success in combating other disease in the future.

A general principle widely accepted by pathologists is that if we understand the life cycle of a disease agent, we can pick out the weak link as a point to apply direct control. Many plant pathologists engaged in basic research on specific pathogens justify their work on this premise. The major blister rust control effort was directed at the weak link in the rust life cycle, the need for two hosts. Eradication of the alternate host *Ribes* spp. should eliminate the blister rust problem, since infection of pines occurs only from basidiospores produced by the fungus growing on *Ribes* plants. The principle, in this case applied to *Cronartium ribicola*, is correct, but in practice it did not work for much of the white pine range. Why did eradication of

Ribes work in only a limited number of cases? Initially, *Ribes* bushes were eliminated by digging and pulling. If any of the root was left, it would sprout to produce a new plant, so eradication often took two or three attempts. Later use of herbicides improved the number of plants killed with one application. Another difficulty in *Ribes* eradication involves the recolonization of a site through seeds. The seeds of *Ribes* plants can remain dormant for many years. Sites that were thought to be free of *Ribes* bushes can be recolonized from dormant seeds, especially following site disturbances caused by logging. *Ribes* eradication works only if it is thoroughly and repeatedly applied to stands of limited size. In the east, control of blister rust is effective if the size of the stand is small enough for thorough *Ribes* eradication.

In the late 1950s and early 1960s a program of blister rust control using antibiotic sprays gained momentum. Two materials, cycloheximide (Actidione) and, later, phytoactin were used. The latter chemical was applied in a mist from an aircraft, much as insecticides are applied to forest areas.

The aircraft method of application is about the only method that can be used for the application of chemicals to forest areas. Even this least expensive method is not entirely satisfactory, because it is difficult to get complete coverage of the trees. Streams and nontarget trees in the treatment area may also get sprayed.

Blister rust control through aerial application of chemicals was not abandoned because of problems associated with aerial application. Chemical control programs were abandoned because the chemicals did not control the disease. Why, then, were they used? Improperly interpreted results of experimental tests with the chemicals showed some control of blister rust. The Forest Service was painfully aware that the *Ribes* eradication programs were costly and not entirely satisfactory. The time was right for any new idea, so the promising chemicals were accepted with open arms. It was very difficult for the people involved to recognize the mistake and reverse the momentum of the control effort. Hopefully, in the future we will avoid premature acceptance of experimental results.

In the early 1960s, another method for the control of blister rust was developed, based upon an understanding of environmental factors affecting production, spread, and infection of pines by basidiospores. The studies of Van Arsdel at the Lake States Forest Experiment Station are a classic example of how an understanding of epidemiology of forest diseases can be used in making management decisions. Beginning with the general knowledge that climatic factors directly affect the development of blister rust, it was found that for successful infection of white pine, a period of 2 weeks of relatively cool late summer weather was needed to cause the rust to produce telia rather than urediniospores on the *Ribes* leaves. For the production of basidiospores from the telia, dissemination of viable spores to pine needles, germination of basidiospores, and infection of needles, 48 hours of continuous 100% relative humidity and temperatures below 68°F are required. One can separate geographical areas into infection hazard zones based on the frequency of these exacting climatic conditions. In the Lake States, the region surrounding the lakes frequently experience these conditions. It is almost impossible in this area to prevent infection by means of *Ribes* eradication control procedures. Lesser levels of blister rust hazard are found in areas where the influence of the lakes is less.

Along the coast of British Columbia and the Pacific northwest, western white pine has been seriously damaged because of the ideal conditions for infection. Rust is also a problem in inland areas where microclimatic conditions are more closely tied to topographic features. Recognition of variation in moisture has lead to development of a blister rust site hazard rating system. Integrated protection strategies have been developed by the U.S. Forest Service by appropriately matching disease resistance to site hazard and early pruning of lower branches where appropriate.

Local climatic conditions vary with topography and aspect. Therefore, region wide hazard ratings must be modified with local features. Prevention of rust disease can be accomplished in medium- to low-hazard zones by avoiding planting in small openings, pruning lower branches, keeping the canopy closed, or using a thin-crowned overstory nurse crop. All of these practices are aimed at reducing microclimatic high humidity caused by dew.

Application of epidemiological principles provides the resource manager with information to help avoid disease problems where possible and to appropriately apply control procedures where they will do the most good. This concept of learning how to live with disease and manipulating disease through environmental awareness is one of the most important contributions that forest pathologists can make to forest management.

Another approach to control involves the use of disease resistance. In the early 1950s, a selection program was initiated to look for disease-resistant white pines in both eastern and western white pines. The fruits of the initial selection programs are now available to western foresters. By 1985 seed orchards were producing F_2 generation seed to reforest more than 4050 ha (10,000 acres) of western white pine lands. The genetic improvement of this seed is such that more than 65% of the seedlings can be expected to develop free of the rust.

Unfortunately, this major achievement with western white pine disease resistance has not been duplicated for eastern white pine. Disease-resistance work started in both areas about the same time, but the eastern work generated some very discouraging results early in the program. When they collected open pollinated seed from disease-free individuals in the field, the seedlings of these selected parents tested as disease susceptible as nonselected seed. These results suggested that there was little genetic control over resistance to the disease. The eastern program shifted from breeding for resistance to vegetative propagation of selected resistant individuals. Hindsight now suggests that if they had made controlled crosses among selected parents they would have identified that breeding for disease resistance was a potentially appropriate means of improving the eastern white pine.

We also recognize today that disease resistance is not necessarily a presence or absence situation. There are a number of resistance mechanisms to white pine blister rust. One type of resistance is expressed as reduced infection spot frequency. Some feature of the host prevents initiation of infection. Another type of resistance involves mortality and shedding of infected needles before the fungus grows into the branch. A third involves restriction of spread within branches, thereby limiting canker appearance and/or mortality. These and other mechanisms reduce the impact of blister rust on white pine populations.

One should recognize that the disease resistance approach to disease control is not without problems (see Chapter 20). There is always the possibility that a more virulent strain of the fungus may arise. There are presently some indications of more virulent strains of *C. ribicola* in the Pacific northwest and British Columbia. In Oregon, they are finding increased cankering in an area that previously had minimal disease. The new fungus strain from the area reduces needle shedding and causes more distinctive stunting of infected leaders than the former strain of the fungus. The importance of these findings to disease-resistant white pine management of the region has yet to be determined.

Benefits of disease resistance can be obtained through regeneration practices as well as resistance breeding. Superior-form, nondiseased seed trees can generally be expected to produce better-form, more-disease-resistant seedlings than can non-selected trees. Careful selection of seed trees for natural regeneration can therefore lead to genetic improvement.

Disease control in forest trees through resistance is a second very important contribution that pathologists can make to forest management. Disease resistance, like epidemiology, is an approach that will require a high degree of sophistication in management. It is not a cure-all for all our problems. Selectively utilized resistance together with epidemiological understanding can be used to reduce future losses for tree disease problems. Concepts, procedures, and expected results of disease resistance breeding of trees are discussed in Chapter 20.

Fusiform Rust of Southern Pines

The fungus *Cronartium quercuum* f. sp. *fusiform*, unlike *C. ribicola*, is a pathogen native to the Gulf and South Atlantic states and represents a serious rust disease because of our utilization of loblolly and slash pines. Prior to 1900 the disease was rare, but because of intensive cultural practices, including planting of infected seedlings, fire control, intensive site preparation, genetic selection of fast-growing trees, and expansion of the range of the susceptible species, the disease has intensified to the point that it is now a serious problem. Recent estimates suggest that $130 million worth of damage is occurring annually and that the disease is increasing at a rate of 2 to 3% per year.

The scientific name of the pathogen may be a bit confusing at first, but it reflects an attempt by taxonomists to group similar organisms properly. The name shows the similarities as well as differences of this fungus and other pine oak gall rust fungi (*C. quercuum* f. sp. *banksianae*, f. sp. *virginianae*, and f. sp. *echinatae*). The four fungi are morphologically similar but are given *forma specialis* designation based on pathogenicity on specific pine hosts. *C. quercuum* f. sp. *fusiforme* occurs on both slash and loblolly pine but is given the f. sp. designation based on the shape of the gall. The other three pathogens, named for their respective hosts, form spherical galls.

An area of very high incidence of fusiform rust occurs from South Carolina through central Georgia, Alabama, and Mississippi to eastern Louisiana. The area of highest rust incidence in loblolly pine is slightly north of the area of highest incidence in slash pine, reflecting the ranges of the two pines. Climate and soil conditions appear to be the most important factors affecting the high incidence, but other factors, such as the presence of oak alternate hosts, seed sources, silvicultural practices, planting practices, and pathogen variation, must be involved. High-rust-hazard sites occur on well-drained sandy soils underlaid by clays. In these high-hazard areas 50 to 100% of the loblolly or slash pine are infected.

The life cycle of *C. quercuum* f. sp. *fusiforme* involves alternation between pines and oaks, principally slash (*Pinus elliottii* var. *elliottii*) or loblolly (*P. taeda*) pines and water (*Quercus nigra*) or willow (*Q. phellos*) oaks. Aeciospores are produced in great quantities, from February to April, on elongate (fusiform)-shaped galls (Fig. 11–11) on infected pines. Although the aeciospore producing period is 3 months for the region, actual spore production in a given area is only about 3 weeks. Aeciospores are wind dispersed to oak leaves.

Infection and invasion of oak leaves results in yellow-orange uredinia and urediniospores (Fig. 11–12) within 7 to 10 days, to infect other oak leaves. Within another week, the brown hair-like telia are formed. Telia remain functional until early June. Four hours of 15 to 27°C temperatures and 100% relative humidity are sufficient to cause the teliospores to produce basidia and basidiospores.

Basidiospores are carried on the air currents to infect recently expanded cotyledons, needles, or succulent shoot tissues of the loblolly or slash pines. Fungus

Figure 11–11 Fusiform-rust-infected slash pine showing multiple fusiform branch galls.

Figure 11-12 Uredinia of *Cronartium quercuum* f.sp. *fusiforme* on water oak.

growth through the needle to the stem or branch eventually results in invasion of stem tissues. Stem or branch tissues respond by swelling or gall formation.

The galls are sites of spermatia production during October to November. Spermatia may occur during the year of infection, but usually they do not occur until the subsequent year. The following spring, aecia are formed on the galls to complete the 2-year life cycle. Galls on the pines expand as the host grows producing annual crops of aeciospores for as long as the stem or branch remains alive. The fungus invasion of the cambium phloem region reduces the growth and may lead to stem breakage but does not generally result in death of the tree.

Activities associated with intensive management have unbalanced this host–pathogen system. One of the first problem areas is the nursery, where the environmental conditions are ideal for infection. Oaks should be eliminated from within and around the nursery, but this is not enough to prevent infection from distant sources of inoculum. As many as 35 to 40 applications of fungicide per growing season have been used to reduce infection of nursery stock. Even with extensive fungicidal control it is necessary to grade the stock carefully. Those seedlings with small galls must be culled out before the stock is outplanted to avoid introduction of the pathogen into new planting areas, for rapid buildup on the remainder of the seedlings.

The nursery manager today has a more effective fungicide, Bayleton, that requires only three or four treatments per season. The earlier fungicides were surface protectants that washed off and needed to be applied to newly emerging foliage. Bayleton is taken up by the plant and systemically translocated to new growth.

The advantages of Bayleton for fusiform rust control are offset by some disadvantages. Bayleton seems to inhibit good mycorrhizal development. There may be some problems with early establishment and survival in outplantings, particularly on the less fertile sites. On good sites this lack of mycorrhizae may not be a serious concern. Resident mycorrhizal fungi quickly replace the mycorrhizae from the nursery in this situation.

Examples of Rusts and Their Control   **173**

Control of fusiform rust in the nursery is an absolute requirement for production of acceptable nursery stock. On the other hand, disease-susceptible seedlings artificially kept alive in the nursery through chemical treatments do not necessarily represent the most appropriate stock for field planting. In the field planting, where chemical control is too costly, these susceptible seedlings rapidly become infected. The ideal nursery-to-field relationship would be to start with disease-resistant seedlings. These should be protected by fungicides in the nursery, where the infection hazards are extreme. These disease-free seedlings should display a degree of natural resistance under inoculum, host development, and environmental conditions of the normal field planting.

A second forest management activity contributing to increased fusiform rust is the implementation of fire control activities. Fires are more damaging to the oaks than to the pines. In the absence of fires, oaks become intermixed with pines, providing an ideal situation for a disease that alternates between oaks and pines.

Intensive site preparation (weed control through chemical and mechanical means) to improve survival and enhance early development planted seedlings is a logical part of intensive forest management. Unfortunately, these activities enhance the intensity of the rust. Obligate parasites such as rusts are generally most destructive on vigorously growing plants. Any activity that improves the vigor and growth of the trees only makes the plants more susceptible to an obligate parasite. Facultative parasites, on the other hand, are usually more destructive on weakened or slow-growing plants.

Genetic improvement is also a logical part of intensive forest management. Unfortunately, genetic selection for fast growth without regard for disease resistance may also result in increased rust disease. The early tree improvement programs made many selections in essentially rust-free areas for planting in high-rust-hazard areas. This activity lead to problems.

Today, genetic improvement programs are geared toward disease resistance. Current annual production of rust-resistant seedlings is estimated at 5 million. Projected production will quickly reach 18 million. The fruits of this disease resistance work have reduced fusiform rust substantially. In one area of southwest Georgia previous plantings averaged 44.4% rust at 5 years. Resistant plantings of loblolly planted in the same area recently average 12.4% at 5 years.

Disease resistance is one of the strongest defenses against fusiform rust, but this is only one component of a developing arsenal of forest management practices. These also include recommendations to:

1. Avoid planting rust-susceptible pines on high-hazard sites and beyond their natural range. Loblolly and slash pines have been extensively planted on sites that were formerly occupied by other species. Where practical, managers should reevaluate planting longleaf, sand, or shortleaf pines. These species have their own special problems that have to be considered carefully. See Chapters 10, 16, and 18.
2. Make sure that nursery stock has been graded to cull all galled seedlings.

3. Increase planting density to compensate for expected losses.
4. Utilize inventory data to make management decisions on replanting, thinning, or harvesting. Growth and yield models provide guidelines for recommendations based on percent of stems infected, stocking density, and age of the stand.

Management of the alternate host, the oaks, does not presently represent a major part of the fusiform rust program. Reduction in oaks should reduce the rust, but the effectiveness (cost/benefit) of oak eradication has not been demonstrated.

In summary, fusiform rust research is an example of a major ongoing contribution of forest pathology to forest management. The application of disease resistance, chemical controls, good nursery management, forest decisions based on site features and inventory data, and the development of yield models for management decisions represent an integration of pathological concepts with many different disciplines to reduce losses in southern pine management.

Pine-Oak Spherical Gall Rust (Eastern Gall Rust)

Pine-oak spherical gall rusts caused by *Cronartium quercuum* f. sp. *echinatae* and f. sp. *virginiana* are very similar to fusiform rust. These occur on shortleaf and Virginia pines, respectively. To make things confusing, all three rusts occur in the south, where the shapes of galls on pines grade from fusiform to spherical. Pine-oak spherical gall rust caused by *C. quercuum* f. sp. *banksiana* can be a problem in the northern states, where jack pine and red oaks occur together. These spherical gall rusts have been called eastern gall rust.

Pine-Pine Gall Rust (Western Gall Rust)

To make things more confusing, another gall rust, caused by *Peridermium* (*Endocronartium*) *harknessii*, produces round galls similar to pine-oak rust on jack and Scots pine. This disease, called pine-pine gall rust or western gall rust, was introduced into many areas of eastern North America from the West, where it occurs on ponderosa and lodgepole pine. Pine-pine gall rust causes extensive galling damage to Scots pine Christmas tree plantings in New York and Pennsylvania (Fig. 11–13). A unique feature of this fungus which aids in the rapid spread of the pathogen is the elimination of an alternate host. The aeciospores produced on pines are capable of reinfecting the pine.

To complete our discussion of the *Cronartium* rusts of pine, we should indicate that there are yet other stem and limb and cone rusts caused by *Cronartium* spp. We have only scratched the surface. We have seen that introduction of a pathogen such as *C. ribicola* to a new area has resulted in serious disease. Also, the modification of cultural practices can cause a pathogen such as *C. quercuum f. sp. fusiforme* to develop into a serious problem. If the *Peridermium harknessii* ever gets

Figure 11-13 Pine-pine gall rust on Scots pine.

into southern pines, we may see another disease significantly negate some of the present fusiform-rust-resistance work. There is a good possibility that a transition from the northeast to the south will take place because of extensive Christmas tree plantings, which provide a corridor of pines from the northeast through the oak-hickory forests of Pennsylvania to the southern pines.

Cedar Apple Rust

Cedar apple rust, caused by various species of *Gymnosporangium*, is the last rust disease that we will discuss. This is really more of a fruit production than a forestry problem, but the disease is common on hawthorns and junipers (Fig. 11-14).

The telial stage (Fig. 11-15) of the fungus causes very little damage to *Juniperus* species. One should note that the disease should really be called juniper apple rust, because *Juniperus* spp., not *Thuja* spp., are the teliospore hosts.

The urediniospore or increase spore stage has been eliminated in this fungus, so that there is an absolute dependence upon the intimate presence of both hosts for survival of the fungus. Spermogonia and aecia form on apple leaves and fruit, reducing the yield and quality of the fruit (Figs. 11-16 and 11-17).

Cedar apple rust is one of the more common problems of ornamental plantings. Hawthorns and flowering crabs regularly used in the northeast are good hosts

Figure 11-14 Cedar apple rust galls on juniper. (Photograph compliments of Dr. Josiah Lowe.)

for the spermogonia and aecial stages of the fungus. These are often associated with upright and prostrate junipers in ornamental plantings. The slimy orange tentacled galls, formed during wet periods in the spring, cause a great deal of apprehension on the part of homeowners who do not understand the disease. Later in the season, unsightly reddish-brown spots on flowering crabs and hawthorns show the next stage of the disease. The best control measure is not to use the two hosts in the same or nearby plantings. (Fig. 11-18).

Figure 11-15 At about the time of leaf flush, telial horns of cedar apple rust expand into yellow jelly-like tendrils.

Examples of Rusts and Their Control

Figure 11-16 Cedar apple rust on the upper surface of hawthorn, showing black spermogonia in the center of yellow infection lesions. (Photograph compliments of Dr. Josiah Lowe.)

Figure 11-17 Close-up of elongated aecial cups on the lower surface of hawthorn leaves. (Photograph compliments of Dr. Josiah Lowe.)

Figure 11-18 The common practice of mixing prostrate junipers and hawthorn in ornamental plantings often results in serious disease in one or both plants.

REFERENCES

ANDERSON, N.A. 1973. Eastern gall rust. USDA For. Serv. For. Pest Leafl. 80. 4 pp.

ANDERSON, R.L. 1973. A summary of white pine blister rust research in the lake states. USDA For. Serv. Gen. Tech. Rep. NC-6. 12 pp.

ANDERSON, R.L., and P.A. MISTRETTA. 1982. Management strategies for reducing losses caused by fusiform rust, annosus root rot and littleleaf disease. USDA For. Serv. Agric. Handb. 597. 30 pp.

ARTHUR, J.C. 1934. Manual of the rusts in the United States and Canada. Purdue Research Foundation, Lafayette, Ind. 438 pp.

BINGHAM, R.T. 1983. Blister rust resistant western white pine for the Inland Empire. USDA For. Serv. Gen. Tech. Rep. INT-146. 45 pp.

BINGHAM, R.T., R.J. HOFF, and G.I. McDONALD. 1972. Biology of rust resistance in forest trees. Proc. NATO-IUFRO Adv. Study Inst. USDA For. Serv. Misc. Publ. 1221. 681 pp.

BLISS, D.E. 1933. The pathogenicity and seasonal development of *Gymnosporangium* in Iowa. Iowa State Coll. Agric. Mech. Arts Res. Bull. 166, pp. 340–392.

BOYCE, J.S. 1957. The fungus causing western gall rust and Woodgate rust of pines. For. Sci. *3*:225–234.

BURDSALL, H.H., JR. 1977. Taxonomy of *Cronartium quercuum* and *C. fusiforme*. Mycologia *69*:503–508.

CHARLTON, J.W. 1963. Relating climate to eastern white pine blister rust infection hazard. USDA For. Serv. Northeast. For. Exp. Stn. 38 pp.

CUMMINS, G.B., and Y. HIRATSUKA. 1983. Illustrated genera of rust fungi. The American Phytopathological Society, St. Paul, Minn. 152 pp.

CZABATOR, F.J. 1971. Fusiform rust of southern pines: a critical review. USDA For. Serv. Res. Pap. SO-65. 39 pp.

DINUS, R.J. 1974. Knowledge about natural ecosystems as a guide to disease control in managed forests. Proc. Am. Phytopathol. Soc. *1*:184–190.

EPSTEIN, L., and M.B. BUURLAGE. 1988. Nuclear division in germinating aeciospores and its taxonomic significance for the western gall rust fungus, *Peridermium harknessii*. Mycologia *80*:235–240.

GODDARD, R.E., G.I. MCDONALD, and R.J. STEINHOFF. 1985. Measurement of field resistance, rust hazard and deployment of blister rust-resistant western white pine. USDA For. Serv. Res. Pap. INT-358. 8 pp.

HAGLE, S.K., G.I. MCDONALD, and E.A. NORBY. 1989. White pine blister rust in northern Idaho and western Montana: alternatives for integrated management. USDA For. Serv. Gen. Tech. Rep. INT-261. 35 pp.

HIRT, R.R. 1964. *Cronartium ribicola*: its growth and reproduction in tissues of eastern white pine. N.Y. State Univ. Coll. For. Syracuse. 30 pp.

HOFF, R.J., and G.I. MCDONALD. 1972. Stem rusts of conifers and the balance of nature. *In* Biology of rust resistance in forest trees. USDA For. Serv. Misc. Publ. 1221, pp. 525–536.

HOFF, R.J., and G.I. MCDONALD. 1977. Selecting western white pine leave-trees. USDA For. Serv. Res. Note INT-218-FR 13.

HOFF, R.J., G.I. MCDONALD, and R.T. BINGHAM. 1976. Mass selection for blister rust resistance: a method for natural regeneration of western white pine. USDA For. Serv. Res. Note INT-202-FR 13.

KETCHAM, D.E., C.A. WELLNER, and S.S. EVANS, JR. 1968. Western white pine management programs realigned on northern Rocky Mountain national forests. J. For. *66*:329–332.

LAIRD, P.P., and W.R. PHELPS. 1975. A rapid method for mass screening of loblolly and slash pine seedlings for resistance to fusiform rust. Plant Dis. Rep. *59*:238–242.

MCDONALD, G.I., E.M. HANSEN, C.A. OSTERHAUS, and S. SAMMAN. 1984. Initial characterization of a new strain of *Cronartium ribicola* from the Cascade Mountains of Oregon. Plant Dis. *68*:800–804.

NANCE, W.L., R.C. FROELICH, T.R. DELL, and E. SHOULDERS. 1983. A growth and yield model for unthinned slash pine plantations infected with rust. *In* Proc. 2nd Biennial Res. Conf. USDA For. Serv. Gen. Rep. 24, pp. 275–282.

PATTON, R.F., and R.N. SPEAR. 1989. Histopathology of colonization in leaf tissue of *Castilleja, Pedicularis, Phaseolus*, and *Ribes* species by *Cronartium ribicola*. Phytopathology *79*:539–547.

PETERSON, R.S., and F.F. JEWELL. 1968. Status of American stem rusts of pine. Annu. Rev. Phytopathol. *6*:23–40.

PHELPS, W.R., and F.L. CZABATOR. 1978. Fusiform rust of southern pines. USDA For. Insect Dis. Leafl. 26. 7 pp.

POWERS, H.R., JR. 1984. Control of fusiform rust of southern pine in the USA. Eur. J. For. Pathol. *14*:426–431.

POWERS, H.R., JR., and J.F. KRAUS. 1988. Rust resistant loblolly pines. Ga. For. Res. Pap. 74. Research Division, Georgia Forestry Commission. 7 pp.

POWERS, H.R., JR., J.P. MCCLURE, H.A. KNIGHT, and G.F. DUTROW. 1974. Incidence and financial impact of fusiform rust in the south. J. For. *72*:398–401.

SAVILE, D.B.O. 1971. Coevolution of the rust fungi and their hosts. Quart. Rev. Biol. 46:211–218.

SNOW, G.A. 1985. A view of resistance to fusiform rust in loblolly pine. 34th Annu. For. Symp. Insects Dis. South. For. Louisiana State University, School of Forestry, Wildlife and Fisheries, La. Agric. Exp. Stn., Baton Rouge, La., pp. 47–51.

VAN ARSDEL, E.P. 1961. Growing white pine in the lake states to avoid blister rust. USDA For. Serv. Lake States For. Exp. Stn. Pap. 92. 11 pp.

VAN ARSDEL, E.P. 1965. Micrometeorology and plant disease epidemiology. Phytopathology 55:945–950.

VAN ARSDEL, E.P. 1967. The nocturnal diffusion and transport of spores. Phytopathology 57:1221–1229.

WILLISTON, H.L., T.J. ROGERS, AND R.L. ANDERSON. 1981. Forest management practices to prevent insect and disease damage to southern pine. USDA For. Serv. Southeastern Area For. Rep. SA-FR 9. 9 pp.

ZILLER, W.G. 1974. The tree rusts of western Canada. Can. For. Serv. Publ. 1329. (Dept. Environ.) 272 pp.

12

FUNGI AS AGENTS OF TREE DISEASE: CANKER DISEASES

- **TYPES OF CANKERS**
- **MODE OF ACTION OF CANKER DISEASES**
- **CANKER DISEASE CYCLE**
- **SYMPTOMS OF CANKER DISEASES**
- **DIAGNOSIS OF CANKER DISEASES**
- **EXAMPLES OF CANKER DISEASES**
- **CONTROL OF CANKERS**

Fungi that cause canker diseases are generally adapted for parasitic existence in the living cells of the phloem, cambium, and outer xylem region of the tree, as well as for saprobic existence upon dead phloem and xylem cells. This type of organism is called a facultative parasite.

Canker fungi cause extremely diverse effects on host trees. They range from rapid killers of host tissue to contributing saprobes that deteriorate branches weakened by other factors.

Canker diseases emphasized in this chapter are all caused by Ascomycotina fungi. There are cankers caused by Oomycetes, such as *Phytophthora cactorum*, which causes bleeding canker of maples and other tree species. Basidiomycotina fungi also may cause cankers. Rust cankers are discussed in Chapter 11. Some Basidiomycete decay fungi, such as *Inonotus obliquus (Poria obliqua)* and *Phellinus igniarius (Fomes igniarius)*, also cause cankers on northern hardwoods. *Inonotus hispidus (Polyporus hispidus), Phellinus spiculosa (Poria spiculosa)* and *Spongipellis pachyodon (Irpex mollis)* are canker rots of southern hardwoods. Basidiospores of these fungi infect through wounds. Mycelial invasion of sapwood results in decay up to 2 meters above and below the canker infection point and cambium death. See Chapter 14 for a more complete discussion of decay.

TYPES OF CANKERS

Canker fungi can be grouped into (1) Saprobes contributing to declines, (2) annual cankers, (3) facultative parasites which produce perennial target cankers, and (4) diffuse cankers. Each of these four types of cankers will be discussed.

MODE OF ACTION OF CANKER DISEASES

Invasion of living tissue of the cambium, phloem, and outer xylem provides canker fungi access to readily utilizable transport and storage nutrients of trees. These fungi may produce toxins that overcome host defensive mechanisms. The host produces callus wound tissue barriers around the invaded area in response. Some canker fungi have been shown to utilize cellulase enzymes for the invasion and decomposition of wood.

Slight differences in mode of action separate the four types of canker fungi. Many saprobic canker fungi are associated with weakened trees. They cannot penetrate the morphological defense barriers of healthy trees, nor can they tolerate chemical defense mechanisms of normal healthy trees. Weakened trees are unable to produce sufficient morphological and chemical barriers to prevent invasion.

An annual canker fungus is an opportunist that infects soon after wounding and invades quickly before the host has a chance to respond. Eventually, the host's response catches up with the invasion to halt further expansion of the canker. A large number of fungi can induce annual cankers.

A perennial canker fungus is much like an annual canker fungus, in that it invades when the host's defenses are slow to respond. Invasion of host tissues during late summer, fall, and winter induces less host response. Perennial canker fungi differ from annual canker fungi in that the former are good saprobic competitors on dead cankered tissues. They also must tolerate some host defense responses better than annual canker fungi, because they appear to break through the host defenses each year.

The interaction of host and fungus in perennial cankers involves more than just what one is led to believe from published research reports. I have personally made thousands of inoculations with many different cultures of *Nectria galligena* and *Ceratocystis fimbriata*. Most of the inoculations produced cankers the first season, a few expanded the second season, but none were still active after 5 years. Some type of balanced relationship must be established between host and pathogen to develop a typical perennial canker. We have little understanding of this relationship.

The fungal pathogen dominates in the diffuse canker system. Host response is minimal or ineffectual. Toxins produced by the fungus and rapid mycelial invasion are characteristics of diffuse cankers.

If one looks back over the described mode of action for the four types of canker pathogens, a gradient should be obvious. At one end of the spectrum the host dominates. At the other the pathogen dominates. In the middle are the opportunistic annual canker fungi which invade through wounds before the host is able to respond. Once the host response occurs, they are held in check. A step further up the pathogenicity ladder are the perennial canker fungi. Perennial canker fungi persist on a host over extended periods of time by limiting invasion to periods when the host response is minimal.

CANKER DISEASE CYCLE

It is generally assumed that canker fungi infect trees through wounds in the stem or through broken branch stubs. These assumptions result from observations of dead branch stubs in the center of many cankers and the development of cankers following artificial inoculation of wounds. Most artificial inoculation techniques involve insertion of massive amounts of mycelial inoculum into wounds. It is erroneously assumed that spores will infect similar wounds.

Those cankers that have been studied in detail show that infection by spores is a different problem, which we do not yet understand. Infection may also be through the foliage, as in the case of Septoria canker of poplars. Infection may occur by means of ascospores or conidia. Information on the importance of each is rather limited.

Invasion involves varying degrees of interaction of pathogen and host as discussed earlier.

Sporulation usually occurs initially as conidia. Ascospores are produced sometime later, presumably after fertilization with a second compatible strain of fungus. The need for fertilization by a compatible strain is also poorly documented.

Long-term survival of perennial and diffuse canker fungi as saprobes on dead tissue of the canker is common. Some of the fungi survive as saprobes in the soil. Others survive as saprobes on dead branches of trees.

Dissemination of spores to infection sites is usually by wind, although splashing rain, insects, rodents, and human beings may also play a role in dissemination.

SYMPTOMS OF CANKER DISEASES

Symptoms of canker diseases range from small irregularities of the stem caused by the callusing of annual cankers, to massive areas of dead bark and the death of trees. Discoloration of the bark of smooth-barked trees is common. Callusing of cankers may close the canker wound or may produce excessive enlargement of the stem around the canker. On thick-barked tree species, cankers may enlarge under normal-appearing bark and be evident only after removal of some of the bark.

DIAGNOSIS OF CANKER DISEASES

The diagnosis of canker diseases is usually straightforward. Characteristic symptoms are often associated with specific fungus signs. Positive diagnosis depends upon identification of the fungus.

The canker-causing Ascomycetes have both a sexual and an asexual spore stage. The two names for the canker-causing fungi are sometimes initially confusing to students but are very convenient to the pathologist, who may encounter either spore stage when diagnosing the cause of a particular canker.

In the absence of signs, it may be necessary to isolate the suspected pathogen in pure culture and follow Koch's postulates to prove pathogenicity. Anyone who tries to isolate from cankers will find a large number of fungi present. It is usually easier to isolate the causal pathogen from the margin of the canker at the interface of diseased and nondiseased tissue. A large number of saprobic fungi can compete and survive in the dead substrate in the center of the canker, but only the pathogen should be able to compete and survive against the host defense mechanisms around the margin of the canker.

EXAMPLES OF CANKER DISEASES

Saprobic Canker Fungi Contributing to Declines

Cytospora (Leucostoma) canker. One of the best examples of a saprobic canker fungus contributing to decline is the Cytospora canker of spruce, particularly Colorado blue spruce (Fig. 12–1). The Cytospora canker problem is characterized by the death of low branches of ornamental blue spruce, usually trees that are 20 years of age or older. The death of branches continues slowly from the bottom upward until the tree is no longer acceptable as an ornamental. Pycnidia of the Imperfect fungus *Leucocytospora* (*Cytospora*) *kunzei* are commonly found on dying branches.

The name of the ascospore stage of the fungus, *Valsa kunzei*, has recently been changed to *Leucostoma kunzei*, and some would change the name of the canker to Leucostoma canker. The imperfect stage has also been changed to *Leucocytospora kunzei*, but no one is suggesting changing the name of the disease to Leucocytospora canker.

Wounding and drought stress seem to be required for infection of Colorado blue spruce by *L. kunzei*. Abundant conidia are oozed from pycnidia particularly in the spring. Pycnidia are very difficult to see in the field because they are immersed in bark tissue that exudes quantities of resin. One can find the pycnidia by placing recently killed branches in a plastic bag with wet paper toweling. The small oval-shaped conidia will ooze from the pycnidia in yellow tendrils under these conditions. In the field the spores are easily dispersed by splashing rain or small water droplets.

Using a decline disease concept to understand this disease would recognize that the age–site relationship is the predisposing factor. The inciting factors are the

Figure 12–1 Mortality of lower branches of Colorado blue spruce. This disease is called Cytospora canker.

drought stress and wounding and the contributing factor is *L. kunzei*. See Chapter 18 for a more extensive development of the decline disease concept.

The commonly accepted method of "control" is to fertilize and water the trees and prune out the dead branches. The spraying of trees with fungicides is recommended by some workers, but there is no experimental evidence to indicate that spraying does more than enrich those being paid to apply the spray.

Beech scale Nectria canker. The most widely accepted explanation for beech bark disease or beech scale Nectria canker (Fig. 12–2) involves the interaction of the scale insect *Cryptococcus fagisuga (C. fagi)* and the canker fungus *Nectria coccinea* var. *faginata*. Even though the disease complex causes extensive destruction of commercial beech, very little work was done on this disease between 1934, when early work was done, and the mid-1970s. This disease, like Cytospora canker, can best be understood as a decline involving predisposing, inciting, and contributing factors.

The scale insect, *C. fagisuga*, was introduced into North America from Europe around the turn of the century and has spread, from the introduction site in Nova Scotia, south westward throughout New England and into New York, Pennsylvania, and West Virginia.

Figure 12-2 Crown death of beech tree associated with beech scale Nectria canker of the main stem.

The wingless young scale insects are attracted to rough places on the smooth beech bark. They feed on the phloem by inserting their stylets through the bark to the phloem. Once they begin to feed, they no longer move.

Female scale insects produce young parthenogenetically (in the absence of fertilization) causing the size of the population to increase rapidly. The bark becomes covered with cottony wax (Figs. 12-3 and 12-4), produced by the sedentary insects.

The insect feeding does not result in the death of trees, but is does seem to numb the natural defenses of the affected tissues. This inability to respond allows the insects to feed in the same area for an extended period of time without the tree producing a necrotic defensive response (wound periderm).

These localized numbed areas are also unable to respond to infection by weak canker pathogens. In North America three different *Nectria* species have been found on the tissues altered by *C. fagisuga*. *N. coccinea* var. *faginata* and *N. galligena* are recognized bark pathogens. The role of *N. ochroleuca*, a third associate, has yet to be clarified. If the scale insect infestation is localized to small patches, the infection and invasion by the fungi results in small circular patches of dead tissue, on which the fungus will eventually fruit (Fig. 12-5). The tree defenses prevent extensive invasion by the pathogen beyond the area affected by the insect. If the insect population becomes densely distributed over large areas of bark (Fig. 12-3), the fungus invasion and sporulation may cover large areas. In this instance the masses of red perithecia (Fig. 12-6) cause the stem to look like it is covered with blood.

As the small wingless scale insect population moved across the northeast, the appearance of the red perithecia of the canker pathogens appeared about 5 to 10 years later. Extensive mortality followed the invasion by the canker pathogens.

Examples of Canker Diseases

187

Figure 12-3 Cottony masses of the scale insect *Cryptococcus fagisuga*.

Following this killing front, there is an aftermath zone of beech bark disease characterized by breakage of residual large living trees, beech snap. Those trees that have some degree of resistance to the massive insect invasion may sustain some attack on localized patches of bark. These patches become invaded by the canker fungi and subsequently by decay fungi and ambrosia beetles. The decay weakens the tree, resulting in breakage. Another characteristic of the aftermath zone is the

Figure 12-4 Close-up of *Cryptococcus fagisuga* scale. The cottony wax has been melted away to show some of the small oval scale insects.

Figure 12–5 Oval patches of *Nectria coccinea* var. *faginata* fruiting on beech.

Figure 12–6 Close-up of clusters of perithecia of *Nectria coccinea* var. *faginata* on beech.

Examples of Canker Diseases

gradual, sometimes, localized infection of small young trees resulting in increasing defect but little mortality.

Our understanding of the biological foundations, the forest management practices, and the ecological relationships of this disease have progressed rapidly in recent years. Trees that are resistant to invasion by the insect have been identified. Some trees have been shown to produce a mechanical response to infection, by the canker fungi, that protects the cambium. A periderm layer causes patches of infected tissue to be sluffed off. Another scale insect *Xylococculus betulae* has been shown to contribute to beech bark disease. Predators and pathogens of the scale insect and the fungi are known.

We now recognize that our recommendations to simply salvage the beech before the disease killed the trees may not have been the most appropriate. The forests often respond to this type of activity by prolific sprouting, particularly if there is extensive wounding to the roots of the trees by the harvesting equipment. Instead of desired maple and birch stands in the aftermath of beech bark disease, harvested stands have turned into beech thickets. More selective harvesting procedures and the use of herbicides to kill stumps and advance regeneration will be required to maintain high-quality productive forests.

Now that the major killing front has passed through most of the northeast, we are left with the havoc of the disease and the management activities in its wake. It is time to apply some of the management recommendations that have been suggested (see Burns and Houston, 1987). These include:

1. Maintain proper stocking by thinning stands. During thinnings discriminate against trees with wounds and other stem irregularities.
2. Remove beech trees with sunken canker lesions and dead patches. Decay will quickly destroy the value of these trees. Many will break off during wind storms.
3. Trees with raised lesions or blocky bark could be left. These may be susceptible to scale but are probably resistant to fungal invasion.
4. Retain trees with smooth bark. These may be resistant to the scale insect. Trees of this type should represent the parent trees of the next-generation stand.
5. Encourage species diversity. Part of our original problem was the overstocking of beech associated with previous selective harvesting activities (high-grading).

Beech bark disease has a major impact on squirrels, bears, turkeys, and other animal populations that rely on beech seeds for their diet. It is therefore not appropriate to eliminate beech from our productive hardwood forests. There appears to be some variation in the amount of mortality and ecological effects of beech bark disease on the species composition and structure of northern hardwood forest in New England. Integration of an understanding of these differences with a selection management system (see Chapter 14) should provide a means for maintaining a beech component to the eastern hardwood forest.

The difficulties encountered in understanding a complex disease association and the economically driven attitudes of timber harvesters have, in the past, limited our responses to beech bark disease. The expeditious course of action was to harvest any marketable beech and let the rest die. The understanding of beech bark disease today as a decline disease involving a number of interchangeable but specifically ordered factors provides a foundation for a significant research effort leading toward sound management strategies. Hopefully, the changing attitudes of an environmentally conscious forest industry will refine and apply the research information to stabilize and improve the situation for beech. Beech bark disease will not go away. We need to develop the know-how to learn to live with the problem.

Annual Cankers

Fusarium canker. Canker fungus–host combinations, in which the fungus is permanently checked by callus development after the first year's invasion, are called annual cankers. Annual cankers are very common but, because of the minimum effect of the host, are often overlooked. Irregularly roughened bark caused by callusing of the wounds is a typical symptom. In species of trees used for lumber, annual canker infections often show up as dark streaks in the wood.

A typical example of an annual canker is Fusarium canker of sugar maple. *Fusarium solani* and a number of other fungi were isolated from annual cankers on sugar maple in Pennsylvania. Inoculation with pure cultures of the fungi demonstrated the pathogenicity of *F. solani*.

Annual cankers have generally received no more than cursory diagnostic study. An exception was an epidemiological study of the problem in Pennsylvania. It was found that infection occurred only over a period of a few years. There appears to be a relatively short period of time in the development of the tree when it is susceptible. Therefore, we see this type of disease develop only once during the life of a stand of trees. Other annual cankers may respond to recurrent injuries in the stand.

Perennial Target Cankers

Nectria canker. Nectria canker, caused by *Nectria galligena*, is a classic perennial target (Figs. 12–7 and 12–8). The fungus, after infection, destroys a minimal amount of the cambium each year. In the spring, the tree produces a callus ridge in the xylem and wound periderm in the phloem to wall off the fungus. During the spring and early summer, the fungus must survive as a saprobe on dead xylem and phloem. During late summer, fall, and winter, the fungus bridges the periderm barrier and moves out and over the callus and invades another minimal portion of the cambium beyond the previous year's development. This alternation of fungus growth and tree growth results in a target type of canker.

If the tree produces sufficient callus, the canker development may be checked. This arresting of fungus growth may occur on part of the canker, resulting in an irregular-looking target.

Perennial target cankers are the result of a balanced interaction of host and fungus and seldom cause death of the tree.

The importance of perennial cankers on the development of the host is really

Figure 12–7 Nectria canker on trembling aspen.

Figure 12–8 Cross section of a Nectria canker on trembling aspen.

Figure 12–10 Eutypella canker of sugar maple.

Figure 12–11 Close-up of Eutypella canker, showing clusters of black projecting perithecial necks and perithecia embedded in the bark as revealed in the knife cut.

Examples of Canker Diseases

These are called diffuse cankers. Chestnut blight is a classic example of a diffuse canker. The fungus destroys the cambium and girdles the tree in a very few years.

Cryphonectria (Endothia) parasitica, the cause of chestnut blight, is a king among fungus pathogens. Introduced into the United States early in this century, within 50 years it had spread so rapidly that the American chestnut was largely destroyed throughout its natural range from Maine to Mississippi. An estimated equivalent of 9 million acres (3.6 million ha) of pure chestnut stands were destroyed by this fungus (Figs. 12–12 to 12–14). A few sprouts and tangled debris of fallen giants remain of what was once the most important hardwood timber species in the United States.

What makes *C. parasitica* the king of pathogens? A look at the disease cycle of chestnut blight will show all the features of a super pathogen.

A super pathogen should produce an abundance of spores on a wide variety of substrates over an extended period of time. Masses of conidia are extruded from pycnidia produced over the entire diseased chestnut bark.

A super pathogen should not be host-specific, but should increase on a number of hosts or dead substrates. The ability to survive on more than one host enhances the survival ability in the absence of the primary host. A number of oaks, chinkapin, shagbark hickory, red maple, and sumac serve as saprobic substrates for *C. parasitica*.

A super pathogen should utilize a number of means to disseminate its spores. *C. parasitica* conidia are produced in sticky tendrils that are ideally suited for dissemination by any insect that comes in contact with the bark of diseased chestnut. Birds that feed on the insects or light on the diseased bark can pick up and

Figure 12–12 Chestnut blight killed the large chestnuts of this stand. Sprouts produced over a long period of time from the stumps result in a brush field.

Figure 12–13 Chestnut-blight-infected sprout.

Figure 12–14 Close-up of Chestnut blight canker, showing pustules on the bark. These are necks of the pycnidia or perithecia of the fungus which are embedded below in the bark.

Examples of Canker Diseases

197

disseminate the spores. Almost anything that moves can carry spores from the surface of diseased bark to wounds on healthy trees. Splashing water can disseminate the spores short distances. Ascospores produced in perithecia on diseased bark are forcibly discharged and wind-disseminated to infect other hosts. To summarize dissemination, wind, water, and many vectors operate to spread the fungus from diseased to healthy trees.

A super pathogen should be capable of infecting through commonly produced infection courts over a variety of environmental conditions. Those pathogens with very exacting environmental requirements and specific infection courts are less capable of expansion throughout a wide environmental range. The climatic conditions from Maine to Mississippi are not at all uniform, yet this fungus expanded over this range. Infection courts are any type of wound produced by insects, birds, man, and wind breakage, to name a few.

A super pathogen should invade the host rapidly enough to avoid induction of resistance mechanisms in the host. One of the distinguishing characteristics of the diffuse canker is the rapid invasion of the host, thereby avoiding the callus defense reaction characteristic of annual and perennial cankers.

To wipe out a species of plant with a pathogen, the virulence, or capacity to cause disease, must exceed all resistance accumulated in the individuals of the host population. The pathogen must kill the host fast enough, and repeatedly, so that the host does not have time to recombine and select individual gene combinations for resistance to the pathogen. The chestnut blight disease, by killing the original population rapidly and killing most sprouts before they have a chance to produce seed, prevents the host from genetic improvement through natural selection. Normally, natural selection for increased resistance would occur in a reproducing population of plants subjected to the pressures of an introduced pathogen.

A super pathogen should be able to survive during extended periods of cold or dry conditions. *C. parasitica* can survive years on dried herbarium specimens as well as a wide range of fluctuating environmental conditions in the field.

Many other very serious disease-causing fungi have some of the characteristics of the ideal super pathogen described but none have combined them all. One can understand differences between different disease-causing fungi by examining or comparing their capacity with the super pathogen.

Much work was done on the chestnut blight disease, particularly during the 1950s, but most of the effort was abandoned, so that by the late 1960s and early 1970s very little was being done. Today, there is a new flurry of activity centered around hypovirulent (nonvirulent) strains of the fungus. In the laboratory, hypovirulent strains have been shown to induce hypovirulence in other strains when grown in the same culture dish. If hypovirulent strains are introduced into wounds around cankers on field-grown chestnuts, sometimes the cankers stop enlarging and the host calluses over the wound. The lack of virulence appears to be associated with a piece of double-stranded RNA which can be transferred through hyphal fusions of virulent and nonvirulent lines of the fungus.

What does all this mean for the future of the American chestnut? At this point it is difficult to say, but the approach is the first glimmer of hope for the species. It is obviously not going to be practical to physically introduce hypovirulent strains into every canker that occurs in a chestnut stand. The ideal situation would be to allow natural spread of hypovirulent strains within a stand. Natural spread of the hypo-virulence is limited due to reduced sporulation of these strains. Another limitation involves the inability of various strains of the fungus to fuse and transfer the hypovirulence factor. Some unknown incompatibility factor is involved. In Europe and in some isolated populations of chestnut is North America, hypovirulence is spreading naturally and reducing the impact of *C. parasitica*. Obviously, much research is needed before we can even determine whether the approach is potentially useful. If something develops from this type of work, it will have major significance not only to chestnut blight but to many other diseases.

Hypoxylon canker. A discussion of diffuse cankers should include a few that are not as devastating as chestnut blight, to show that it is possible to manage tree species with diffuse cankers. This is not to minimize the overall importance of diffuse cankers, since they are very destructive, but to show that we can learn to live with most of them.

Hypoxylon canker caused by *Hypoxylon mammatum*, is a disease of bigtooth and trembling aspens that was first reported in 1921 (Figs. 12-15 and 12-16). In contrast to *C. parasitica*, this is a pathogen native to North America, which has just recently been recognized in Europe. European forest pathologists are concerned with the potential for this disease, as it affects native *Populus tremula* and hybrids with *P. tremula, P. tremuloides*, and *P. grandidentata* parentage. The disease is also a problem on introduced *P. trichocarpa* lines.

In the Lake states, where aspen is the most important commercial wood

Figure 12-15 Hypoxylon canker of aspen.

Examples of Canker Diseases

Figure 12-16 Hypoxylon canker often results in the stem breaking off at the canker.

species, up to 10% of the trees are affected. In individual stands up to 90% may be infected. An annual loss equivalent to one-fourth of the annual increment for the species is reported.

The disease is readily diagnosed by the characteristic black and white mottle it produces in infected bark. Bark of young cankers and margins of older cankers become initially yellow to orange rather than the normal green color. With time, diseased bark and wood tissue becomes blackened and cracked in a checkerboard-like fashion. The asexual stage of the fungus appears, the year following infection, on peg-like pillars that cause the outer layer of bark to blister and peel back (Fig. 12-17). The sexual stage appears 2 to 3 years after infection. Ascospores are produced in perithecia embedded in fungus stroma (Fig. 12-18). The stroma and perithecia are initially white but become black with age.

A long and eminent list of forest pathologists have contributed to our understanding of Hypoxylon canker, yet we still have many questionable and unanswered steps in the life cycle to unravel. One notable gap in our understanding is the exact site and conditions under which infection takes place. No reproducible demonstration of ascospore infection of trees in the field has yet been made. Many workers have attempted indirectly to determine the site and conditions of infection, so we have a number of reports on this topic. Insect wounds, branch stubs, small twigs, bark saprophytes, bark fungal-inhibitory toxins, fungus host-killing toxins, bark moisture, and branch axils are all implicated by different groups of researchers.

Another area that has produced a number of concepts involves the role of environmental conditions on disease development. Both poor and good sites are

Figure 12-17 Asexual spores of the *Hypoxylon mammatum* fungus develop on peg-like pillars below the outer bark.

Figure 12-18 Close-up of Hypoxylon canker, showing clusters of *Hypoxylon mammatum* perithecia embedded in fungus stroma.

Examples of Canker Diseases

implicated with disease. Reduced amounts of disease in dense stands implies a stand-density relationship.

Moisture stress is one key environmental factor that appears to be related to disease incidence. This interpretation is based on field surveys and controlled inoculation tests. A fundamental understanding of how moisture stress affects this and other facultative pathogens may provide a foundation for proper selection and management of trees with these types of problems.

Another advancement in our understanding of this disease comes from recent work on genetics of resistance in aspen to Hypoxylon canker. Three mechanisms of resistance were found. Resistance due to callus formation appears to be controlled by a few major genes, but pathogen variation may overcome this type of resistance. Resistance due to branch death is low and the heritability for resistance is also low. Retardation of canker enlargement is one of the best forms of resistance but is difficult to evaluate in short-term experiments. A more complete discussion of the genetics of resistance is included in Chapter 20.

A third contribution to our understanding of this disease involves understanding how the disease increases within the stands of aspen. This topic is developed more fully in Chapter 19, but the important point to recognize with Hypoxylon canker is that intensification within an even-aged stand is not dependent upon the amount of disease in the stand. Diseased trees within even-aged stands do not contribute inoculum for infection of other trees. Trees are most susceptible to infection by Hypoxylon canker when they are seedlings. Three years between infection and sporulation allow most of the trees surrounding an infected tree to grow out of the susceptible stage before sporulation of the fungus occurs.

As we move to more controlled management systems, the selection of genetic resistance in aspen can be used to reduce losses due to Hypoxylon. Today, we generally use natural reproduction to generate aspen stands.

Scleroderris canker. Scleroderris canker, caused by *Gremmeniella abietina* (formerly known as *Scleroderris lagerbergii*) and currently also known as *Ascocalyx abietina*, is an excellent example to show that even highly sophisticated mid-twentieth-century forest pathologists can have a new disease introduced under their noses twice and not recognize its importance until it is well established.

The first detailed description of this most important disease of red and jack pines was presented in 1966. Failure of plantings to survive had been noticed for at least 10 years, but the cause could not be determined until a visiting European forest pathologist provided the idea that the disease was the same as a well-known disease in Europe. Death of lower branches and trees is recognized as Scleroderris canker by culturing or recognition of pycnidia of *Brunchorstia pinea*, the imperfect stage, or the presence of apothecia of *G. abietina*. A characteristic symptom of the disease is the yellowing of needle bases on infected branches and the yellowish-green color of cambium and wood tissues (Figs. 12–19 and 12–20). The branch death phase of this disease is the primary symptom. Stem cankering occurs but is sufficiently uncommon to cause some to describe the disease as a shoot dieback rather than a canker.

Today, Scleroderris canker occurs in plantations in the states of Michigan, Wisconsin, Minnesota, Vermont, New Hampshire, Maine, and New York. In

Figure 12-19 Necrosis of needle base associating with Scleroderris canker in the branch.

Figure 12-20 Red pine branch tips killed by Scleroderris canker as seen in early summer.

Examples of Canker Diseases

Canada, work in the provinces of Ontario, Quebec, and New Brunswick has shown the wide distribution and importance of the disease. When people know what to look for, the disease is shown to be present and accounts for plantation failures formerly thought to be due to some attribute of the site.

Usually, infected plantations occur in areas where cold air accumulates, such as topographic depressions or openings in the forest canopy. A strong dependence on cool, damp, environments is seen for this disease. The fungus also appears to grow and compete better at lower temperatures.

The disease is initially distributed to the plantations on infected nursery stock. Recognition of nurseries that produce infected stock is an important step toward prevention of additional problems. It is very difficult to detect infection in nursery stock until it is planted out, so one has to use records of sources of stock for known infected plantations to determine which nurseries are the source of the problem.

Thorough distribution of inoculum throughout vast areas of natural regeneration of jack pine in northern Ontario was also observed. In this case, inoculum, came from a distant source, because no remnants of the former infected stand were apparent. The spores were probably carried to the area on a weather front from the south and deposited in the young stand by rain.

Epidemic disease development following long-range spore dispersal of a massive amount of inoculum is rarely observed for tree diseases. It may be more common than we think, though, because long-range dispersal is well documented for agricultural crops.

After about 10 years of intensive study in the United States and Canada, Scleroderris canker was reasonably well understood and controlled through fungicide sprays in the nursery. Of course, it was also important to avoid replanting where the disease was well established.

In 1975, a new Scleroderris canker disease was reported from New York. This disease occurred on large trees up to 18 m in height. Mortality was extensive on red and Scots pine (Fig. 12–21). It was suggested that the *Gremmeniella* fungus causing death of large pine was different from the *Gremmeniella* fungus occurring on small trees in the Lake states. The disease in New York was caused by a strain of the fungus that closely resembles some European strains of the fungus. By 1977, approximately 14,000 ha was known to be infected.

The new Scleroderris canker disease in New York spawned and focused a massive highly coordinated research effort by Canadian and U.S. pathologists. Quarantines were established to prevent movement of infected materials. Interactions among North American, European, and Asian scientists provided a framework for generating and testing a number of hypotheses to explain the outbreak. The ideas that a new more virulent race of the fungus was causing the problem, that unique environmental factors had set the stage, that management practices using nonlocal seed sources were involved, that this new pathogen represents a major threat to the pine forests of the central, western, and southern forests, that other species besides pine can serve as hosts, and that acid rain had weakened the trees or favored the pathogen were considered and tested. Extensive work on the fundamental biology of the pathogen was also accomplished.

At the present time the New York epidemic has subsided, but flair-ups of the

Figure 12-21 Mortality of large red pine caused by Scleroderris canker.

disease have occurred recently in eastern Canada. Many of the quarantines have been dropped or modified. It would be convenient for forest pathologists to take credit for bringing the disease under control but this is not the case. Human intervention may contribute to the development of an epidemic but usually natural factors are the driving force that returns the system to equilibrium.

What have we learned about Scleroderris canker? Although the idea of a new more virulent race of the pathogen was the primary impetus for the major research effort, the research to demonstrate new or more virulent characteristics for the pathogen were not very convincing. Environmental factors were again demonstrated to be the primary driving variables for the epidemic. The 1970 period was one of the wettest periods of this century for New York. One interpretation of the epidemics suggests that conditions for invasion of tissues by the pathogen is a key factor. If snow covers branches or air temperatures remain around the freezing point for at least 44 days, infections will spread to produce disease symptoms. In the absence of 44 conducive days no symptoms develop. There was a strong association of conducive periods and recent outbreaks. The role of nonlocal seed sources has been shown to be a major factor in Scleroderris disease epidemics in Europe, but no one seriously tested this question for North America. It is interesting to note that the most seriously affected red pine plantations were part of a massive planting project during the 1930s using nonlocal seed sources. The potential for spread of this disease is still not fully understood. There seems to be some limit to the southern spread based on climatic variables. The susceptibility of pines, spruce, firs, hemlocks, Douglas-fir, and larches to infection was demonstrated but pines represent the primary hosts in North America. There appear to be races of the fungus that specialize on specific hosts. The role of acid rain may have been a politically

Examples of Canker Diseases

expedient thing to test during this time of major concern about acid rain, but the data do not suggest any major impact of acid rain on Scleroderris canker. On the practical side, research results indicate that pruning lower branches from infected plantations reduced the level of inoculum in the stand.

Pitch canker. Pitch canker caused by *Fusarium moniliforme* var. *subglutinans* is a disease complex of loblolly, slash, and other southern pines that is difficult to classify within the canker diseases. The disease is a major problem in seed orchards but is also causing damage to plantations. The pathogen infects stems causing bleeding, resinous cankers with no callus swelling. It infects shoots producing a crown dieback. It infects female flowers, thereby reducing the seed production. It also infects and causes mortality to seedlings in the nursery. Growth losses in slash pine are estimated at 13 to 30 million cubic feet annually. Mortality of up to 25% occurs, but annual mortality of 2% is more common. The importance of injuries for infection is well documented but the number of other interacting factors is extremely variable from one location to another. Each site needs to be considered individually.

There is no specific control program at this time, but investigations leading to an integrated management strategy are under way. These include aspects of chemical control for both the fungus and insects (possible wounding agents), the use of an antagonistic bacterium and other avirulent competitors and procedures for careful harvesting and storing of seed.

CONTROL OF CANKERS

Control measures, such as keeping trees healthy by fertilizing and watering, avoiding wounding, and selectively removing diseased trees, are sometimes recommended, but little evidence is available for measurable reduction in disease infection using these methods. Sound control recommendations are based on epidemiological understanding of environmental, host, and pathogen interaction. Application of this understanding to the control of cankers awaits changes in forest management practices. The forest manager must recognize the potential to manipulate the disease before it develops. At the present time, pathological understanding is most often requested after the problem develops, and then the pathologist is able to do little more than name the disease and predict the loss.

REFERENCES

ANAGNOSTAKIS, S.L. 1987. Chestnut blight: the classical problem of an introduced pathogen. Mycologia *79*:23–37.

ANONYMOUS. 1954. Chestnut blight and resistant chestnuts. USDA Farmers' Bull. 2068. 21 pp.

BEATTIE, R.K., and J.D. DILLER. 1954. Fifty years of chestnut blight in America. J. For. *52*:323–329.

BELANGER, R.R., P.D. MANION, and D.H. GRIFFIN. 1989. *Hypoxylon mammatum* ascospore infection of *Populus tremuloides* clones: effects of moisture stress in tissue culture. Phytopathology *79*:315–317.

BIER, J.E. 1940. Studies in forest pathology. III. Hyoxylon canker of poplar. Can. Dep. Agric. Publ. 691 Tech. Bull. 27. 40 pp.

BLAKESLEE, G.M., L.D. DWINELL, and R.L. ANDERSON. 1980. Pitch canker of southern pines. USDA For. Serv. For. Rep. SA-FR 11. 15 pp.

BRUCK, R.I., and P.D. MANION. 1980. Interacting environmental factors associated with incidence of Hypoxylon canker on trembling aspen. Can. For. Res. *10:17–24*.

BURNS, B.S., and D.R. HOUSTON. 1987. Managing beech bark disease. North. J. Appl. For. *4*:28–33.

CROWDY, S.H. 1952. Observations on apple canker. IV. The infection of leaf scars. Ann. Appl. Biol. *39*:569–587.

DONAUBAUER, E., and B.R. STEPHAN, eds. 1986. Recent research on scleroderris canker of conifers. Forstliche Bundesversuchsanstalt, Vienna. 167 pp.

DORWORTH, C.E. 1970. *Scleroderris lagerbergii* Gremmen and the pine replant problem in central Ontario. Can. For. Serv. Ont. Reg. For. Res. Lab., Sault Ste. Marie, Ont. Inf. Rep. O-X-139. 12 pp.

DWINELL, L.D., J.B. BARROWS-BROADDUS, and E.G. KUHLMAN. Pitch canker: a disease complex of southern pines. Plant Dis. *69*:270–276.

EHRLICH, J. 1934. The beech bark disease, a *Nectria* disease of *Fagus*, following *Cryptococcus fagi* (Baer). Can. J. Res. *10*:593–692.

FILIP, S.M. 1978. Impact of beech bark disease on even-age management of a northern hardwood forest (1952 to 1976). USDA For. Serv. Gen. Tech. Rep. NE-45. 7 pp.

FRENCH, W.J. 1969. Eutypella canker on *Acer* in New York. N.Y. State Univ. Coll. For. Syracuse Tech. Publ. 94. 56 pp.

GROSS, H.L. 1984. Defect associated with Eutypella canker of maple. For. Chron. *60*:15–17.

HOUSTON, D.R. and J.T. O'BRIEN. 1983. Beech bark disease. USDA For. Serv. For. Insect Dis. Leafl. 75. 8 pp.

HOUSTON, D.R., and D. WAINHOUSE, eds. 1983. Proc. I.U.F.R.O. Beech Bark Dis. Working Party Conf. USDA For. Serv. Gen. Tech. Rep. WO-37. 140 pp.

HOUSTON, D.R., E.J. PARKER, R. PERRIN, and K.J. LANG. 1979. Beech bark disease: a comparison of the disease in North America, Great Britain, France, and Germany. Eur. J. For. Pathol. *9*:199–211.

HUBBES, M. 1964. New facts on host-parasite relationships in Hypoxylon canker of aspen. Can. J. Bot. *42*:1489–1494.

JORGENSEN, E., and J.D. CUTLEY. 1961. Branch and stem cankers of white and Norway spruce in Ontario. For. Chron. *37*:394–404.

KAMIRI, L.K., and F.F. LAEMMLEN. 1981. Effects of drought-stress and wounding on cytospora canker development on Colorado blue spruce. J. Arboric. *7*:113–116.

MACDONALD, W.L., F.C. CECH, J. LUCHOK, and C. SMITH eds. 1978. Proc. Am. Chestnut Symp. W. Va. Univ. Morgantown, W. Va. 122 pp.

MANION, P.D. 1975. Two infection sites of *Hypoxylon mammatum* in trembling aspen (*Populus tremuloides*). Can. Bot. *53*:2621–2624.

MANION, P.D., ed. 1984. Scleroderris canker of conifers. Martinus Nijhoff/Dr. W. Junk Publishers. Dordrecht, The Netherlands. 273 pp.

MANION, P.D. and M. BLUME. 1975. Epidemiology of Hyoxylon canker of aspen. Proc. Am. Phytopathol. Soc. *2*:101.

MANION, P.D., and D.W. FRENCH. 1967. *Nectria galligena* and *Ceratocystis fimbriata* cankers of aspen in Minnesota. For. Sci. *13*:23–28.

MANION, P.D., and D.H. GRIFFIN. 1986. Sixty-five years of research on hypoxylon canker of aspen. Plant Dis. *70*:803–808.

MAROSY, M., R.F. PATTON, AND C.D. UPPER. 1989. A conducive day concept to explain the effect of low temperature on the development of scleroderris shoot blight. Phytopathology *79*:1293–1301.

McCRACKEN, F.I., and E.R. TOOLE. 1974. Canker-rots in southern hardwoods. USDA For. Serv. For. Pest Leafl. 33. 4 pp.

MILLER-WEEKS, M., and J.T. O'BRIEN. 1978. Beech bark evaluation survey. USDA For. Serv. Northeast. Area, State Private For. Eval. Rep P-78-2-4.

ROANE, M.K., G.J. GRIFFIN, and J.R. ELKINS. 1986. Chestnut blight, other Endothia diseases, and the genus *Endothia*. The American Phytopathological Society, St. Paul, Minn. 53 pp.

SCHOENEWEISS, D.F. 1983. Drought predisposition to cytospora canker in blue spruce. Plant Dis. *67:383–385.*

SHIGO, A.L. 1972. The beech bark disease today in the northeastern U.S. J. For. *70*:286–289.

SKILLING, D.D. 1977. The development of a more virulent strain of *Scleroderris lagerbergii* in New York state. Eur. J. For. Pathol. *7*:297–302.

SPAULDING, P. T.J. GRANT, and T.T. AYERS. 1936. Investigations of Nectria diseases in hardwoods of New England. J. For. *34*:169–179.

THOMAS, C.S., and J.H. HART. 1986. Relationship between year of infection, tree age, tree growth and nectria canker of black walnut in Michigan. Plant Dis. *70*:1121–1124.

TWERY, M.J., and W.A. PATTERSON III. 1984. Variation in beech bark disease and its effects on species composition and structure of northern hardwood stands in central New England. Can. J. For. Res. *14*:565–574.

VALENTINE, F.A., P.D. MANION, and K.E. MOORE. 1976. Genetic control of resistance to Hypoxylon infection and canker development in *Populus tremuloides*. Proc. 12th Lake States For. Tree Improv. Conf. USDA For. Serv. Gen. Tech. Rep. NC-26, pp. 132–146.

VAN ALFEN, N.K., R.A. JAYNES, S.L. ANAGNOSTAKIS, and P.R. DAY. 1975. Chestnut blight: biological control by transmissible hypovirulence in *Endothia parasitica*. Science *189*:890–891.

WATERMAN, A.M. 1955. The relation of *Valsa kunzei* to cankers on conifers. Phytopathology *45*:686–692.

WEIDENSAUL, T.C., and F.A. WOOD. 1974. Analysis of a maple canker epidemic in Pennsylvania. Phytopathology *64*:1024–1027.

WOOD, F.A., and J.M. SKELLY. 1964. The etiology of an annual canker on maple. Phytopathology *54*:269–272.

13

FUNGI AS AGENTS OF TREE DISEASES: VASCULAR WILT DISEASES

- *TYPES OF WILTS*
- *MODE OF ACTION OF WILT DISEASES*
- *WILT DISEASE CYCLE*
- *SYMPTOMS OF WILT DISEASES*
- *DIAGNOSIS OF WILT DISEASES*
- *EXAMPLES OF WILT DISEASES AND CONTROLS*
- *ORIGIN OF WILT DISEASES*

Photosynthesis and transpiration by leaves require large amounts of water. In vascular plants, water is taken up from the soil by the roots and transported via the vessels or tracheids of the xylem to the leaves.

A continuous unbroken column of water from the roots to the leaves allows water to be lifted hundreds of feet. Root pressure is part of the upward pumping mechanism. Some energy for upward pumping may be supplied by living parenchyma cells around the vessels. However, the basic mechanism of water transport relies on the cohesive tension of water molecules. Molecule-by-molecule replacement of transpired water in the leaves with molecules of water picked up in the roots occurs to keep a closed water system. Any breaks in the closed system immediately induce an interruption or void space in the column of water, that expands instantaneously because of the intense vacuum. A vacuum can enlarge the void space but cannot lift water above 9.75 m, because this is the maximum vertical displacement of a column of water under atmospheric pressure.

Transpiration of individual trees varies from zero on rainy nights to 200 to 400 liters per day on sunny days. Any disruption in the transport of water therefore quickly induces wilting and death.

Rapid wilting and death of "normal"-looking trees during hot, dry periods is a dramatic thing to see, and therefore wilt disease epidemics produce a high degree of emotional concern by individuals and communities.

TYPES OF WILTS

Wilting of leaves can be caused by bacterial or fungal colonization of plant vessels. It can be caused by organisms that disrupt or destroy uptake of water by feeder roots. It can be caused by drought. It can also be caused by canker fungi girdling branches.

The types of wilts discussed in this chapter are caused by the fungal invasion of xylem vessels.

MODE OF ACTION OF WILT DISEASES

Wilt-disease-causing organisms invade the vessels of plants, disrupt water movement, and subsequently cause wilting.

Wilt diseases can occur in all plants with vessels, but are especially serious in ring-porous trees. Upward water movement in ring-porous trees takes place primarily in the current year's vessels. Therefore, plugging of the current year's vessels causes rapid wilting. Diffuse porous trees transport water in vessels of a number of annual rings, so many small vessels need to be plugged before wilting occurs.

The mechanisms of disruption are varied and not completely understood. One basic mechanism involves the production of tyloses by the plant. Tyloses are a defense reaction plugging of vessels produced by disruption of the plasma membrane of parenchyma cells adjoining the vessels. Membranes and the contents of parenchyma cells expand through the connecting pits into the vessels. Another mechanism of wilting involves plugging of vessel end-wall perforation plates by the fungus spores and hyphae caught on the end-wall plate. A third mechanism involves production of toxins by wilt-causing organisms that disrupt the water-pumping mechanisms either at the leaves or along the vessels. A fourth mechanism involves changes in viscosity of vessel fluids as the result of enzymatic degradation and solubilization of cell-wall material, thereby slowing down the flow rate of water to the leaves. A fifth mechanism involves parasitism of parenchyma cells, with dysfunction of the vessels as a secondary reaction. A sixth mechanism involves embolism or introduction of air bubbles into the vessels through fungus disruption of cell walls.

The effect of any of these six mechanisms is to damage or slow down water transport within the tree. Air is introduced into the vessels and the affected part of the system stops functioning.

WILT DISEASE CYCLE

The inoculation point for wilt pathogens is a wound in either the stem or the root system, depending upon the disease. The intense vacuum of the vessel system often draws spores well within the host during wounding.

Germination and subsequent mycelial or budding-type sporulation occurs. Plugging of vessels and disruption of vessel integrity by the fungus causes additional breakage and suction columns. The fungus may be moved about rapidly. Invasion of parenchyma cells and movement between vessels occurs through pits, expanding the sphere of influence of the fungus.

Sporulation and dissemination of wilt pathogens generally occur after the tree dies. Sporulation is affected by how fast the tree dries out. Different types of relationships have evolved between the wilt pathogens and insect vectors. These are more fully developed with specific examples.

SYMPTOMS OF WILT DISEASES

Initially, limp or drooping normal-colored leaves may be observed. Within a few days, the wilted leaves change color to yellow brown and die. They may be shed or remain on the tree.

Vascular discoloration is usually evident. Cross or diagonal sections of infected branches will show discoloration of the current year's vessels.

DIAGNOSIS OF WILT DISEASES

The diagnosis of wilt diseases is initially based on symptoms. Verification of a specific causal agent is accomplished by isolation and identification of the fungus on agar media. Branches collected from recently dead areas are surface-sterilized, and chips of outer xylem are aseptically transferred to culture media. About 1 week later, the fungi are identified based on characteristic asexual fruiting structures.

Proof of pathogenicity of a new suspected pathogen is accomplished using Koch's postulates.

EXAMPLES OF WILT DISEASES AND CONTROLS

Dutch Elm Disease

Infection by *Ceratocystis ulmi* (Figs. 13-1 to 13-5) takes place as a result of feeding wounds in small branch crotches by *Scolytus multistriatus* (smaller European elm bark beetle). Recent arguments suggest that the proper name of the pathogen should be *Ophiostoma ulmi*.

Figure 13–1 Dutch-elm-diseased American elm, showing branches with wilting leaves in the lower crown and dead as well as normal-looking branches in the upper crown.

Figure 13–2 Close-up of wilting leaves of American elm.

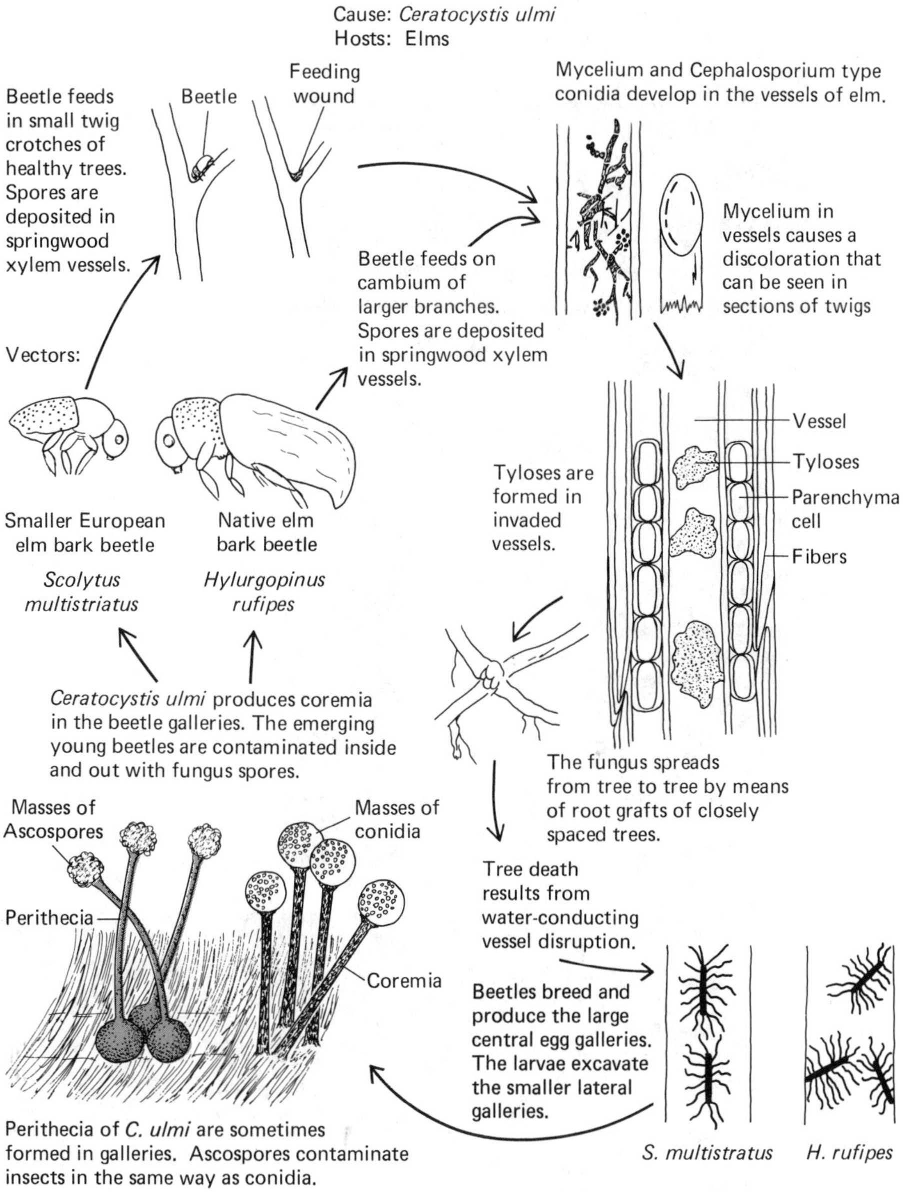

Cause: *Ceratocystis ulmi*
Hosts: Elms

Beetle feeds in small twig crotches of healthy trees. Spores are deposited in springwood xylem vessels.

Beetle

Feeding wound

Mycelium and Cephalosporium type conidia develop in the vessels of elm.

Beetle feeds on cambium of larger branches. Spores are deposited in springwood xylem vessels.

Mycelium in vessels causes a discoloration that can be seen in sections of twigs

Vectors:

Smaller European elm bark beetle

Scolytus multistriatus

Native elm bark beetle

Hylurgopinus rufipes

Tyloses are formed in invaded vessels.

Vessel
Tyloses
Parenchyma cell
Fibers

Ceratocystis ulmi produces coremia in the beetle galleries. The emerging young beetles are contaminated inside and out with fungus spores.

The fungus spreads from tree to tree by means of root grafts of closely spaced trees.

Masses of Ascospores

Masses of conidia

Perithecia

Coremia

Tree death results from water-conducting vessel disruption.

Beetles breed and produce the large central egg galleries. The larvae excavate the smaller lateral galleries.

S. multistratus   *H. rufipes*

Perithecia of *C. ulmi* are sometimes formed in galleries. Ascospores contaminate insects in the same way as conidia.

Figure 13-3   Disease cycle of Dutch elm disease.

The infection takes place in the spring of the year, when large springwood vessels are near the surface. Elm trees that do not produce large springwood vessels have been shown to be somewhat resistant to the disease. Infection can also take place in large stems as a result of feeding activity of the native elm bark beetle (*Hylurgopinus rufipes*). In both cases, the beetle feeds on the phloem cambium region. Some chewing into the xylem exposes recently formed vessels. Spores of *C.*

Figure 13-4 Smaller European elm bark beetle galleries in American elm.

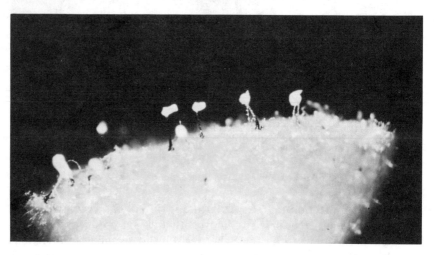

Figure 13-5 Coremia of *Ceratocystis ulmi* are produced in bark beetle galleries. Elm bark beetles become contaminated by the conidia produced in white sticky masses at the tips of the coremia.

ulmi on the surface of the feeding beetle may be dislodged from the beetle and introduced into injured vessels. The fungus probably moves passively down the tree as suction columns develop in disrupted vessels. It moves around the circumference of the tree in the root region, and eventually plugs vessels around the whole tree. The fungus gains access to vessels of adjacent trees through root grafts between trees.

Fungus sporulation in beetle pupal chambers and galleries is ideally coordinated with the activity of the two types of bark beetles. Spores are passively carried as contaminants on the surface of beetles which emerge from infected dead trees and fly to healthy trees to feed.

Control of Dutch elm disease (DED) is more a political problem than a biological problem. The importance of the elimination of insect-breeding places in dead logs and the severing of root grafts has been known for a long time. Spraying with insecticides has never been completely satisfactory and is definitely not sufficient as a single means of control. Extensive losses of street elms result because political bodies are unwilling to properly care for the trees. The cost to remove an elm that has been dead for 2 years is similar to the cost to remove a recently killed tree. Removal of a few living trees to improve spacing is probably less expensive than removal after they die, because of root grafts to DED-infected trees. The cost is the same, but early removal of dead trees and improvement of spacing or severing root grafts reduces subsequent losses in the remaining population of elms.

Control of DED must be centered around a good sanitation program to remove all dead elm material. Dead branches in standing live trees as well as dead elm trees must be removed from the control area and burned or buried. In the absence of careful elimination of all dead elm material, it is almost impossible to manipulate DED effectively.

In some communities it is almost impossible to manage all the elms because of wild populations around and within the community. Beetle management through sanitation can be supplemented, in this instance, with mass trapping using an aggregation pheromone for the European elm bark beetle. One approach uses mass trapping of beetles with pheromone samples placed in the center of sticky pieces of paper attached to utility poles. Another approach attaches the pheromone sample to trees that are dying of Dutch elm disease. These trap trees draw in large numbers of beetles that breed and lay eggs. If the trees are injected with cacodylic acid, the larval stages of the insect do not develop properly, thereby reducing the insect population.

Can Dutch elm disease be controlled? Many communities try. Many fail. Very few are satisfied. One has to consider the economics of the system. There is no question that Dutch elm disease control programs cost money. But costs to the community are 37 to 76% less with control programs than without. A community can expect to maintain most of its elm population only with a sustained high-performance control program. Any suspension in effort cannot be reversed. Losses will occur. But costs and losses will occur in any mature tree population. Therefore, the answer is yes—Dutch elm disease can be controlled. Elms can be maintained as cost effectively as can any other species.

Overenthusiasm today regarding the use of systemic fungicides for the control of Dutch elm disease is somewhat misplaced. The cost of treatment per tree is rather high. How much more effectively could the money be utilized for the destruction of root grafts and beetle breeding places? The questionably effective "control" by the fungicide is only good for 1 to 3 years, so the high cost of application must be repeated. The method of application involves pressure injection through wounds made in the stem.

Wounds produced for injection are extensively invaded by stain and decay fungi. All of these drawbacks make fungicidal injections practical for a limited number of very important trees.

My note of sarcastic pessimism in relation to DED is intended. Hopefully, people like you, who have an appreciation of the biologists' ways of manipulating nature, will provide a political base to help avoid such problems in the future. There will be many new problems involving ornamental trees in the future no matter what species is planted.

A point to keep in mind when evaluating or considering control is what level of disease incidence you are willing to accept, because absolute control is not possible. One must also recognize that trees are not immortal and that disease prevention or control is more difficult with older trees. Expectations for saving the original overmature population of elms in our eastern cities were overly optimistic. Many of these trees were dying in the absence of Dutch elm disease. A young, vigorously growing, well-cared-for population of elms, if planted today, could be expected to do better than most of the other tree species presently being planted in our cities. Today there is available a Dutch elm disease–resistant American elm that is marketed under the name American Liberty. A single resistant elm line is not necessarily the answer to the Dutch elm disease problem, but such a tree could encourage some communities to try a few elms. It is also interesting to note that Washington, D.C., has mandated an effective commitment to maintaining its elm population. The role of elms in urban tree management is discussed further in Chapter 22.

Oak Wilt

Oak wilt caused by *Ceratocystis fagacearum* is a serious disease, particularly of red and black oaks. The disease is distributed from Minnesota to Pennsylvania south to North Carolina, Arkansas, and Texas. Pathologists have been concerned over the effect this disease would have on oaks of the deep south, where it has not been reported until recently. Inoculations of oaks in southern Arkansas caused death of the inoculated trees, but the pathogen did not sporulate and spread to additional trees. Recent reports of *C. fagacearum* associated with the serious decline of live oak in Texas has changed our concepts of this disease. The pathogen does not occur on the European continent, and there is a strong effort to prevent its introduction. For some unknown reason, the pathogen also does not occur in New York or New Jersey.

The disease cycle of the oak wilt pathogen is similar to Dutch elm disease in some ways but is very different in others (Figs. 13-6 to 13-8). The first difference is in the means of dissemination and infection. The oak wilt pathogen is rather passively disseminated by sap feeding Nitidulid beetles. These insects are attracted to the sweet-smelling odor of the fungus associated with fungus mats (Fig. 13-9) that develop below the bark of recently killed trees. Pressure pads, produced with the

Cause: *Ceratocystis fagacearum*

Hosts: Species of Quercus (all oaks)

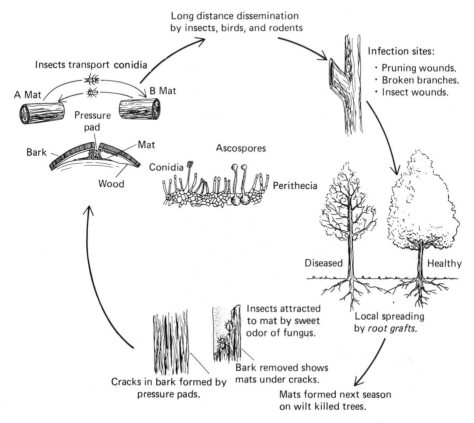

Figure 13-6 Oak wilt disease cycle.

mats, crack the bark and allow access by insects. The fungus sporulating on the mat is passively picked up by the Nitidulids and any other visitors to the mats, including squirrels and birds. The greater importance for Nitidulid beetles in vectoring the fungus is their attraction not only to the mats but also to wounds in healthy trees.

Within vessels, the *C. fagacearum* pathogen is similar to the *C. ulmi* pathogen. Root grafts are also a major means of spread from one tree to another.

Control of oak wilt has traditionally involved sanitation similar to Dutch elm disease, but the reason for the removal of the dead trees was to eliminate material on which fungus mats were formed, rather than to eliminate beetle breeding areas. Methods of deep girdling to cause rapid death and drying of infected trees, as well as cutting of infected trees and surrounding trees, were used in major control efforts in West Virginia and Pennsylvania, respectively. Evaluation of the effectiveness of control efforts compared to noncontrolled areas demonstrated the ineffectiveness of both treatments in preventing spread of the disease. The control procedures reduced mat production but failed to influence the disease increase.

Examples of Wilt Diseases and Controls

Figure 13-7 Oak wilt infection center in Wisconsin.

At the present time, one can recommend that pruning be avoided during the growing season in areas where oak wilt occurs. This is because it appears that wounds attract a number of insects that may be vectors of the pathogen. Root grafts should be severed between infected and healthy trees. Diseased trees should be cut and the wood disposed of to prevent the production of pressure pads. To prevent the exportation of oak wilt to Europe, pathologists have developed a methyl bromide fumigation protocol to kill the pathogen and any possible insect vectors in logs before they are loaded onto ships. Another effective procedure is to debark the logs.

Verticillium Wilt

Verticillium wilt, caused by *Verticillium dahliae* or *V. albo-atrum,* is a vascular wilt disease that could also be considered a root disease (see Chapter 16) because infection occurs through root contact with soilborne inoculum.

The *Verticillium* fungus causes wilts on a number of agricultural crops as well as on trees. The most seriously affected trees are maples and elms, but the symptoms are slow to develop and not as striking as the last two wilts. If Dutch elm disease is present in an area, we very seldom recognize a Verticillium-wilt-affected elm tree. In

Figure 13-8 Close-up of wilted leaves.

Figure 13-9 *Ceratocystis fagacearum* pressure pads produced below the bark. The sweet odor of the fungus mats attracts insects, birds, and rodents, which become vectors of the fungus spores produced on the fungus mat. (Photograph compliments of Dr. Savel Silverborg.)

Examples of Wilt Diseases and Controls

maples, the death of branches and the tree is slow enough that the condition is often diagnosed as part of the maple decline problem (Figs. 13-10 and 13-11).

Verticillium wilt causes a gradual wilting and death of branches, and eventually the whole tree. One characteristic symptom of the disease is a green or brown streaking in the xylem of affected branches, stems, and roots. Positive identification is possible only by culturing the fungus. Infection with this disease is completely different from Dutch elm disease or oak wilt. In this instance, infection occurs through small wounds in the roots. As the roots grow, they may pass near some inoculum which is maintained as dormant sclerotia in the soil. The presence of susceptible host roots causes germination of the inoculum leading to infection. Invasion of the vascular system can be rather irregular. Some branches are invaded while others are not. Wilt symptoms and death can occur to individual branches and eventually the total tree.

Sclerotia produced by the fungus within the roots of the infected host are released to the soil as the dead host tissue decomposes. The sclerotia can remain dormant for extended periods. Replanting trees in infested soil results in a high probability of infection of the new plantings. Therefore, the only recommendation that can be made with regard to this disease is to cut the diseased trees and not to replant with another susceptible species.

Verticillium wilt may become a more serious problem in the future for urban populations of trees, because of the present lack of concern for the problem. Many shade-tree nurseries today are established on former agricultural land. If the former agricultural crop included tomatoes or potatoes, the level of *Verticillium* sclerotia in the soil will be very high, because these crops and agricultural practices are ideally suited for the development of the pathogen. At the present time, trees showing

Figure 13-10   Verticillium-wilt-infected branch on Norway maple.

Figure 13-11 Close-up of Verticillium wilt, showing necrotic blotch symptoms on Norway maple leaves and green discoloration in the xylem of the infected branch.

symptoms of Verticillium wilt are not discriminated against. The nursery operator cuts off the branches showing the symptoms and sells these trees along with the healthy ones. They have a good chance of surviving the guarantee period, after which the nursery operation is off the hook. We do not know the long-term effect of this practice or the conditions that cause the disease to flare up again in infected trees. We can predict that there will be future problems due to this practice.

We do not fully understand the possible consequences of many of our actions today. Future generations will have to absorb the costs of our mistakes with Verticillium wilt of maples, just as this generation had to absorb the cost of the mistake involved in planting too many elms.

ORIGIN OF WILT DISEASES

It is interesting to consider the origin of the wilt pathogens we are presently plagued with. The suggestion has been made that for agricultural crops the application of plant breeding and intensive monoculture undoubtedly contributed to the emergence of some fungal wilt pathogens.

The origin of the oak wilt pathogen is a mystery. Some have suggested that it was introduced, based on a high rate of disease increase, but further data have not substantiated the high rates of disease increase. Oak wilt does not occur on any other continent, so it must be native to North America. We might speculate that *C. fagacearum* originated from a saprobic stain fungus.

The hypothesis of evolution from saprobic stain fungus is not unique to oak wilt. The idea has been proposed for the Dutch elm disease pathogen and is backed up by some observations. The *C. ulmi* pathogen as a serious disease agent originated on the European continent around the turn of the century. All introductions to other continents can be traced to this center of origin. The pathogen spreads like an introduced pathogen in Europe as well as anywhere it becomes established.

An interesting finding with the *C. ulmi* fungus is that some strains or cultures of the fungus are not pathogenic. These survive as saprobic stain fungi on dead elm logs.

The association of bark beetles and species of *Ceratocystis* is more than coincidence. The long-necked perithecia with an accumulation of sticky ascospores at the top of the perithecia neck, as well as the long coremia, with sticky spores in a mass at the top, appear to be an evolutionary adaptation for insect dissemination and association. Both spore stages are found in insect galleries. The genus *Ceratocystis* is well known for its association with insects and wood stain.

To make the transition from a saprobic existence in association with bark beetles to a parasite of elm trees would require a mutation that allows the fungus to survive within the vessels of living trees. It must be able to tolerate the defense mechanisms or bypass the activation of defense mechanisms by the tree to survive in the living tree. We do not know enough about the mechanism of interaction between the tree and the fungus to speculate as to what type of mutation occurred to allow survival within the tree. Once such a mutant occurred, it would by chance be deposited in the vessels during normal feeding by beetles. The fungus that could survive in the tree and cause its death would probably predominate in the staining of the wood and invasion of beetle galleries, so a strong selection pressure for the pathogenic strain would develop in the fungus population.

A postscript to the foregoing speculation regarding the origin of *C. ulmi* should indicate that additional mutations and selection for pathogenic capacity can occur. Pathologists on the European continent are now finding that the elms they developed and planted widely because of resistance to the Dutch elm disease are now dying of a wilt disease they would like to call American elm disease. They speculate that we exported a more virulent strain of the fungus *C. ulmi* back to Europe on elm logs presently being exported to Europe for manufacturing into veneers. The aggressive strain of the fungus appears to be spreading in North America also.

There is one feature common to these three wilt diseases. They all probably are of recent serious pathological origin. If this is the case, we can expect wilt diseases of other species to develop along similar lines—we have not seen the last of the wilt disease epidemics. Hopefully, we can learn from the ones we now have, so as to avoid extensive losses from future wilt epidemics.

REFERENCES

APPLE, D.N., R. PETERS, and R. LEWIS. 1987. Tree susceptibility, inoculum availability and potential vectors in a Texas oak wilt center. J. Arboric. *13*:169–173.

CAMPANA, R.J. 1974. DED controls: will systemics work? Weed Trees Turf, *74* (May):16–17.

CANNON, W.N., JR., and D.P. WORLEY. 1976. Dutch elm disease control: performance and costs. USDA For. Serv. Res. Pap. NE 345. 7 pp.

CANNON, W.N., JR., J.H. BARGER, and D.P. WORLEY. 1977. Dutch elm disease control: intensive sanitation and survey economics. USDA For. Serv. Res. Pap. NE 387. 9 pp.

DIMOND, A.E. 1970. Biophysics and biochemistry of the vascular wilt syndrome. Annu. Rev. Phytopathol. *8*:301–322.

ELGERSMA, D.M. 1969. Resistance mechanisms of elms to *Ceratocystis ulmi*. Phytopathol. Lab. "Willie Commelin Scholten" 77, Baarn, The Netherlands. 84 pp.

HIMELICK, E.B. 1969. Tree and shrub hosts of *Verticillium albo-atrum*. Ill. Nat. Hist. Surv. Biol. Notes 66. 8 pp.

HOUSTON, D.R. 1985. Spread and increase of *Ceratocystis ulmi* with cultural characteristics of the aggressive strain in northeastern North America. Plant Dis. *69*:677–680.

Jewell, F.F.1956. Insect transmission of oak wilt. Phytopathology *46*:244–257.

JONES, T.W. 1971. An appraisal of oak wilt control programs in Pennsylvania and West Virginia. USDA For. Serv. Res. Pap. NE-204. 15 pp.

JONES, T.W. 1972. Oak wilt. USDA For. Serv. For. Pest Leafl. 29. 7 pp.

JUZWIK, J., D.W. FRENCH, and J. JERESEK. 1985. Overland spread of the oak wilt fungus in Minnesota. J. Arboric. *11*:323–327.

KRAMER, P.J., and T.T. KOZLOWSKI. 1960. Physiology of trees. McGraw-Hill Book Company, New York. 642 pp.

LANIER, G.N., D.C. SCHUBERT, and P.D. MANION. 1988. Dutch elm disease and elm yellows in central New York. Plant Dis. *72*:189–194.

MACDONALD, W.L., E.L. SCHMIDT, and E.J. HARNER. 1985. Methyl bromide eradication of the oak wilt fungus from red and white oak logs. For. Prod. J. *35*:11–16.

MACE, M.E., A.A. BELL, and C.H. BECKMAN. 1981. Fungal wilt diseases of plants. Academic Press, Inc., New York. 640 pp.

MERRILL, W. 1967. The oak wilt epidemics in Pennsylvania and West Virginia: an analysis. Phytopathology *57*:1206–1210.

MERRILL, W. 1968. Effect of control programs on development of epidemics of Dutch elm disease. Phytopathology *58*:1060.

MILLER, H.C., S.B. SILVERBORG, and R. J. CAMPANA. 1969. Dutch elm disease: relation of spread and intensification to control by sanitation in Syracuse, New York. Plant Dis. Rep. *53*:551–555.

NEWBANKS, D., A. BOSCH, and M.H. ZIMMERMAN. 1983. Evidence of xylem dysfunction by embolization in Dutch elm disease. Phytopathology *73*:1060–1063.

SINCLAIR, W.A., and R.J. CAMPANA, eds. 1978. Dutch elm disease perspectives after 60 years. Northeast. Reg. Res. Publ. Search. Vol. 8, No. 5. 52 pp.

SINCLAIR, W.A., J.L. SAUNDERS, and E.J. BRAUN. 1975. Dutch elm disease and phloem necrosis. Cornell Tree Pest Leafl. A-9. 20 pp.

SINCLAIR, W.A., K.L. SMITH, and A.O. LARSEN. 1981. Verticillium wilt of maples: symptoms related to movement of the pathogen in stems. Phytopathology *71*:340–345.

SMITH, L.D., and D. NEELY. 1979. Relative susceptibility of tree species to *Verticillium dahliae*. Plant Dis. Rep. *63*:328–32.

STIPES, R.J., and R.J. CAMPANA. 1981. Compendium of elm diseases. The American Phytopathological Society, St. Paul, Minn. 96 pp.

WILSON, C.L. 1965. *Ceratocystis ulmi* in elm wood. Phytopathology *55*:447.

ZIMMERMAN, M.M. 1963. How sap moves in trees. Sci. Am. *208*(3):132–142.

14

FUNGI AS AGENTS
OF TREE DISEASES:
WOOD DECAY

- **TYPES OF WOOD DECAY**
- **MODE OF ACTION OF WOOD DECAY FUNGI**
- **WOOD DECAY "DISEASE" CYCLE**
- **EVIDENCE AND EFFECTS OF DECAY**
- **METHODS FOR DETECTION OF DECAY**
- **IDENTIFICATION OF DECAY FUNGI BASED ON FRUIT BODIES**
- **EXAMPLES OF HEART ROT AND SAPROBIC DECAYS**
- **ROLE OF HEART ROT IN FOREST MANAGEMENT AND PLANT SUCCESSIONS**
- **CONTROL OF HEART ROT DECAY**
- **EXAMPLES OF WOOD DECAY IN THE HOME**
- **CONTROL OF WOOD DECAY IN PRODUCTS**

Wood decay is a very broad subject involving at least two major subdivisions. One aspect of wood decay deals with decay of living trees. Another deals with decay of dead trees and wood products. Decay of living trees, heart rot, accounts for more loss in saw timber than fire, insects, weather, or any other disease agent. About one-third of all losses in saw timber are caused by heart rot. Decay of dead trees is absolutely required to properly recycle carbon, nitrogen, and other elements that are tied up in the tree. Decay of wood products is also significant. Ten percent of the annual cut of timber is utilized to replace wood that has deteriorated because of decay fungi.

TYPES OF WOOD DECAY

Wood decay in its broadest sense, wood deterioration, is caused by a wide variety of insects, marine animals, fungi, bacteria, and physical factors of the environment. This chapter deals with the major decay causing organisms, the fungi that chemically and enzymatically digest woody cell walls.

Even the limited area of fungi that enzymatically digest woody cell walls is too broad, because fungi that cause cankers and fungi that cause wilts may enzymatically digest cell-wall material. In an attempt to narrow the subject, we limit our discussion to those fungi that live and derive nutrients primarily by digestion of woody-cell-wall materials. Those fungi that utilize living cell contents as major source of nutrients but also digest wood are discussed in other chapters.

Heart Rotters and Saprobic Decayers

A basic separation of wood decay can be made on the basis of whether the fungus survives primarily in living trees or in dead trees and wood products. The decay fungi of living trees are heart rots. The decay fungi of dead trees and wood products are saprobic decomposers. Heart rot fungi are not confined to the heartwood of living trees. These fungi usually infect through wounds in the sapwood of living trees. Through a tree defense process, termed compartmentalization (see heart rot disease cycle), the heart rot fungus is generally restricted in its invasion to only those tissues present at the time of wounding. Over the years additional wood is produced by the tree to produce eventually a cylinder of sound decay-free wood surrounding the decay column in the center (heartwood) of the tree.

Saprobic decay fungi rapidly invade the sapwood of dead trees. A large wound on a living tree may result in a large area of dead tissue with the same properties as a dead tree. Therefore, saprobic decayers, like *Fomes fomentarius,* may be seen on living beech trees. Saprobic decay fungi invade sapwood first but will eventually also decompose the more decay-resistant heartwood of dead trees and wood products.

We like to categorize the decay fungi into these two groups, but they may not recognize our system. For example, *Ganoderma applanatum* (*Fomes applanatus*) operates very much like a saprobic decay fungus in the colonization and decomposition of American elm stumps, but operates like a heart rot fungus by causing a root and butt rot of living trembling aspen. One cannot, therefore, totally categorize the fungi into these two groups without considering the host.

Top Rot and Root and Butt Rot

The heart rots are further separated into top rots and root or butt rots, depending upon which portion of the tree is affected. Top rot decay fungi are found in the wood of upper portions of trees. The fungi seldom progress very far into the roots and therefore do not spread from one tree to another via roots or from the stumps of a harvested crop to the next generation of sprouts or suckers.

Root and butt rot fungi colonize the lower stem and roots of trees. They may parasitize the cambium of roots or may remain in the central xylem tissues. Those that parasitize cambium are discussed more fully in Chapter 16.

Root and butt rot fungi can cause serious problems with stands regenerated from stump sprouts and root suckers. Decay in the parent trees and infection of cut stumps may inoculate the new stand with root rot fungi. Fire and logging scars are also common points of infection for root and butt rotters.

Slash Rot and Products Decay

The saprobic decay fungi are artificially separated into slash rotters and products decay fungi. Slash rotters decompose branches, stems, and roots of dead trees to recycle the carbon, minerals, and other basic elements tied up in the complex organic structure of wood. The products decay fungi are a group of slash rotters commonly associated with the deterioration of wood products.

White Rot, Brown Rot, and Soft Rot

Another method of classifying decay types is based on the chemistry of the decay process and the modified appearance of the decayed wood (Fig. 14-1). Some fungi, called brown rot fungi, decompose primarily the carbohydrate component of the wood. Preferential decomposition of the carbohydrates, cellulose, and hemicellulose, would leave much of the brown-colored lignin to give the wood a brown appearance. White rot fungi, on the other hand, decompose all cell wall components. However, they may attack the lignin, cellulose, or hemicellulose in different orders. In some advanced white rot decay most of the lignin has been removed, leaving a small residue of light-colored cellulose, often in localized pockets.

It is interesting to speculate on why these fungi leave some of the cellulose. It is further interesting to speculate on why many other fungi that can decompose cellulose do not use up this available material. The current thinking is that the wood substrate runs out of nitrogen. All fungi need both a carbon and a nitrogen source. If the nitrogen is used up, there is no further decomposition even if there is a readily available carbon source.

The white rots are further subdivided into stringy, spongy, mottled, or pocket rots. Sometimes the decayed wood of white rot fungi is more yellow or yellow-brown than white. But the decayed wood is rather fibrous in appearance. The white rot fungi, decomposing the lignin of the middle lamella, cause the cells to separate into the fibrous material. The brown rot decay is readily differentiated from these yellow colored white rots because brown rot fungi do not decompose the middle lamella sufficiently to cause separation of the cells into a fibrous material.

Brown rot in its later stages generally causes fractures across the grain that break the wood up into small or large brown-colored cubes. The mycelium of the fungus may develop white felts in the brown rot fractures, but the cross-grain fractures distinctively identify the brown rot fungi. Examples of these types of rots are shown in Fig. 14-1.

(a) (b) (c)

(d) (e) (f)

Figure 14-1 Appearance of decayed wood. A, Brown cubical rot; B, Brown cubical rot with mycelial felts in fractures; C, Surface checking associated with soft rot; D, White Stingy rot; E, White Spongy rot; F, White pocket or mottled rot.

A third type of decay fungus, called a soft rotter, is separated from the others primarily on the selective attack of only a portion of the cell wall. They decompose largely cellulose and hemicellulose in localized pockets in the secondary cell wall. Soft rot occurs commonly in wood that is saturated with water or in wood that is in direct contact with soil. Soft rot decay is characterized by surface softness and formation of shallow surface cross checking in the wood upon drying, as seen in Fig. 14-1. Soft rots are caused by fungi in the subdivisions Ascomycotina and Fungi Imperfecti, while the other decays are usually caused by Hymenomycetes of the subdivision Basidiomycotina.

MODE OF ACTION OF WOOD DECAY FUNGI

The separation of wood decay fungi into white and brown rot is better understood once one develops an understanding of the basic chemistry and structure of wood and the basic enzymatic reactions of decay fungi. A wood decay model will tie together the chemistry of wood structure and enzymatic activity of fungi.

Wood Chemistry and Structure

Wood is a unique, strong, durable material that is decomposed by a limited number of microorganisms. This may sound like a contradiction to the main subject of this chapter, but it is important to recognize that wood is a rather durable material. Durability and strength are the reasons woody plants are able to survive and compete for long periods of time. Durability, strength, ease of shaping and fastening, economy, and abundance are reasons wood is one of our most important structural materials.

The durability and strength of wood are associated with specialized structural cells of the xylem, tracheids in conifers, and fibers in hardwoods. The xylem is made up of other specialized cells for transport and storage, but it is the tracheids and fibers that represent the main structural elements.

Structural woody cells have thick walls composed primarily of cellulose, hemicellulose and lignin. The holocellulose or total carbohydrate fraction of wood is about 70 to 75% of the extractive-free weight of softwoods and 75 to 82% of the extractive-free weight of hardwoods. Lignin represents 18 to 30% of the extractive-free weight or most of the noncarbohydrate weight.

The term "extractive" refers to a diverse group of chemical materials in wood that are readily extracted from wood with water or organic solvents. The extractives are not part of the cell wall. They are storage products within living cells, metabolic end products of normal cell death, or metabolic reaction products of wound-response mechanisms. Storage products are sugars and starches. The other compounds are a complex of phenolic materials.

Cellulose is a long-chain polymer of about 10,000 glucose units held together with ß-1, 4 glycosidic bonds (Fig. 14-2). Cellulose chains may aggregate into microfibrils in which a number of cellulose chains are cross-linked by hydrogen bonding into crystalline cellulose. Other portions of the cellulose chains are more loosely associated. There are number of theories on the structure of microfibrils, but all

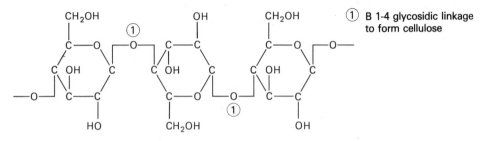

Figure 14–2 Portion of the cellulose macromolecule.

have crystalline, highly organized cellulose regions and noncrystalline, amorphous, cellulose regions.

Lignin is impregnated within the noncrystalline, amorphous portions of the cellulose matrix (Fig. 14-3). Lignin is also a major constituent of the cementing material between cells called the middle lamella. Lignin is a complex polymer with many cross-linkages between the basic building blocks, phenylpropane. In Fig. 14-3, six phenylpropane units are joined together by five different linkages. These five plus a few other linkages polymerize large numbers of phenylpropane units into a heterogeneous highly branched three dimensional structure.

Part of the holocellulose fraction of wood is made up of something other than ß-linked glucose polymers. This other fraction, hemicellulose, is made up of glucose, galactose, mannose, glucuronic acid, arabinose, and xylose units linked by glycosidic bonds between sugars. The hemicellulose molecules are tightly intermixed with the lignin polymer in the amorphous cellulose, around the crystalline cellulose regions.

Cellulose is essentially the same in all trees. However, major differences in trees occurs in the hemicelluloses and lignins. In hardwoods hemicelluloses are made of a polymer rich in xylose, a five-carbon sugar. In conifers hemicelluloses are rich in the six-carbon sugars, glucose, and mannose. The building blocks of conifer lignin generally have one methyl (OCH_3) group attached to each benzene ring, while the hardwood lignin has both one and two methyl group attachments.

True cellulose, lignin, and hemicellulose account for about 95% of the chemical constituents of the woody cell. The remaining fraction consists of pectins and

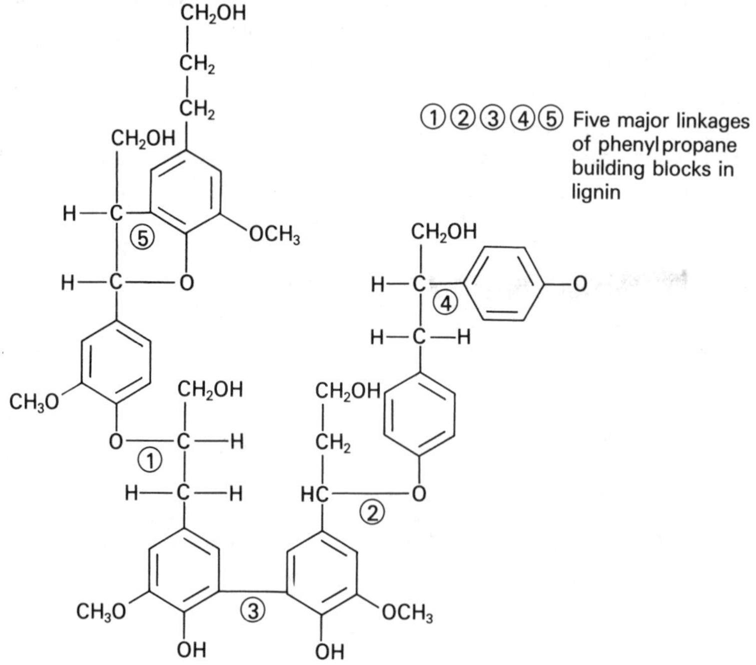

Figure 14-3   Lignin chemical structure.

extractives such as starch, complex phenols, and minerals. Pectins are glycosidically linked polymers such as hemicellulose, made up primarily of glucuronic and galacturonic acids. Starch is very much like cellulose in that it is made up of glucose units, but the glycosidic linkages between the glucose molecules in starch are α-1,4 and α-1,6 linkages. What seems like a small difference between the cellulose beta linkages and the starch alpha linkages is significant enough to prevent most animals, including man, from utilizing cellulose for food. Phenols are hydroxylated aromatic compounds of numerous configurations. The odors, colors, and natural decay resistance of wood are often the result of specific phenols.

The basic chemical building blocks of wood just described are organized into the woody cell (Fig. 14-4). Cellulose, lignin, and hemicellulose are aggregated into three dimensional substructures, which are further organized into fibrils with specific orientation within the various layers of the cell wall. The fibrilar structure of the cell wall can sometimes be seen with the light microscope, but more detailed observations require the electron microscope. A thin inner portion of the secondary wall (S_3) is made up of fibrils with an oblique angular direction in relation to the long axis of the cell. The middle portion of the secondary wall (S_2), where most of the structure and mass of the cell wall is found, is made up of fibrils with an orientation very much along the long axis of the cell. The outer layer of the secondary wall (S_1) has fibrils oriented at almost right angles to the main axis of the cell.

The secondary cell wall is surrounded on the outside by a primary wall (P), composed of cellulose, hemicellulose, and pectin. A small fraction of the primary wall is protein. The protein is bound within the wall and is not readily available to fungi, which find the wood substrate rather deficient in nitrogen.

The lumen surface or inner portion of the secondary wall is covered by a warty layer composed of debris left over when the cell died, as well as deposition products like phenols impregnated into the cells during the transition from sapwood to heartwood.

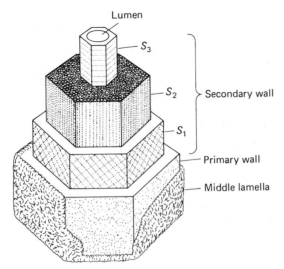

Figure 14-4 Woody-cell-wal structure.

The middle lamella, composed primarily of lignin and pectin, cements the cells together into a woody structural material. During chemical pulping of wood, the middle lamella is chemically degraded, thereby releasing the fibers, which are then regrouped and pressed together into paper.

Now that we have a general understanding of the chemical composition and basic structure of wood cells, we will consider how fiber and tracheid structural cells are formed in the tree. These cells are formed from a layer of cells between the bark and the wood, called cambium or lateral meristem. The living cambium cells divide longitudinally, successively to form phloem (bark) on the outside and xylem (wood) on the inside. The newly formed xylem cells are surrounded by a primary wall but lack the structural secondary wall. The newly formed cells expand into long spindle-shaped cells. The secondary wall is then synthesized to give them rigidity and durability. Fiber and tracheid cells die soon after the secondary wall is formed.

The sapwood or outer portion of the stem is composed of (1) dead fibers and vessels in hardwoods, or dead tracheids in conifers; and (2) living parenchyma cells. The parenchyma cells store extra starch which the tree can draw upon as energy for growth or to produce chemical defense compounds in the event of injury. In some species, such as oak or walnut, the parenchyma and ray cells eventually die, so that the central portion of the stem is entirely composed of dead cells. During the process of dying, these cells synthesize the phenolic compounds that make up the distinctive color, odor, and decay resistance of the heartwood. Other species, such as aspen, maple, and birch, do not form true heartwood. A few living parenchyma cells can be found well within the central xylem. Any color changes in these species are due to the host chemical reactions to wounds and microorganism invasion.

Wood Decay Model

Free moisture is necessary for decay. Recall from the discussion of the requirements for growth of fungi in Chapter 8 that free moisture is necessary for most fungi. Fungi decay wood only if the moisture content is above fiber saturation.

If one allows a dry piece of wood to absorb moisture in an atmosphere of 100% relative humidity, the water molecules will be absorbed and bind, by hydrogen bonds, to hydroxyl groups, primarily in the amorphous regions of the cellulose (Fig. 14-3). The absorption of water causes swelling of the cellulose fibrils and dimensional changes to the wood. Once all the potential bonds are fulfilled, the wood is said to be at fiber saturation. The actual percentage of water at this point is variable, depending upon the density of the wood, but an average figure for fiber saturation is around 28% moisture content. Theoretically, wood maintained just at fiber saturation does not have any free water within the lumens of the cells, but slight temperature fluctuations can cause moisture to condense and form a film of water in the lumen of the cell, allowing fungal growth and decay to commence.

The wood of living trees, containing in excess of 100% moisture based on oven-dry weight, can be decayed by fungi. Wood maintained in water will accumulate water until all of the cell lumens are filled with water. At or near this total saturation point, the white and brown rot decay fungi do not operate very well because of lack of oxygen. Therefore, for decay to take place, the wood must be above fiber saturation but not totally saturated.

Soft rot fungi can decay wood over a wide range of moisture conditions. Some decay wood that is totally saturated with water. Others operate closer to fiber saturation conditions. These fungi degrade the surface layers of submerged or continually wetted wood of greenhouse benches, boats, dock posts, and other structures. Only the surface layers of the wood are degraded, because of lack of oxygen in the interior of the wood. Oxygen dissolved in the water supplies the fungus with its oxygen requirements. Strings of diamond-shaped microscopic cavities in the secondary wall are often produced by soft rot fungi. On other occasions, the soft rot produces major erosions or trenches in the secondary wall.

The hyphae of white and brown rot decay fungi generally grow in the lumen of the cell. A gelatinous sheath surrounds the fungus hyphae to provide intimate contact of the hyphae and the cell wall. Enzymes and nonenzymatic depolymerizing agents produced by the fungus diffuse through this sheath to the woody cell wall surface. Cell wall breakdown products similarly diffuse through the sheath to be absorbed through the wall of the fungus. In this process it is important to recognize that free water in the cell lumen is critical for diffusion of the enzymes and breakdown products. The fungus hyphae pass from one cell to another through pits between the cells, or they may enzymatically digest bore holes through the wall.

The complex spatial arrangement of cellulose and hemicellulose within the lignin barrier or shield make wood indigestible to most organisms (Fig. 14-5). A few fungi, wood decayers, utilize specialized extracellular enzymes and nonenzymatic depolymerizing agents to overcome the lignin barrier and break these polymers into soluble fragments that can subsequently be absorbed into the fungus and utilized for energy. The mechanisms by which wood decay fungi convert wood to chemical energy have been intensively investigated for many years. There have been a number of hypotheses to describe the process. Recent research is demonstrating the interrelationships of enzymatic and nonenzymatic depolymerizing agents as well as the role of nitrogen and other factors in the breakup of the complex three-polymer system of wood.

Enzymes that degrade cellulose and hemicelluloses have been investigated for a number of years. These enzymes generally fall into two categories. Some cellulase enzymes (endocellulases) split the cellulose chain randomly into fragments of various lengths. Others (exocellulases) start at one end of the long cellulose chain and split off single-glucose units or two-glucose units, diemers (cellobiose).

In addition to the cellulase enzymes, some type of nonenzymatic cellulose depolymerizing agent produced by fungi has been suspected, since the known cellulase enzymes are too large to penetrate deeply into the crystalline regions of the secondary wall. The present theories suggest that radical-generating oxidases within the fungus produces something like a superoxide anion that is released from the fungus and diffuses into the cell wall to initiate the primary attack on the amorphous regions between the crystalline cellulose. This type of nonenzymatic activity is particularly characteristic of brown rot fungi.

It has been very difficult to work out the degradation process for the very

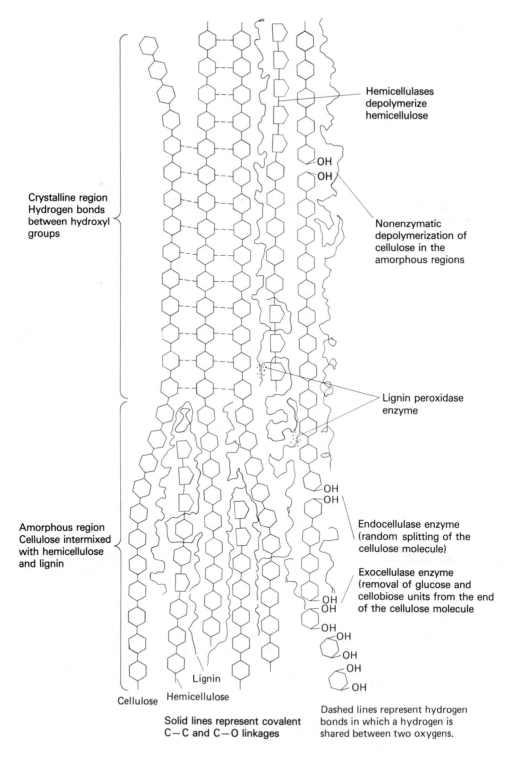

Hemicellulases
depolymerize
hemicellulose

OH
OH

Crystalline region
Hydrogen bonds
between hydroxyl
groups

Nonenzymatic
depolymerization of
cellulose in the
amorphous regions

Lignin peroxidase
enzyme

OH
OH

Amorphous region
Cellulose intermixed
with hemicellulose
and lignin

Endocellulase enzyme
(random splitting of the
cellulose molecule)

Exocellulase enzyme
(removal of glucose and
cellobiose units from the end
of the cellulose molecule

OH
OH

OH
OH
OH
OH
OH

Lignin

Cellulose   Hemicellulose

Solid lines represent covalent
C—C and C—O linkages

Dashed lines represent hydrogen
bonds in which a hydrogen is
shared between two oxygens.

Figure 14-5 Organization of cellulose, hemicellulose, and lignin into wood and the
enzymes involved in its decomposition.

complex lignin molecule. Only recently have researchers identified a lignin-depolymerizing enzyme, lignin peroxidase. Associated with this enzyme is another enzyme, manganese peroxidase, that requires manganese. The functioning of these two enzymes requires hydrogen peroxide, and therefore the enzymes involved in production of hydrogen peroxide are also required for lignin degredation.

Lignin peroxidase is produced when the soluble nitrogen, carbon, or sulfur have been essentially exhausted. The starved fungus initiates maintenance or secondary metabolism. Under these conditions, the enzymes diffuse from the hyphae to the woody cell wall, resulting in depolymerization of many different bonds within the lignin molecule.

The mode of action for lignin peroxidase involves removal of an electron from the aromatic ring. The instability of the aromatic radical spontaneously breaks the molecule at many possible points, including the ring or linkages between rings. Lignin peroxidase is therefore a rather nonspecific enzyme.

This nonspecificity of action is somewhat unusual for enzymes. The nonspecificity results in a number of different breakdown products, complicating the initial studies. However, nonspecificity has turned out to be a rather interesting feature. Lignin peroxidase is now being studied intensively for possible application in the breakdown of many different aromatic pollutants.

The specific features of the wood decay model vary somewhat among the white rot, brown rot, and soft rot fungi. They also differ within groups. In general, the white rot fungi degrade both cellulose and lignin at the same time. These fungi also preferentially utilize exocellulase enzymes that chop away at the cellulose molecules from the ends. Under some conditions certain white rot fungi may selectively delignify the wood. The white pocket rots and the white mottled rots are examples of selective delignification. It is not known what causes the selective delignification. The same fungus may cause selective delignification in a localized area and a few millimeters away cause simultaneous degradation of cellulose and lignin. There is serious interest in understanding selective delignification as a potential biological pulping process to augment or replace the current chemical processes for pulping wood for paper production.

The brown rot fungi, on the other hand, do not utilize very much of the lignin molecule. They also contrast to the white rots in that the cellulose molecule, in the amorphous regions, seems to be rapidly broken up into various length fragments with nonenzymatic depolymerizing agents. These fragments are further broken down, by randomly splitting endocellulases, ultimately to cellobiose and glucose.

The soft rot fungi utilize both lignin and cellulose. Sometimes the soft rot fungi grow from the lumen of the cell into the secondary wall, where they produce strings of diamond shaped cavities (Fig. 14-25). Other soft rot fungi or the same fungus on a different substrate may digest the secondary wall from the lumen surface.

The white and brown rot fungi stay in the cell lumen except to bore holes between cells. Nonselective white rot fungi generally degrade the wall uniformly from the inside, enlarging the lumen, or they erode troughs in the secondary wall adjacent to the hyphae. Brown rot fungi generally diffuse the nonenzymatic depolymerizing agents deeply into the wall and therefore do not produce erosion troughs.

WOOD DECAY "DISEASE" CYCLE

It is somewhat inappropriate to refer to all decays as disease, because we generally expect that, to be diseased, an organism should be living. Saprobic and products decay are problems of nonliving tree products. Heart rot, on the other hand, is a problem of living trees involving a dynamic interaction of the host and the pathogen. Therefore, in some instances we use the qualification "disease" when describing a disease cycle.

Heart Rot Disease Cycle

The disease cycle of a heart rot fungus (Fig. 14-6) begins when a basidiospore, randomly disseminated by the wind, comes in contact with a wound in a tree. If the temperature is right and the wound supplies the proper condition of moisture,

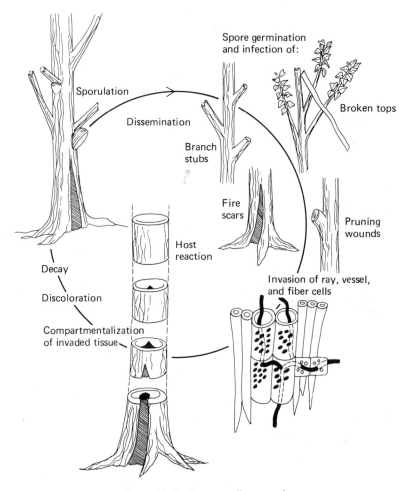

Figure 14-6 Heart rot disease cycle.

nutrients, and absence of inhibitors, the spore germinates to form a germ tube which further expands into a hypha (Fig. 14-7). The hypha releases enzymes and non-enzymatic depolymerizing agents for the digestion of cell-wall material. The hyphae branch and grow in the lumens of the cells. Movement from one cell to another is generally through pits in cell walls but, as decay progresses, bore holes are formed by enzymatic digestion of the cell wall (Fig. 14-8).

Trees respond to wounding and pathogen invasion. A tree can be visualized as a number of subcompartments. The most obvious is the new cone of cells laid down each year over the last in the form of an annual ring. Vital functions of trees, in temperate climates, take place in the recent cones. The older cones are compartmentalized from the new cone by a series of small-diameter, thick-walled, highly lignified cells produced during the summer (latewood).

The stem is further compartmentalized loosely into wedge-shaped segments by rays, which provide living cell communication between the older inner xylem and the recent outer xylem and phloem.

The rays form a screen of small groups of living cells capable of a short-term response to invasion. If the row of ray cells maintain contact with the living cambium of a vigorous tree, the ray cells may be in a better position to utilize nutrients for extensive chemical response and containment of invasion by the decay fungus.

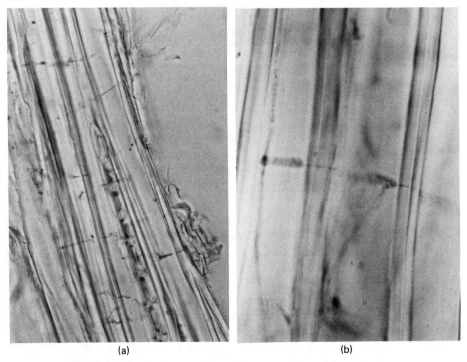

(a)   (b)

Figure 14-7   (a) Photomicrograph of *Phellinus tremulae* hyphae in aspen fibers. (b) Note the reduction in the size of the hyphae as it penetrates the cell wall in this enlargement.

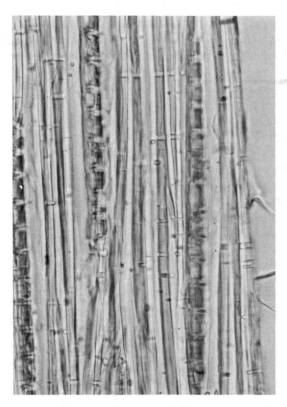

Figure 14-8 Photomicrograph of bore holes in aspen fibers caused by *Phellinus tremulae* decay.

The upper and lower ends of the wound response compartment are formed by tyloses and extractives plugging vessels and tracheids. The tyloses of hardwoods are discussed in Chapter 13. Tyloses are also produced in vessels in response to wounding and invasion by decay fungi. Conifers and hardwoods utilize a massive accumulation of extractives to fill the lumens of tracheid cells. Resinous accumulation of extractives are readily seen in the dead branch stubs of both hardwoods and conifers. The parenchyma cells provide the energy and materials for production of tyloses. Parenchyma cells and resin ducts also provide the energy and materials for resin accumulations.

The outer surface of the response compartment is the strongest side of the defense-response mechanism. Living cells of the cambium produce a wall of defense which generally prevents expansion of the invading fungus into wood laid down after the initial wound. But some fungi (canker rot fungi) interact parasitically with the cambium wound response to produce a canker which overcomes the outer response wall.

The cambium response involves the production of a sheath of chemically modified cells that form a barrier zone between the wounded tissue and subsequent tissue formed by future growth of the tree. This barrier zone is strongest near the wound but extends around the circumference of the tree as well as up and down. The wounded cone is essentially separated from future cones by the barrier zone.

In summary, wounding a tree induces a series of oxidative reactions. Further chemical reactions and morphological responses follow to compartmentalize the invading microorganisms to maintain structural integrity of the tree stem or branch (Fig. 14-9). Tyloses form in vessels above and below the wound to seal the upper and lower ends of the compartment. Parenchyma cells of the rays produce toxic phenolic compounds to impregnate cells around the sides of the wound. The summer wood cells of each annual ring slow down the penetration of fungi into the interior of the stem. The cambium on the margin of the wound produces future cones. Over time the cambium produces callus and eventually normal cells to close the wound.

If the wound is closed, the invasion process is essentially stopped, but the discoloration caused by microbial activity and host response remains. Additional wounding may reactivate the process and initiate new barrier zones.

If the wound remains open because of frost cracking or because an invaded branch stub is too large to be totally encased in new tissue, the decay may continue to progress and eventually break down all the compartment walls except the barrier zone. The decay will theoretically decompose the entire woody volume of the tree that was present at the time of wounding. This total invasion and decomposition takes many decades or centuries in long-lived trees. While the decay is progressing

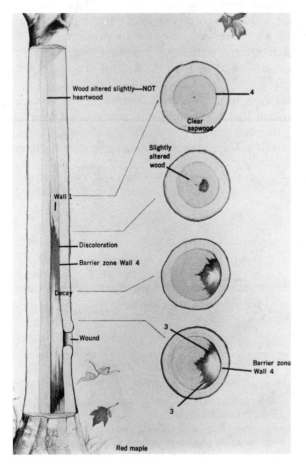

Figure 14-9 Compartmentalization of a decay column following a wound in red maple. (From a drawing by Dave Carrol, compliments of Dr. Alex Shigo, U.S. Forest Service.)

Wood Decay "Disease" Cycle

through the wood inside the barrier zone, the tree produces new healthy wood outside the barrier zone.

Some fungi are specifically adapted to tolerate the chemical-response compounds of the tree. These fungi would represent the limited number of heart rot decay fungi, at least one per tree species. These fungi utilize the chemically modified ecological niche to their advantage. By tolerating the chemicals produced by the tree, they do not have to compete with other microorganisms which cannot tolerate the toxic chemicals.

Most infecting basidiospores have a single genome ($1n$), so it is necessary in the life cycle of some of the decay fungi for a second basidiospore to infect the same substrate. The second spore initiates an infection colony. The two genomes are brought together by fusion (anastamosis) of the hyphae of the two colonies.

The dikaryotic mycelium ($n + n$) continues to develop within the stem until some triggering mechanism, which is still not known, causes the fungus to move out of the stem along paths of least resistance to fruit. Dead branch stubs commonly provide ideal paths for the fungus to progress out to the surface to fruit, as well as expansion beyond the original compartment as just described. Fruit bodies are produced under or on the broken dead branches. Nutrients generated by the enzymatic activity of the hyphae within the wood supply the fruiting body with large amounts of energy, nutrients, and water necessary for sporulation.

Recall the nuclear fusion necessary to form a diploid nucleus and meiotic division within the basidium, discussed in Chapter 8. This stage completes the disease cycle.

It is interesting to consider the huge number of spores that are produced by a fruit body. I was amazed to find that a *Phellinus tremulae* (*Fomes igniarius* var. *populinus*) fruit body could produce in excess of 100,000 spores per square millimeter of pore surface (Fig. 14-10). If you recall this example from Chapter 8, you will

Figure 14-10 Note the white spore casts on the slides. The naturally occurring *Phellinus tremulae* fruit body inside the plastic box produced 100,000 spores per square millimeter of slide surface in less than 12 hours.

remember that this number is a daily estimate of the spore-producing capabilities of the fungus.

If one measures the amount of nutrients the fungus must derive from wood for synthesis of the vast number of spores, a serious inconsistency is evident. The digestion of wood does not supply enough nutrients. The total nitrogen in the spores produced is in excess of the total nitrogen available in the decay column area. We do not know where the fungus derives the nitrogen it needs. Some investigators are developing a hypothesis of nitrogen fixation by bacteria in association with fungus hyphae in wood or in the fungus fruit body. Others would speculate that the fungus actually parasitizes living cells and transport cells around the continually festering branch stub wound.

Concepts of Infection and Invasion by Heart Rot Fungi

The science of forest pathology began in 1874 when Robert Hartig described the association of fungus hyphae decaying wood with the fruit bodies found on the outside of trees. Hartig observed the causal relationship between the Basidiomycete fungus fruiting on the surface and the decay of wood. These observations expanded into the classic concepts of decay. Infection of wounds by basidiospores eventually results in decay. Wounds, dead branches, broken tops, and fire scars are commonly seen as points where heart rot decay is continuous with the outside of the stem, so they are obviously points of infection. The concept was expanded by others to assume that for infection to take place, the wound must expose heartwood.

In 1938, Haddow published a comprehensive study of heart rot in eastern white pine caused by *Phellinus pini* (*Fomes pini*). His observations indicated that the fungus infected small twigs, as small as a few millimeters in diameter. Dead twigs of very young trees were associated with infection. White pine weevil-killed leaders were also points of infection. His work emphasized the infection by heart rot fungi of very small branches. The Haddow concept of infection was later further developed by Etheridge, who worked with the Indian paint fungus, *Echinodontium tinctorium,* on western firs. He found numerous points of infection associated with small dead branches. Etheridge claims that after initial invasion of the small twig, the fungus becomes dormant until some point in the future when the tree is stressed or injured. Wounds made during logging do not represent points of infection but rather points where host response to the wound causes activation and invasion of already existing infections.

A third concept of decay infection and invasion is the Shigo concept of succession leading to decay. The concept involves initially a host response to wounding, next invasion of the modified wood by pioneer organisms such as bacteria and nonhymenomycete fungi, and finally the infection and invasion by the Hymenomycete fungi of the wood that has been modified by the pioneer organisms. Nonhymenomycete fungi are primarily Fungi Imperfecti. Hymenomycete fungi are the Basidiomycotina decay fungi usually given credit for the decay. The central feature of the Shigo concept is that wood decay fungi are secondary organisms that follow pioneer organisms that modify the wood substrate. The pioneer organisms are supposedly responsible for detoxifying host-response phenols and modifying the pH of the wood.

A fourth concept suggests that the development of decay can best be explained in terms of the moisture content of sapwood. Excess moisture in normal functioning sapwood prevents the establishment or expansion of decay. Decay occurs in areas where normal functioning of sapwood is disrupted (for example, with wounds, or in the heartwood) (Boddy and Raymer, 1983).

The reader may think that the whole subject of infection and invasion must be thoroughly confusing to specialists, since we have four distinct concepts of the process. Actually, each concept is probably correct for some types of decay and may therefore be useful in categorizing types of decay in trees.

Before we accept any one or all four of these hypotheses, it may be informative to look at the evidence for each. The classic Hartig concept is founded on the observations of continuity of the decayed central column of the tree with the fruit body emerging from the stem at the base of decayed branches or from wounds. The validity of pathogenic relationship is verified by inoculation of trees with mycelium of pure cultures of decay fungi and the development of decay. Re-isolation from the decay-induced tissue completes Koch's normally accepted proof of pathogenicity.

The Haddow–Etheridge modification of the basic concept is in the area of infection courts. Culturing from large numbers of locations within the stem, branches, and twigs of firs recovered *Echinodontium tinctorium* from the obvious decayed central portion of the tree but also from small twigs and branches that were not obviously decayed. Also, a determination of the nuclear condition of the various isolates indicated that many of those compartmentalized in the twigs were haploid cultures. Presumably, they were recent single basidiospore infections that had not developed sufficiently to make contact and anastomose with another infection, resulting from a second spore.

A limited number of studies on basidiospore germination in the infection process of decay fungi have shown that basidiospores will germinate and infect wounds made in trees. Studies with *Chondrostereum purpureum* (*Stereum purpureum*), *Heterobasidion annosum* (*Fomes annosus*), and *Phellinus tremulae* (*Fomes igniarius* var. *populinus*) show that the spores will germinate and infect freshly made wounds. No one has demonstrated that the basidiospores of heart rot fungi will germinate and invade wood previously invaded by pioneer organisms.

It would be reasonable to assume that some fungi utilize stem or branch wounds as infection courts and that other initiate infection through the death of small twigs.

The Shigo succession concept is based primarily on evidence that fungi are compartmentalized during the invasion process. The concept further develops from many unsuccessful attempts to produce decay following inoculation with decay fungi. The inoculation wounds of many trees are invaded by a specific group of nondecay fungi. Note the difference between the terms "infection" and "invasion," because part of the difference between the Shigo and Haddow–Etheridge concepts relates to the difference between these two terms. Compartmentalization appears to be so general with invasion columns that it is assumed that compartmentalization is also the dominant factor influencing infection.

Isolations from large numbers of decay columns in northern hardwoods and mapping of locations where specific types of organisms were recovered led to the compartmentalization concept. On the margins of the decay and discolored zones, bacteria and Fungi Imperfecti were recovered. In the decayed wood behind the discoloration, other decay fungi were recovered. In the sound tissue outside the discolored wood, microorganisms were seldom recovered.

Isolations from large numbers of discoloration columns, resulting from inoculations with decay fungi, seldom recovered the decay fungus but commonly recovered *Phialophora* and *Trichocladium,* the same Fungi Imperfecti recovered from margins of decay columns of naturally infected stems. The lack of success in establishment of the decay fungus is assumed to be because the decay fungi cannot readily become established in freshly wounded wood. The wound response and pioneer invaders require a certain amount of time to prepare the substrate for invasion by the decay fungus. Others have inoculated trees with heart rot fungi and recovered the same fungi later, therefore showing the lack of need for the wound response and pioneer organisms.

An additional bit of evidence for prior modification of the substrate for development of decay fungi was provided by laboratory studies using blocks of wood from sound, discolored, and decayed tissues. Blocks of discolored and decayed tissue lost more weight when exposed to decay by specific decay fungi than did the sound block. The conclusion that decay progresses more rapidly in the modified wood substrate suggests a progressive succession of microorganisms in decay.

A note of caution in interpretation of isolation results is necessary to place some of the foregoing evidence in perspective. If an analogy can be made between the microorganism associations in wood and the macroorganism associations in a circus train, the roles of the specific organisms may be clearer. In the circus train, a large number of animals are compartmentalized into small cages. If one were to look at total numbers of organisms in portions of the train, it would appear that the lions, tigers, and other wild animals dominate but these animals are kept in their places by just a few dominant human beings. The real controllers are the human beings, even though they may be outnumbered by the caged animals. How would some nonbiased entity envision the dominant organism and the roles of the specific organisms if he or she were to come upon the circus train after a major derailment in the middle of the night? The nonbiased entity would probably easily recognize that the lions and tigers are now in control of the organisms that make up the train.

The placing of a chip of wood on agar media is like breaking up the circus train. Fungi (Fungi Imperfecti) that are aggressive and able to utilize the abundant sugar provided by the media rapidly overgrow decay fungi that have more specialized enzymes for slow decomposition of relatively inert substrates such as cellulose and lignin. Is the *Phialophora* a lion or a lion tamer in the original wood? We have no way of knowing at the present time, so that we (as nonbiased entities) must exercise care in assigning importance to one organism rather than another.

The moisture content theory is based on an understanding of the relationship of water content and oxygen tension. The normal functioning sapwood around decay columns is sufficiently high in moisture and low in oxygen to account for the restriction in expansion of the decay fungi. The decay organism may expand only into tissues in which normal sapwood function has been disrupted.

It is my suggestion that, with our present state of understanding of the infection and invasion phenomenon, we accept the possibility that any one of the four concepts may be correct, at least with regard to certain decay phenomena. Additional sophistication of our techniques and further studies of many more decay processes are necessary before we can make judgments regarding which concept should take the dominant place in introductory forest pathology teaching.

Why spend so much time on the concepts of infection and invasion? Because we do not understand these processes, we are very limited in our recommendations for prevention and control of losses due to heart rot. Avoiding wounding, for example, may or may not really make any difference. At any rate, it is sometimes impossible to avoid infection. Every tree is wounded many times. Why do some have heart rot at an early age and others seem to be free of decay? Once we can answer these questions, we will be in a much better position to make control recommendations.

Saprobic Decay "Disease" Cycle

More is known about the decay process in saprobic systems because it is a simpler system, in which the wood substrate is not responding as in the case of heart rot. Infection takes place when basidiospores come in contact with substrate, provided that there is sufficient moisture (above fiber saturation), an absence of spore germination inhibitors, moderate temperatures, and oxygen. The rate of invasion and enzymatic digestion of wood is dependent upon these same variables. Sporulation may require previous dikaryotization with a second compatible strain of the fungus as with some of the heart rot fungi. Spore dispersal by air currents completes the cycle.

It is interesting to note that, if appropriate moisture, temperature, lack of inhibitors, and oxygen requirements are met, decay will take place. The dispersal of spores must be very efficient, as spores are almost universally present.

EVIDENCE AND EFFECTS OF DECAY

Weight Loss White loses more weight than brown

Extracellular cellulase and lignin-degrading enzymes of the fungus break up the woody cell wall into fragments which can be further degraded inside the fungus to CO_2 and H_2O. This metabolism of wall material results in a weight loss to the wood.

Table 14-1 presents the changes in cell-wall chemical composition resulting from decay by a white rot fungus, *Trametes versicolor* (*Polyporus versicolor*), and a brown rot fungus, *Poria placenta* (*Poria monticola*). Note that in the case of the

TABLE 14-1 COMPOSITION OF SWEETGUM SAPWOOD IN PROGRESSIVE STAGES OF DECAY, BASED ON THE MOISTURE-FREE WEIGHT OF THE ORIGINAL SOUND WOOD

Type of decay-causal organism	Average weight loss (%)	Percent of original cell wall		
		Cellulose	Hemicellulose	Lignin
White rot,				
Trametes versicolor	0[a]	52	25	23
	25	40	19	17
	55	23	11	11
Brown rot,				
Poria placenta	0[a]	52	25	23
	20	40	17	23
	45	22	10	23

[a]Soundwood.

Source: Condensed from Cowling (1961).

white rot, when the test specimen has lost about 50% of its original weight, the lignin as well as the cellulose is reduced by about 50%. With brown rot, lignin is almost unchanged from the original.

Strength Loss   *white stronger (less loss) than brown*

The cellulolytic activities of the brown and white rot fungi operate somewhat differently, as can be seen by measuring the degree of polymerization of the cellulose at various periods of exposure to white and brown rot fungi (Fig. 14-11). The brown rot fungi utilize nonenzymatic agents and endocellulases to randomly split the cellulose chains, thereby producing many short fragments. The white rot fungi separate a limited number of cellulose chains and degrade chains primarily from the ends by exocellulases.

The splitting of many of the long chains of cellulose by brown rot fungi produce very dramatic strength losses within a very short period of exposure. White rot fungi take much longer to degrade enough structural elements to produce comparable strength losses. Decay fungi cause strength losses which may cause a living tree to break or a branch to break and fall. In buildings and wood products, strength losses may result in breakage of utility poles, failure of bridges, collapse of rafter or floor joists, and many other structural failures.

Reduced Standing Timber Volume

Heart rot fungi produce serious defect and volume reductions while a tree is still growing. If one is measuring volume increment added to trees yearly by growth, one should subtract the increasing increment of defect caused by heart rot fungi to get a realistic estimate of growth. There is a point in the life of trees when the amount of increment produced by growth is exactly equal to the amount of wood destroyed by heart rot fungi. This point is a theoretical age of maximum volume of usable wood

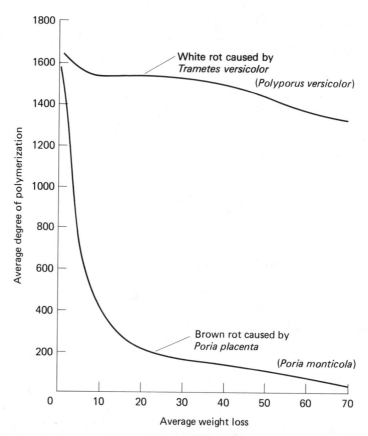

Figure 14–11 Changes in the degree of polymerization of holocellulose from sweetgum sapwood with progressive stages of decay. (From Cowling, 1961.)

and is referred to as the pathological rotation age. For obvious reasons, the forest manager does not allow timber to reach pathological rotation, and realistically would harvest the timber long before this point was reached. Intensive forest management involving short rotations obviously does not need to be concerned with pathological rotation, because harvesting occurs well before decay becomes a major problem. But trees that are not harvested for timber, but left for recreational or aesthetic reasons, will go past pathological rotation. At or near the pathological rotation age, the trees are becoming more and more weakened, as decay consumes more wood than is being laid down by growth. The manager of such forests must recognize the increasing hazard resulting from the weakening of the trees growing beyond pathological rotation.

Saprobic Recycling of Wood Residues

Saprobic fungi can be used as indicators of trees that are in the process of dying. Weakened trees or trees with large wounds that do not readily callus over are invaded by sap rot fungi. They do not progress much beyond the area of dead or dying tissues, but will cause weakening of the invaded tissue and possibly contribute to the overall weakening of the tree, climaxing in the breakage of the stem or a branch.

Saprobic fungi are important contributors to the decomposition of logging slash. Cleanup activities of this type should be recognized and encouraged. If logging debris is broken up and limbs are cut and scattered so that the debris is close to the ground, the environment is ideally suited for decomposition by saprobic fungi. Large stems, and limbs that stick up into the air are more slowly decomposed by fungi.

Saprobic fungi, in their eagerness to clean up wood residues, become a problem to foresters attempting to salvage fire- or insect-killed trees. A large amount of research has been conducted and is presently under way to evaluate the rate of deterioration of balsam fir killed by the defoliation of the spruce bud worm. Saprobic fungi such as *Stereum sanquinolentum* and *Trichaptum abietinum* (*Polyporus abietinus*) are serious decomposers of the killed fir.

Sabrobic fungi also exceed their usefulness by invading logs of timber stored for extended periods of time and by invading wood products if the moisture content is maintained above fiber saturation. The saprobic fungi of wood products will be discussed later in the chapter.

Increased Permeability of Wood

Wood decay fungi increase the permeability of the wood to water and other liquids. Increased permeability to water enhances the wettability of wood used in construction, thereby enhancing the potential development of future decay. Increased permeability increases the amount of paint needed if the wood is used in siding and house trim.

Increased Electrical Conductivity of Wood

Increased electrical conductivity is associated with wood decay. Decay appears to cause accumulation of ions in the discolored wood around the decaying region. It is based upon this increased electrical conductivity that an electrical measuring device, a Shigometer, is able to detect decay and discoloration in living trees (Fig. 14-12). The Shigometer measures resistance to a pulsed electric current. A twisted two-wire insulated probe is inserted into a 3/32-inch hole drilled into the tree. The tips of the wires are uninsulated and therefore one conducts a pulsed current into the tree while the other conducts the return flow of current back to a portable voltmeter for the measurement of resistance.

The probe is slowly inserted into the hole while the meter is being monitored. A rapid decrease in electrical resistance followed by an increase occurs whenever the probe contacts move from an area of sound tissue into a discolored zone and then into a decay column.

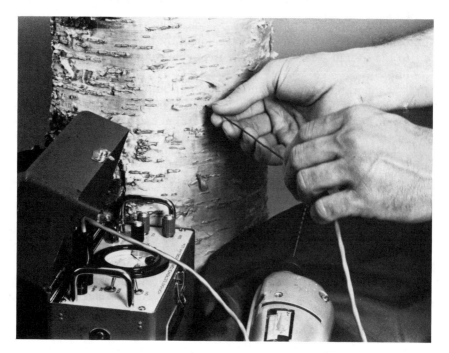

Figure 14–12 Shigometer used to detect decay in trees by measuring the resistance to a pulsed electric current.

Changes in Wood Volume

Wood decay causes changes in the volume of wood, particularly with brown rot fungi (Fig. 14-13). Irregular volume changes upon drying result in the warping and bending of wood or collapse of localized pockets on wood surfaces.

Changes in Pulping Quality

The pulping quality of wood is affected most by brown rot fungi. The increased lignin residue and rapid cellulose depolymerization of brown rot fungi make pulping of brown rotted wood very impractical. Wood degraded by white rot fungi is not seriously objectionable for pulping purposes unless the wood is bought on a volume basis. A given volume of wood decayed by white rot fungi has less weight and will yield less pulp than the same volume of sound wood. If wood is bought on a weight basis, there is very little difference between the weight and quality of pulp produced from sound or from white-rot-decayed wood. Selective white rot fungi that delignify wood may eventually be used in wood pulping processes to remove lignin from wood, to bleach pulps, or to remove toxic waste products (chlorinated phenols) in effluents.

Figure 14-13 Note changes in wood volume with decay in these experimental test blocks. The upper row of blocks decayed by a brown rotter were initially the same size as the lower row of blocks decayed by a white rotter. (Photograph compliments of Dr. Robert Zabel.)

Discoloration of Wood

An objectionable feature of decayed wood is the discoloration associated with decay. Wood that is to be used for furniture or decorative purposes is generally of higher value if discolorations and decay are absent. The term "generally" must be included because in some instances defects due to decay are highly prized. One example is the use of pecky cypress for paneling. The brown rot fungus causes cavities in cypress that are of decorative value. Another example of the use of decayed wood in preference to sound is the use of wood decayed by *Phellinus pini* (*Fomes pini*). This white rot fungus produces white pockets in the wood which give a pleasing appearance to paneling. To sell this type of decayed wood, it is important not to call it rot. *Phellinus pini*-decayed woods sells under trade names like "driftwood" or other more emotionally appealing titles.

Reduced Caloric Value

The last property of wood affected by decay that will be mentioned is the caloric value. Anyone who has cut or purchased firewood recognizes the desirability of sound wood over decayed. A cord of sound wood will produce more heat than a cord of decayed wood. Actually, if the wood were bought on a weight basis rather than by volume, there would not be much difference between the caloric value of sound wood and that of decayed wood.

Evidences of decay were discussed in the preceding section. We expand on these in this section and categorize them as visual macroscopic indicators, microscopic indicators, cultural isolation, and physical test indicators used for recognition of decay in trees, logs, lumber, and wooden structures.

The ability to recognize decay in trees, logs, lumber, and wooden structures can be an important means of avoiding economic losses. The buyer of timber must be able to recognize decay in trees, because the amount of decay will affect the value of the logs. The park manager or city forester must be able to recognize decay, to prevent damage to life and property as a result of breakage of defective trees. The log buyer at the sawmill or pulp mill must recognize decay, because of the effects on yield and quality of the products produced from the logs. The lumber dealer and building contractors must recognize decay, because decayed lumber will have less structural strength or require excessive amounts of paint to finish. The homeowner should be able to recognize decay when buying a house and in maintaining it, to avoid expensive replacement costs of defective parts of the structure.

Usually, decay can be detected by visual macroscopic characteristics but under special circumstances it may be necessary to verify the macroscopic characters with microscopic features, cultural isolation and identification of the causal agent, and physical tests of the wood. Wood used in ladders or in locations where freedom from defects is very important are examples of situations in which all of the characteristics used to detect decay might be used.

Detecting decay has two major difficulties. One is that in its incipient or early stages some decays display only minor changes in macroscopic properties even though serious strength losses have already developed. This is the case for some of the brown rots. A second difficulty is the internal or hidden decay situation in trees, utility poles, and so on. Special sampling techniques may be required to identify decay in such cases.

Macroscopic Characteristics of Decay

Decay of trees. Decay in trees is associated with heart rot fungi, which decompose both the heartwood and sapwood portions of the living stem, and saprobic fungi, which decompose dead branches and dead portions of the stem.

The fruiting bodies of decay fungi, referred to as conks by field foresters, are positive indicators of advanced decay of both saprobic and heart rot fungi. If the conk can be identified as to species, it is possible to predict approximately how much of the tree is defective as a result of the decay (Fig. 14-14). For that reason, it is desirable for the forester to be able to recognize the fruit bodies of the common species of decay fungi present in a region.

The specific identification of decay fungi can sometimes be confusing to a beginner. You will note that two names have been used for most of the decay fungi mentioned in this book. The first name is the most currently accepted name for the fungus and the name in parentheses is the name utilized in much of the older literature. Unfortunately, it is necessary to be familiar with both.

Figure 14-14 Conk or fruit body of *Phellinus igniarius* on beech indicates a 10- to 14-ft (3- to 4.2-m) decay column. (Photograph compliments of Dr. S. B. Silverborg.)

Swollen punk knots and resinosis are indications of decay (Fig. 14-15). Decay fungi do not always fruit on the surface of every stem being decayed. During the early stages of invasion, fruiting is uncommon and certain decay fungi, such as *Phellinus pini* (*Fomes pini*) on eastern white pine, seldom fruit except on wind-thrown trees or logging slash. The presence of swollen knots around former branch stubs and the punky soft tissue behind the swelling are good symptoms. These knots do not heal over and often resin will exude from the wound.

A decayed branch stub is a good indicator of some decay within the stem (Fig. 14-16). Branch stubs that are not points of infection for decay fungi are resin-impregnated and very hard.

Remnants of conks may be present. Many decay fungi produce annual fruit bodies that are consumed by insects and other fungi, so that only remnants of the conk remain. Careful examination of the stem and exposed roots is necessary to detect these remnants.

Basal fire scars are common points of infection by decay fungi and can always be suspected as indicators of decay (Fig. 14-17). The amount of decay associated with the fire scar is directly proportional to the length of time since the fire.

Decay in logs. Decay in logs may be from heart rot which developed in the living tree or may result from saprobic fungi becoming established in the logs during extended storage.

Heart rot in logs can be recognized by the same indicators as were used for the recognition of decay in trees.

Methods for Detection of Decay

Figure 14-15 Punk knot and resinosis on white pine, indicating internal discoloration and decay.

Figure 14-16 Decayed branch stub on beech, indicating internal discoloration and decay.

Figure 14-17 Basal scar indicator of internal discoloration and decay in sugar maple. (Photograph compliments of Dr. S. B. Silverborg.)

Saprobic decay during extended storage can usually be recognized by the presence of fruit bodies of the decay fungus (Fig. 14-18). The fruit bodies are more numerous than those formed by heart rotters. They are formed at places other than branch stubs and old wounds, and the fruit body orientation shows that they were formed after harvest.

On the sawed ends, decay can be recognized by hollow cores, zone lines, discoloration patterns, and roughened saw cuts (Fig. 14-19).

Decay in lumber and primary wood products. Decay of primary wood products may be the remnants of heart rot from the living tree (Fig. 14-20), or the result of infection of stored logs or poorly seasoned lumber by saprobic fungi. Heart rot fungi do not survive long in lumber, so no further development of heart rot decay will occur in the lumber. Saprobic fungi can continue to decay the lumber as long as the moisture content is above fiber saturation. The moisture may result from moisture not dried out of the green lumber or may result from rain or groundwater being picked up because of improper piling.

Irregular color variations reduce the value of lumber. Lumber grading is based on the number and size of knots as well as color variations caused by decay and stain fungi (Fig. 14-21).

The zone lines that mark the border between decayed and sound wood or two antagonistic fungi are accumulations of thick-walled pigmented hyphae and other pigmented materials of both the fungus and the wood and are excellent indicators of decay.

Methods for Detection of Decay

Figure 14–18 Saprobic decay of stored hardwood logs caused by *Trametes versicolor*.

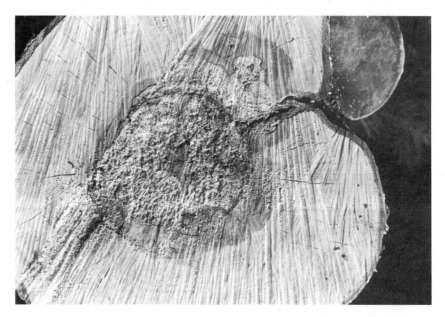

Figure 14–19 Indicators of decay in logs. Zone lines, discoloration, and roughened saw cut on the end of a log. The fruit body of *Phellinus igniarius* is also present on this birch log.

Figure 14-20 Defective board, showing white pocket rot caused in the living tree by the heart rot fungus *Phellinus pini*.

The punk knots caused by heart rot fungi and decayed branch stubs of living trees are observed as decayed knots in lumber.

Mycelial fans may occur (Fig. 14-22). Tightly stacked lumber does not dry out. The spaces between the boards are excellent places for decay fungi to produce sheets of mycelium called mycelial fans. The growth of the mycelium between the boards rapidly spreads the fungus throughout the stacked lumber.

Early stages of decay by both heart rot and saprobic fungi are not easily detected but may be very important concerns, depending upon the final use of the product. Roughened saw cuts and slight fiber pulling during planing (brashness) are indicators of incipient (early) decay.

Decay in buildings and wooden structures. The most serious decay of wooden structures results from saprobic fungi invading wood that becomes wetted by rain, water-vapor condensation, or plumbing leaks.

Paint peeling is a good indicator of decay potential. Moisture may occur in walls because of condensation associated with improper moisture barriers or because of leakage from the roof. Decay fungi will become established and cause structural damage whenever the moisture content of the wood rises above fiber

Methods for Detection of Decay   **255**

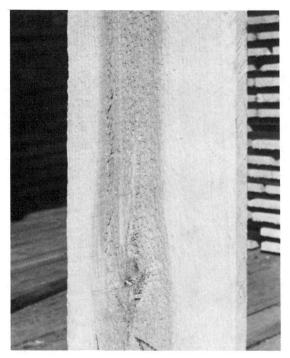

Figure 14–21 Red discoloration to this white pine board is associated with the early stages of *Phellinus pini* decay in the tree.

saturation. Paint peeling is not the result of decay but an indicator of conditions ideal for decay.

A sagging roof line is a good indicator of improper construction and early phases of decay. If moisture condenses in the attic because of an improper moisture barrier between the attic and moisture-laden, warm rooms, or if moisture gets into the attic due to leakage around chimneys or because of defective roofing, decay will slowly weaken the roof joists and cause them to sag.

Figure 14–22 Mycelial fan which developed on the surface of this board. Moisture conditions may become favorable for decay fungi in tightly stacked lumber.

An early evidence of decay is discoloration of a window frame. Condensation of moisture on the inside of windows, or moisture seeping into the window casings because of improper construction of a drip cap on the outside, wets the wood above fiber saturation.

Carpenter ants are good indicators of conditions favorable for decay (Fig. 14-23). Carpenter ants need wood with a moisture content just above fiber saturation to maintain the proper atmospheric moisture for the development of their eggs (Fig. 14-24). They do not eat wood. Wood for the carpenter ant is just an ideal place to maintain a constant environment. Ants will excavate wood that is decayed in preference to sound wood. Chemical control of the ants may be satisfying to the homeowner, but it does not change the development of the decay.

Where the floor joists sit on the concrete wall, moisture is sometimes diffused into the wood from the ground through the concrete. Discoloration of the joists is early evidence of the problem. Decay and structural failure will follow if steps such as use of a plastic barrier to prevent the wicking of moisture are not taken.

Failure of steps or a deck is an evidence of advanced decay. Untreated wood, in contact with the ground or exposed to the elements, will eventually decay. If one does not recognize the problem before failure occurs, a major replacement cost will develop. Use properly treated wood when ground contact or exposure to moisture cannot be prevented.

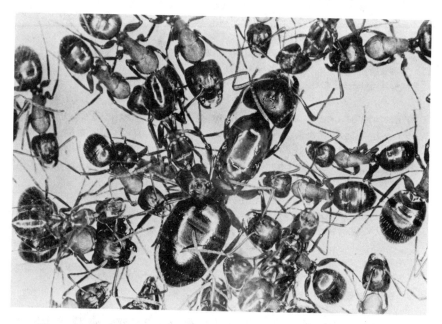

Figure 14-23 Carpenter ants are good indicators of conditions favorable for decay. Shown are a queen and numerous worker ants. (Photograph compliments of Dr. John Simeone.)

Methods for Detection of Decay

Figure 14–24 Samples of carpenter ant activity and decay of wood in a house with a serious moisture problem. (Photograph compliments of Dr. John Simeone.)

Microscopic Characteristics of Decay

Presence of characteristic hyphae of decay fungi. Fungus hyphae can often be observed in thin sections of wood examined under the microscope (Fig. 14-7). Some decay fungi have specific characteristics to their hyphae. Others cannot be distinguished from hyphae of other fungi that may invade the wood to utilize storage products of the ray and parenchyma cells.

Hyphae with clamp connections are characteristic of many Hymenomycete fungi. If the hyphae have clamps, there is a good possibility that the fungus can cause decay. If it does not have clamps, it may or may not be a decay fungus.

Hyphae associated with cell-wall degradation are excellent diagnostic characteristics of decay.

Hyphae in zone lines are usually thick-walled and pigmented. These hyphae are relatively persistent indicators of decay.

Cell-wall degradation. Hyphae are seldom seen in thoroughly decayed wood, so the effects of former hyphae on cell walls are used for diagnostic purposes. The fungi that decay wood appear to recycle the limited nitrogen available in the wood. Hyphae in tissues that are thoroughly decayed are probably reabsorbed and the nutrients transported out to the periphery of the colony, where invasion of new tissues is in progress.

Bore holes (Fig. 14-8) are produced by hyphae to penetrate from one cell to another. The initial hole is smaller than the diameter of the hypha, which has to restrict its diameter to pass from one cell to another. Enzymatic activity of the hyphae later enlarge bore holes to many times the original diameter.

Cell-wall erosion or thinning occurs as a result of enzymatic activity of fungus hyphae within the lumen of the cells. Brown rot fungi produce a generalized erosion and cracks of the S_2 layer of the secondary wall. White rot fungi produce erosion troughs in the secondary cell wall where the hyphae make contact with the wall, or they may cause more general erosion, enlarging the lumen. Soft rot fungi usually produce diamond-shaped cavities in the S_2 layer or trench-like cavities in the S_3 and S_2 layers (Fig. 14-25).

(a)

(b)

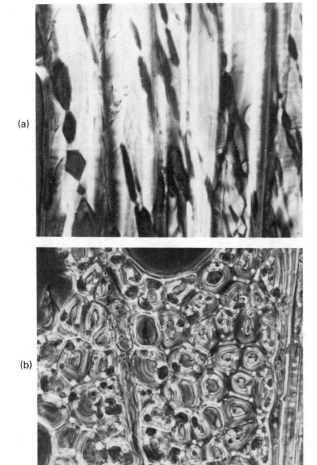

Figure 14-25 Soft rot of beech fibers caused by *Chaetomium globosum:* (a) longitudinal section, showing diamond-shaped cavities in the secondary wall; (b) cross section showing cavities in the secondary wall. (Photograph compliments of of Alirio Perez-Mogollon.)

Methods for Detection of Decay

Isolation from Wood and Identification of Decay Fungi in Pure Culture

Isolation procedures. Procedures for the isolation of fungi from wood are relatively simple. Small chips of wood are aseptically removed from wood and transferred to agar media. Aseptic techniques and media vary depending upon the training and preferences of the researcher. Large numbers of fungi will be recovered from almost any decayed wood. Variations in procedures attempt to selectively isolate decay fungi in preference to others.

Identification of decay fungi in culture. Decay fungi are identified in culture based on a number of cultural characteristics. A publication by Nobles (1965) provides a reference for the cultural identification of decay fungi. A more recent reference by Stalpers (1978) can also be used. The growth rate is used as one of the characteristics for identification of the fungi. The color of the mycelial mat in culture is often characteristic but in some instances may be rather variable. Microscopic characteristics of the fungus are important. The presence or absence of clamp connections, asexual spores (oidia), specialized hyphae, and other mycelial characteristics are used for identification. A phenol oxidase reaction on media containing an oxidizable phenolic derivative such as gallic or tannic acid is used to separate white rot fungi from brown rot fungi (Fig. 14-26). The odor of the culture is sometimes characteristic. Sometimes the fungus will fruit in culture, so that fruit body characteristics can be utilized to identify the fungus.

Physical Tests for Recognition of Decay

Limitations. A serious limitation of physical tests for the verification of decay is that normal nondecayed wood is rather variable in its physical characteristics.

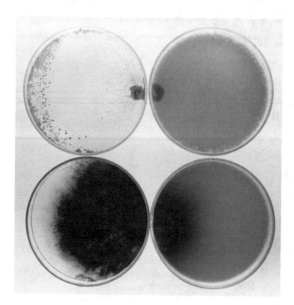

Figure 14-26 Phenol oxidase test. Note the extensive darkening of the medium caused by the white rot fungus in the lower right plate containing gallic acid. The upper right plate is a brown rot fungus on gallic acid medium. Note the minimal reaction in this plate as compared to the white rot fungus in the lower right plate. The left plates show the same fungi growing on malt extract agar.

Therefore, for a physical test to be valid, the wood must be significantly weakened below the expected range of variation for sound wood.

Specific tests. Specific weight loss, resistance to penetration, resistance to impact loading, resistance to fracture, and many other tests are used. Changes in electrical properties can be used for both trees and wood products such as poles. Use of the Shigometer for this purpose has already been discussed. X-ray and fluorescent microscopic analysis of fibers will show the effects of decay fungal enzymes on crystalline regions of the cellulose molecule and decomposition of lignin, respectively.

IDENTIFICATION OF DECAY FUNGI BASED ON FRUIT BODIES

Aphyllophorales

Early taxonomists of Hymenomycetes observed differences in the shapes of fruit body spore-producing surfaces. They classified the fungi into genera based on the shape of the spore-producing surface (hymenium) and the overall shape and texture of the fruit body. Mushrooms with a hymenium covering gill-like plates were grouped into many genera in the order Agaricales. Other Hymenomycetes with hymenial surfaces lining pores or covering smooth or spiny surfaces were placed in the order Polyporales. The Polyporales were subdivided into these common genera: *Polyporus,* with the hymenial surface lining pores in fleshy or leathery annual bracket fruit bodies; *Fomes,* with the hymenial surface lining pores of woody perennial bracket fruit bodies; *Daedalea,* with the hymenial surface lining short slit-like pores; *Poria,* with the hymenium lining pores but the fruit body resupinate; *Hydnum,* with the hymenial surface covering teeth-like projections from annual fruit bodies; *Echinodontium,* with hymenial surface covering teeth-like projections from perennial fruit bodies; and *Thelephora* and *Stereum,* with a flat hymenium occurring on a smooth leathery fruit body.

Most of the forest pathology literature relating to decay fungi has utilized this original Friesian classification system. Overholts' *Polyporaceae of the United States, Alaska and Canada* (1953) is an excellent reference using this system.

In recent years, mycologists have recognized that the former hymenial and annual or perennial criteria used for the classification of genera of the Polyporales do not identify mutually exclusive groups nor do they group the fungi phylogenetically. Some very similar fungi are grouped into different genera because the hymenial surfaces are different.

Microscopic examination of fruit bodies reveals that three types of hyphae make up some fruit bodies (trimitic); see Fig. 14-27. Others are made up of two types (dimitic), and a third group are made up of one type (monomitic). Other microscopic characters, such as color and shape of spores and the presence of sterile structures in the hymenium, are now considered important. European mycologists have led the movement toward a new classification of the Polyporales into many more genera than were originally recognized. The system is still in a state of flux, but many pathologists have shifted over to the new system. Keys by Pegler (1967),

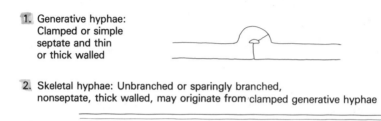

1. Generative hyphae:
Clamped or simple
septate and thin
or thick walled

2. Skeletal hyphae: Unbranched or sparingly branched,
nonseptate, thick walled, may originate from clamped generative hyphae

3. Binding hyphae: Highly branched and thick walled

Figure 14-27 Hyphal types in fruit bodies of decay fungi.

Domanski et al. (1973), Shaffer (1968), and Gilbertson and Ryvarden (1986) utilize larger numbers of genera for classifying the Polyporales.

It will be necessary to recognize both systems. The decay fungi described in this book will be listed using both names.

EXAMPLES OF HEART ROT AND SAPROBIC DECAYS

Every tree species has at least one, but a very limited number of decay fungi that can cause heart rot in the living tree. Once the tree dies, or on dead branches of the living tree, a large number of saprobic decay fungi can develop. This type of observation, as well as the observation that the fungi that cause heart rot do not persist very long once the tree dies, lead to the conclusion that there is some specialization in the type of interaction that goes on between a heart rot fungus and its tree host.

The nature of the specialized interaction between heart rot fungi and host is not known, but one can speculate that tolerance of phenolic defense compounds produced by the tree is involved. Another aspect of the specialized interaction may involve the sources of the limiting nitrogen. In the living tree, heart rot fungi may parasitize living cells of the sapwood and cambium to derive sufficient nitrogen for hyphal growth and sporulation. Nitrogen in the living tree is replenished by the transport system of the tree. Nitrogen in the dead tree is static or replenished by diffusion of nitrogen from the soil or in rainwater.

A relatively complete discussion of many of the decays caused by fungi is provided in the classic textbook *Forest Pathology* by Boyce (1961). Therefore, rather than repeat Boyce's treatment, a few short statements will be made for the most common decays.

Hepting (1971) provides a very complete list of the decay fungi associated with forest and shade trees in the United States. By using Hepting's book to determine a list of possible names of fungi, the amateur can then use technical mycological books to identify fruit bodies. If more detail is necessary, one can refer to Boyce or to the references listed by Hepting.

Phellinus is one of the most important genera of heart rot fungi. The genus is characterized by fungi causing white rot. The fruit body is leathery to woody, usually perennial, with a brown context (interior) and pore layer. The hyphal system is dimitic and without clamps.

Phellinus igniarius (*Fomes igniarius*) is a very common heart rot pathogen of poplars, birches, oaks, maples, and beech. *Phellinus tremulae* (*Fomes igniarius* var. *populinus*) is a rather similar fungus with a wintergreen odor (Fig. 14-28). *Phellinus tremulae* is particularly serious as a decay fungus of trembling and bigtooth aspens. A black zone line around the stringy white rot decay column is usually characteristic of both these decays. The decayed wood loses most of its strength but is usually acceptable as pulpwood.

Phellinus pini (*Fomes pini*) is a major heart rot pathogen of most conifers (Fig. 14-29). The fruit bodies range from large bracket structures, to aggregates of small brackets, to resupinate (flat against the surface). The early stages of decay produce a distinctive red discoloration in the wood. At this stage the disease is known as red ring rot. Advanced decay forms white pockets of rot. Construction-grade lumber and lower grades of plywood can be made from the decayed wood. As already mentioned, decorative paneling can be made from *P. pini*-decayed wood.

Phaeolus schweinitzii (*Polyporus schweinitzii*) is one of the more common root and butt rot fungi of conifers (Fig. 14-30). The genus contains only one fungus, which is a brown rotter, with a spongy annual pored fruit body. The fruit body

Figure 14–28 *Phellinus tremulae* on aspen.

Examples of Heart Rot and Saprobic Decays

Figure 14-29 *Phellinus pini* on Douglas-fir.

Figure 14-30 *Phaeolus schweinitzii* on spruce root.

usually arises from the main roots of infected trees by means of a stipe (stalk). The context of the fruit body is rusty yellow to brown and is made up of a dimitic hyphal system. The upper surface is tomentose (velvety), yellow to purple brown to black in color. *Phaeolus schweinitzii*-infected conifer stands are very seriously damaged during high winds, and caution should therefore be exercised when developing recreation sites in infected stands. The volume lost to *P. schweinitzii* is large if the trees are windthrown. If the trees can be harvested before they fall, the volume lost is usually minimal and restricted to the lower portion, usually less than 8 feet (2.4 m), of the bole of smaller trees and somewhat more in larger trees.

Three other root and butt rotters, *Heterobasidion annosum* (*Fomes annosus*), *Armillaria* spp. and *Phellinus weirii* (*Poria weirii*), are discussed more fully in Chapter 16.

Inonotus is a genus with fleshy, brown, annual, pored fruit bodies. The hyphal system is monomitic. Basidiospores are distinctly rust-colored. The decay is usually white rot. Two species of interest in the northeast are *Inonotus obliquus* (*Poria obliqua*) on birches (Fig. 14-31) and *Inonotus glomeratus* (*Polyporus glomeratus*) on maples and beech (Fig. 14-32). Decay by these canker rot fungi is usually identified by the presence of a sterile conk on the tree. The cinder-like growth forms at a branch stub. Fertile forms of the fungi develop after the trees die. The decayed wood is acceptable for pulping purposes but is too weak to use for lumber.

Oxyporus populinus (*Fomes connatus*) causes a white rot of maples and other hardwoods (Fig. 14-33). The fruit bodies are white, pored, and perennial. The hyphal system is monomitic. The fruit body is often found associated with Eutypella cankers. It is often covered on the upper surface with moss, giving the fruit body a green appearance. Decay of *O. populinus* often results in a hollow tree of no merchantable value.

Figure 14–31 *Inonotus obliquus* sterile conk on birch.

Examples of Heart Rot and Saprobic Decays

Figure 14-32 *Inonotus glomeratus* sterile conk and canker on sugar maple.

Laetiporus sulphureus (*Polyporus sulphureus*) causes a brown heart rot of many hardwoods, including oak, cherry, locust, and others (Fig. 14-34). The poroid fruit body is annual, yellow, and consists of numerous large brackets. The context is fleshy and made up of dimitic hyphal system. Decay by this fungus rapidly destroys the wood, which eventually disintegrates into a brown powder.

In the Pacific Northwest, *Echinodontium tinctorium* is one of the more destructive heart rotters of firs and hemlock (Fig. 14-35). The toothed perennial fruit body is reddish brown in color. The yellow stringy rot caused by this white rot fungus extends up to 16 ft (4.9 m) in either direction from the fruit body. In many instances, multiple infections render the infected tree useless as a forest product.

Ganoderma applanatum (*Fomes applanatus*) is an extremely common saprobic decayer and heart rot fungus of hardwoods (Fig. 14-36). The large perennial fruit bodies have a white pore surface and a smooth dark upper surface. The context is red-brown and consists of a trimitic hyphal system. The white mottled rot is usually restricted to the lower portions of the tree and, in living trees, is often responsible for weakening the tree enough to cause it to fall over in a heavy wind.

Many additional heart rots could be mentioned, but the above are some of the more common ones.

It is sometimes difficult for a beginner to distinguish heart rots from saprobic decayers. As a general rule, heart rot fungi fruit at the base of branch stubs, whereas saprobic fungi produce fruit bodies anywhere on the stem. The saprobic fruit bodies

Figure 14-33 *Oxyporus populinus* on sugar maple.

Figure 14-34 *Laetiporus sulphureus* on cherry.

Examples of Heart Rot and Saprobic Decays   267

Figure 14–35 *Echinodontium tinctorium* on western hemlock.

Figure 14–36 *Ganoderma applanatum* on elm stump.

are often numerous, whereas the heart rot fruit bodies usually occur as one or very few per tree.

The number of saprobic decay fungi is much larger than the number of heart rot fungi. Therefore, a limited presentation on a few of them requires some arbitrary selection criteria. I will comment briefly on a few common ones from the northeast. These are also present in other regions but may not represent the most common fungi in those regions.

Trametes versicolor (*Coriolus versicolor*) (*Polyporus versicolor*) is one of a large group of white rot decay fungi (Fig. 14-37). The genus is characterized by annual, pored, leathery, corky or woody fruit bodies. The hyphal system is trimitic. The thin (less than 5 mm) fruit bodies have a light-colored context. *T. versicolor* is a small, thin, aggregating, bracket fungus with concentric zonation of gray, green, brown, and black bands on the upper surface and a white pore surface. It occurs as a saprobic fungus primarily on hardwoods, and is an important decay fungus of hardwood products.

Gloeophyllum saepiarium (*Lenzites saepiaria*) and *Gloeophyllum trabeum* (*L. trabea*) are two common, very similar looking, saprobic fungi primarily of conifers (Fig. 14-38). The brown rot decay by these fungi often occurs on fenceposts, poles, and other wood structures. The brown fruit bodies are corky, flat, somewhat zonate and tomentose above, with a plate-like hymenial surface. These fungi are very important products decay fungi.

Fomitopsis pinicola (*Fomes pinicola*) is a brown rot fungus of both conifers and hardwoods (Fig. 14-39). It is primarily a saprobic decayer but can act as a weak

Figure 14-37 *Trametes versicolor* on dead maple.

Figure 14-38 *Gloeophyllum trabeum* on discarded pole.

Figure 14-39 *Fomitopsis pinicola* on dead spruce.

Fungi as Agents of Tree Diseases: Wood Decay   Chap. 14

heart rot pathogen. The perennial, pored, woody fruit body is made up of a trimitic hyphal system. The context is a straw-yellow in color. The lower surface is yellow-cream and the upper surface is black with a yellow to red outer band. The brown rot decay caused by this fungus breaks the wood into large blocks. White hyphal mats often occur between the blocks.

Fomes fomentarius is the only species retained in the genus *Fomes* in the new classification system. *Fomes fomentarius* is a white rotter, with a thick, woody, perennial fruit body (Fig. 14-40). Pores are very distinctive and extend 2 to 7 mm into the fruit body. The hyphal system is trimitic. The upper surface is gray to black, the lower surface is a gray to brown, and the context is brown. The hoof-like shape of the fruit body is somewhat characteristic. This white rotter is common on birch and beech, but also occurs on many other hardwoods. As with other saprobic fungi, the presence of fruit bodies of *F. fomentarius* generally indicates that the tree is dead and extensively decayed.

The last saprobic fungus presented in this limited group of examples is an Ascomycotina decay fungus. *Hypoxylon deustum* (*Ustulina vulgaris*) is a common decay fungus of maple and beech (Fig. 14-41). The most characteristic symptom of the decay is the extensive zone-line development in the infected wood. In the field, advanced decay will disintegrate most of the wood leaving the zone lines intact. The decayed stump section of the tree looks fire-charred. The fungus fruits by forming perithecia in black stroma on the surface of the decayed stump.

Figure 14–40 *Fomes fomentarius* on dead beech.

Figure 14-41 *Hypoxylon deustum* on maple.

ROLE OF HEART ROT IN FOREST MANAGEMENT AND PLANT SUCCESSIONS

In rapidly growing healthy trees, heart rot decay is not a serious threat to structure, survival, and normal activities. But discoloration and decay may cause economic losses in forest products produced from such trees. Old trees, suppressed trees, and unhealthy trees are not capable of strong response to invasion by stain and decay fungi and do not grow fast enough to maintain structural integrity. Decay is therefore a major factor in selectively removing the old and inferior trees from the forest population.

It is important to understand the role of heart rot decay in the maintenance and perpetuation of the trees of superior vigor in the forest. If selective logging practices remove the nondecayed, high-quality, and most vigorous trees from the forest, what is left is the low-vigor, poorly formed trees that are already being invaded by decay fungi. Once the vigorous superior trees are removed, there is no serious competition for the poor-quality trees. These may then take over the site and eventually provide the genetically inferior seed for the next forest generation.

Those who advocate or utilize selective logging practices should recognize the impact of their activities on the future potential of the forest. I have heard it suggested that selective logging is like shooting the winners of a horse race and using the losers to breed the future generations. The understanding of these processes in northeastern hardwoods has lead to management recommendations that recognize the importance of maintaining the vigorous high-quality trees in the population long enough to ensure that these will provide the seed for the next generation. The "selection system" involves harvesting or killing of the inferior trees during intermediate cuts to provide appropriate openings and seedbed for regeneration of seed from the superior trees. The key to the system is the perpetuation of stand productivity by maintenance of comparable groundspace in each age class. The cuts are set up to remove most of the inferior trees and enough high-quality trees to produce a desired diameter distribution and to make it economically profitable. The ideal layout involves removal of trees in small groups to reduce logging and skidding damage to remaining trees. The small group removal also provides enough disturbance and opening to stimulate regeneration.

Another topic worth considering is the role of heart rots in natural plant community successions. If nature is left unchecked, the heart rot fungi are the main means by which mature trees are finally brought down. The heart rot fungi of aspen cause mature trees to die and break off, allowing other species to come through. Everyone recognizes aspens as pioneer species, but the same can be said for many pines and redwoods. These species become established on fire-disturbed sites. The few giants that survive the fire are the source of seed as well as the source of fungus inoculum for infection of the next generation by heart rot fungi. The infection of young saplings guarantees that eventually the heart rot fungi of the lower stem will decay more wood than is produced by the cambium (recall the pathological rotation concept). The only known biological agent other than man that is capable of bringing down the giant redwoods is a heart rot fungus, *Poria sequoiae*. A fruitful area for a pathologically-oriented ecologist is to investigate the relative importance of heart rot fungi in the development of present-day stand composition, as compared to the more traditionally important agents, such as logging and fire.

CONTROL OF HEART ROT DECAY

Our limited knowledge of infection courts and infection processes of heart rot fungi limits our ability to make good control recommendations.

The association of decay with wounds suggests avoiding wounds as a possible recommendation. Conscious efforts to prevent wounds in selection or partial cut-logging operations can avoid a great many future decay problems. But all wounds cannot be avoided.

Although tree wound dressings are sold which allegedly prevent infection by stain and decay fungi, there is serious doubt as to the benefits of these materials.

Many trees do a good job of natural healing of wounds. It would appear that some of this ability is genetically controlled, because in clonal species such as aspen some clones are more infected than others. If we could determine precisely how the tree prevents infection, we could perhaps help the tree by some type of topical dressing when wounding cannot be avoided.

Proper pruning practices and shaping up of irregular wounds enhances healing and presumably reduces infection by decay fungi. Recent research on pruning suggest that the previous concepts of cutting the branch flush with the stem is not a good practice. Present recommendations suggest that the pruning cuts should be close to but not injure the bark ridge that surrounds the attachment of the branch to the main stem. If one cuts only the branch and not the ridge of tissue at the attachment to the stem, the wound response is confined to the branch tissues. If the wound injures the branch bark ridge, the wound response involves the stem tissues. Less discoloration and more confined wound response occurs if the branch bark ridge is not injured. Tearing of the bark below the pruned branch should be avoided by first undercutting the branch. Callus tissues around a dead branch should not be cut into when pruning dead branches. Loose bark should be removed from around an irregular wound to form a smooth diamond-shaped wound.

The shaping of irregular wounds is best done immediately after the wound occurs. If callus tissue has already formed, the further aggravation by shaping the wound probably does more harm than good.

Selection of which trees to cut and which trees to leave when thinning a young vegetatively propagated (coppice) stand can avoid decay later. Cut all high stump sprouts and save the low ones. In companion sprouts, cut one only if it is small; if the sprouts are larger than 4 inches, either cut both or leave both.

As described in the preceding section, an increase in decay can be expected if one selectively logs the dominant trees in the stand. The formerly suppressed trees may respond to increased light with rapid growth, but they have a high probability of having been infected with decay. As suppressed trees, they were less capable of providing an adequate defense against infection and invasion by decay fungi. The established decay in suppressed trees limits the production of high-quality wood.

Rapidly growing young trees can better resist infection and invasion by decay fungi, so that overmature, slow-growing trees should be selected against in both forest and urban tree populations.

EXAMPLES OF WOOD DECAY IN THE HOME

Decay can cause expensive repair bills for the homeowner. It is paradoxical that decay is so common in the home because simple means of prevention have been well-known for years. That decay need not destroy a home is attested by the fact that well-designed buildings of early settlers of this country are still standing hundreds of years later, because the wood that was properly used 200 or more years ago is still sound.

Why, then, is decay so common? One reason is that modern builders and homeowners fail to recognize how to prevent decay. An understanding of decay in the home and its prevention by builders and the general public would go a long way toward reducing expensive repair bills and conservation of our wood resource.

Evidences of Decay in Buildings

Some of the indicators of decay in buildings were covered with the general treatment of recognition of decay earlier in this chapter. These will now be reviewed and expanded.

A common symptom of decay is the musty smell often associated with basements and in other cases with rooms in the upper part of the house. The musty smell may come from fungi which are working over an old pair of boots sitting in the corner, or from fungi degrading the floor joists (Fig. 14-42). One cannot tell from the smell, but if the moisture is sufficient for fungi to degrade the boots, it is also sufficient for fungi to decay the wood.

Black carpenter ants have the same environmental requirements as decay fungi. Instead of killing the ants with chemical insecticides, appreciate the ants for showing you that you have a decay problem. If you eliminate the environmental conditions that caused the ants to take up residence within your home, they will leave to find a better place to live. The carpenter ant is a predator of other insects and should be applauded for his good work rather than condemned for pointing out your decay hazards.

Peeling and discoloration of painted siding can be covered up with aluminum or fiberglass siding. This treatment, like the killing of the ants, is a cosmetic treatment of the symptoms and not control of the problem. The paint is peeling because there is water in the walls. Determine the cause and take care of the moisture before hiding it behind aluminum.

Sagging of floors and roof occur before final breakage. Recognize the cause and take steps to correct the problem before the whole building collapses.

Figure 14-42   Decay of floor joists in the crawl space below a house.

Examples of Wood Decay in the Home

275

The last indicator of decay in the home, which appears well after the musty smell, ants, paint peeling, and sagging of floors, are the fruit bodies of the decay fungi. By the time you have fruit bodies of fungi, you have an expensive repair problem. I have seen mushrooms growing up through a plush wall-to-wall carpeted bathroom and have had to tell the homeowner that it would not make any difference if he kept the mushrooms picked—that he should, instead, have the leaky toilet fixed and replace the floor.

Decay Hazard Zones and Control Treatments (Fig. 14-43)

Attic. The attic is a decay hazard zone because of the condensation of water vapor from heated rooms on the cold roof rafters and roof boards. Obviously, the condensation problem is associated with homes in northern climates, but leaks around chimneys and because of improper flashing of the valleys of the roof are common to all climates. Condensation can be controlled by a vapor barrier in the ceiling of heated rooms to prevent water vapor from escaping. Aluminum-foil-backed insulation or plastic sheeting between the ceiling joists and the ceiling boards are the best methods. A good coat of oil-base paint is almost as effective.

All leaks in the roof should be immediately repaired. The old adage that you cannot repair the roof when it is raining and do not need to when it is not does not take into account that fungi continue to operate rain or shine.

Another preventative measure associated with attic problems is to provide adequate ventilation to dissipate any moisture that might build up because of the first two problems, as well as from condensation of water vapor from trapped moisture-laden air during cooling. Adequate ventilation requires placement of louvers on the side walls near the roof ridge and under the eaves of the overhang.

Walls. Moisture in the siding and framing may result from (1) condensation of water vapor, as in the attic problem; (2) water leaking into the walls from ice backed up on the roof overhang; (3) rain wetting of the siding; and (4) moisture trapped in the walls during the construction of a new home.

Control of condensation in the walls is by use of foil-backed insulation, plastic sheeting, and oil-base paint. The first two are preferred. Paint is somewhat less efficient, but better than nothing.

Water leaking into the walls from ice dams on the overhang of the roof occurs during winters when heavy snowfall is followed by a period of very cold weather. In poorly insulated homes, heat from the rooms rising into the attic causes the snow to melt and water to run off. The roof overhang does not get warmed by the escaping heat, so water freezes in the gutters and overhang. As more and more ice is built up, the water may work back under the shingles and into the walls. Rain during a warming trend would accentuate the amount of water and may cause leakage far enough back on the roof to have moisture leaking through the ceiling. This ice problem does not develop every year, so following the winters when it does occur, there is a rash of house painting and application of aluminum siding. Neither of these are control measures. The best control measures involve good insulation

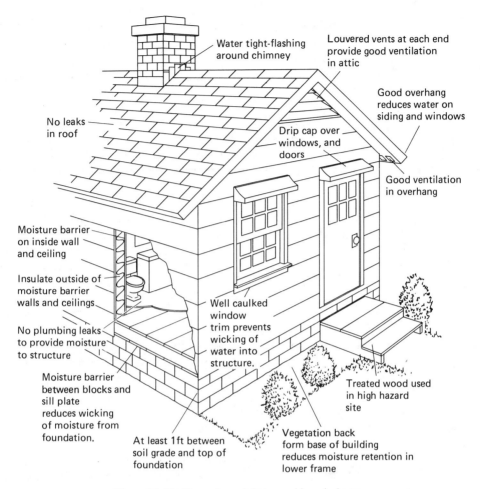

Water tight-flashing
around chimney

Louvered vents at each end
provide good ventilation
in attic

Good overhang
reduces water on
siding and windows

No leaks
in roof

Drip cap over
windows, and
doors

Good ventilation
in overhang

Moisture barrier
on inside wall
and ceiling

Insulate outside of
moisture barrier
walls and ceilings.

No plumbing leaks
to provide moisture
to structure

Well caulked
window
trim prevents
wicking of
water into
structure.

Moisture barrier
between blocks and
sill plate
reduces wicking
of moisture from
foundation.

Treated wood used
in high hazard
site

At least 1ft between
soil grade and top of
foundation

Vegetation back
form base of building
reduces moisture retention in
lower frame

Figure 14-43 Prevention of decay problems in homes.

between the rooms and the attic (a minimum of 8 inches (20 cm) of fiberglass) and good attic ventilation with ridge and overhang louvers. If the roof is maintained at the same temperature throughout, water will not form in one location and freeze in another.

The wetting of siding by rain is generally not a serious source of water in the walls unless the siding is not tight or has cracks that wick the moisture into the walls. Wetting of untreated siding usually does not cause problems. Many old unpainted barns and farm houses continue to stand. Mechanical weathering of the wood takes place but very little decay occurs. Roof overhangs of 18 in. (46 cm) will prevent much of the wetting by rain. Painting will help moisture to run off rather than penetrate, but is effective only in the absence of cracking or peeling.

Moisture trapped in walls because of unseasoned lumber or wet lumber used during construction is a problem of short duration, because the material will eventually dry out. Such practices should be avoided, though, because the period of drying time may be long enough for decay and weakening of the structure to occur.

Examples of Wood Decay in the Home

It is important to recognize that water enters cracks in the structure rapidly in the liquid form. Once inside the wall it very slowly leaves in the vapor form. Gravity and condensation may permit water to accumulate in substantial amounts, sometimes at considerable distances from the openings where water vapor can escape. Water trapped in walls can therefore be a serious problem.

Windows. Window frames are especially vulnerable to moisture buildup. For that reason, wood used in window frames is usually treated with fungicide preservatives. The wood is treated during the manufacture of the windows. It is an effective preservative if the window frames are not exposed to excessive amounts of moisture. Any preservative will, with time, leach out of the treated piece, and then the material is vulnerable to decay. On the outside, proper construction of a moisture drip cap over windows is important. The window trim and frame should be sealed into the siding by caulking of all cracks where water might leak in. On the inside of the house, it is important to adjust or control the level of humidification. Present-day humidifiers are capable of producing enough humidity to almost peel the wallpaper from the walls. It is important to reduce the humidity as the outside temperature decreases. If moisture consistently condenses on the windows, it is too high.

Plates, joists, and siding. Plates, joists, and siding are sometimes decayed because water is wicked into the material from the soil through the concrete foundation. A minimum distance of exposed foundation above the soil grade of 1 ft, (30 cm), and treatment of the contact between the wood and the foundation with a waterproof substance such as asphalt, will prevent the problem.

Moisture can also accumulate in the joist plates and lower siding because vegetation close to buildings prevents adequate ventilation and drying. Caution should be exercised in planting dense decorative vegetation around the foundation. Usually, this type or problem occurs because the plantings are allowed to grow too large, so that a little maintenance and pruning of foundation vegetation is all that is necessary to avoid decay of the lower walls.

Subflooring and joists. The subflooring and floor joists are often seriously decayed because moisture seeps into the material because of faulty plumbing. Early repair of faulty plumbing will avoid costly replacement repairs to the floor and structural members.

Moisture can accumulate in subflooring and floor joists in structures with a crawl space rather than a basement. Evaporation of water from the soil, followed by condensation on the floor, can be avoided by placing a moisture barrier such as asphalt paper or plastic on top of the soil. Another good precaution is to provide ventilation louvers to the crawl space to allow the moisture to escape.

Decorative wood. Wooden steps, patios, and decorative wood used in the exterior of homes represent a special decay hazard because one cannot prevent wetting of the material. Construction methods that allow rapid drying of the wetted wood are necessary. The flooring of patios or steps should be loose enough to allow ventilation to dry it out. Avoid small cracks or joints between various members where moisture will wick into the wood and not easily dry out. Any wood that is in direct contact with the ground will decay rapidly if it is not treated with a preservative. Therefore, use treated wood any place where moisture will be a problem.

CONTROL OF WOOD DECAY IN PRODUCTS

Control of wood decay in products involves two simple rules: First, design and use structures in ways to minimize accumulation and retention of water. Shed external water away from the structure and ventilate the structure to dissipate water produced within the structure. Second, in cases where it is impossible to keep wood dry, use naturally durable or chemically treated wood.

Wood Preservatives

A number of excellent preservatives are available for different types of wood use. Railroad ties are exposed to extreme hazards and have generally been pressure treated with creosote. Creosote was once the unusable residue from coal coking and was quite variable in its toxicity to decay fungi. Today, good control over creosote production results in a very fungitoxic compound with very good resistance to leaching from treated wood.

Creosote has an objectionable smell, cannot be painted over, and will produce burns on your skin. Today there are major restrictions with the use of creosote. At one time pentachlorophenol was a commonly used preservative in many products from utility poles to decks and decorative wood. Some of the impurities in the production of pentachlorophenol have proven to be very toxic and therefore the use of this product is now restricted for use on only utility poles and similar products.

Today the treatment of choice for many applications is to pressure treat with chromated copper arsenate (CCA). The treated product has a green color but can be painted or stained as desired. A number of other chemicals, such as copper naphthenate, chromated zinc chloride, and boron are also available for treating wood.

Wood is preferably treated with a preservative using a pressure system for maximum impregnation and distribution. In the absence of a pressure treatment, one can achieve some lesser level of durability by dipping or brushing the preservative onto the pieces of wood. Nonpressure treatments result in less retention and poorer distribution of preservatives. Microorganisms in the soil or water leaching will eventually remove the protective material more rapidly, leaving the wood vulnerable to decay.

It is not possible to effectively pressure treat all wood. For example, the western spruce and fir lumber presently sold as construction-grade lumber in the Northeast is almost impossible to properly treat. The preservative does not penetrate the wood. A good preservative treatment should have deep penetration of the preservative. Southern yellow pine sapwood is readily penetrated by preservatives and is therefore often specified where maximum durability is desired.

Natural Resistance of Wood to Decay

The heartwood of a number of species of trees is impregnated with toxic compounds that provide natural decay resistance to sarobic fungi.

Table 14-2 lists the natural decay resistance of a number of our domestic woods. The list of nonresistant woods contains those without heartwood, such as poplars, birches, and maples. The sapwood of all species is equally susceptible to decay by saprobic fungi. Therefore, it is important to specify heartwood when buying naturally durable wood.

It is important to emphasize the lack of durability of sapwood of naturally durable woods. I have often seen people buying or using cedar fenceposts or redwood boards that have only a very small central core of durable heartwood. Occasionally I have noted fruit bodies of sap rot fungi on the posts as they are stacked in the retail sales yard. Why do they waste their money on such a product?

TABLE 14-2 NATURAL DECAY RESISTANCE OF DOMESTIC WOODS

Exceptionally resistant[a]	Resistant	Moderately resistant	Nonresistant
Baldcypress (old growth)	Catalpa	Baldcypress	Alder
Black locust	Cedar	Douglas-fir	Ashes
Red mulberry	Black cherry	Honeylocust	Basswood
Osage-orange	Chestnut	Western larch	Beech
Pacific yew	Arizona cypress	Swamp chestnut oak	Birches
	Junipers	Eastern white pine	Buckeye
	White and bur oaks	Longleaf pine	Butternut
	Redwood	Slash pine	Elms
	Sassafras	Tamarack	Hackberry
	Black walnut		Hemlocks
			Hickories
			Magnolia
			Maples
			Red and black oaks
			Most pines
			Poplars
			Spruces
			Sweetgum
			Sycamore
			Willows
			Yellow-poplar

[a]Sapwood of even the most resistant woods is equivalent to nonresistant woods, so categories are based on heartwood only.

Source: Sheffer and Cowling, (1966).

REFERENCES

Amburgey, T.L. 1974. Wood-inhabiting fungi prevention and control. Pest Control *42*:22–25.

ANDERSON, L.O. 1972. Condensation problems: their prevention and solution. USDA For. Serv. Res. Pap. FPL 132. 37 pp.

ANONYMOUS. 1974. Wood handbook: wood as an engineering material. USDA For. Serv. Agric. Handb. 72.

BAMBER, R.K., and K. FUKAZAWA. 1985. Sapwood and heartwood: a review. For. Abs. *46*:567–580.

BIESTERTELDT, R.C., T.L. AMBURGEY, and L.H. WILLIAMS. 1973. Finding and keeping a healthy house. USDA For. Serv. Gen. Tech. Rep. SO-1. 19 pp.

BLANCHETTE, R.A. 1982. Progressive stages of discoloration and decay associated with the canker-rot fungus, *Inonotus obliquus,* in birch. Phytopathology *72*:1272–1277.

BLANCHETTE, R.A. 1982. Decay and canker formation by *Phellinus pini* in white and balsam fir. Can. J. For. Res. *12*:538–544.

BLANCHETTE, R.A. 1984. Screening wood decayed by white rot fungi for preferential lignin degradation. Appl. Environ. Microbiol. *48*:647–653.

BODDY, L., and A.D.M. RAYNER. 1983. Origins of decay in living deciduous trees: the role of moisture content and a reappraisal of the expanded concept of tree decay. New Phytol. *94*:623–641.

BOYCE, J.S. 1961. Forest pathology, 3rd ed. McGraw-Hill Book Company, New York. 572 pp.

BROOKS, F.T., and W.C. MORE. 1923. On the invasion of woody tissues by wound parasites. Proc. Camb. Philos. Soc. (Biol. Sci.) *1*:56–58.

COWLING, E.B. 1961. Comparative biochemistry of the decay of sweetgum sapwood by white rot and brown rot fungi. USDA For. Serv. Agric. Tech. Bull. 1258. 79 pp.

COWLING, E.C., and W. BROWN. 1969. Structural features of cellulosic materials in relation to enzymatic hydrolysis. *In* Cellulases and their applications. Adv. Chem. Ser. 95. American Chemical Society, Washington, D.C., pp. 152–187.

DILL, I., and G. KRAEPELIN. 1986. Palo podrido: model for extensive delignification of wood by *Ganoderma applanatum.* Appl. Environ. Microbiol. *52*:1305–1312.

DOMANSKI, S. 1972. Fungi, Polyporaceae I (resupinate), Mucronoporaceae I (resupinate). Translation of the book "Grzyby," published in 1965 by Panstwowe Wydawnictwo Naukowe. Available from the U.S. Department of Commerce, National Technical Information Service, Springfield, VA.

DOMANSKI, S., H. ORTOS, and A. SKIRGIELLO. 1973. Fungi, Polyporaceae II (pileatae), Mucronoporaceae II (pileatae). Translation of the book "Grzyby," Vol. 3, published in 1967 by Panstwowe Wydawnictwo Naukowe. Available from the U.S. Department of Commerce, National Technical Information Service, Springfield, Va.

ETHERIDGE, D.E., and H.M. CRAIG. 1976. Factors influencing infection and initiation of decay by the Indian paint fungus (*Echinodontium tinctorium*) in western hemlock. Can. J. For. Res. *6*:299–318.

FERGUS, C.L. 1960. Illustrated genera of wood decay fungi. Burgess Publishing Company, Minneapolis, Minn.

GILBERTSON, R.L. 1980. Wood-rotting fungi of North America. Mycologia *72*:1–49.

GILBERTSON, R.L., and L. RYVARDEN. 1986. North American Polypores. Fungiflora, Oslo, Norway.

GJOVIK, L.K., and R.H. BAECHLER. 1977. Selection, production, procurement, and use of preservative-treated wood, supplementing federal specification TT-W-571. USDA For. Serv. Gen. Tech. Rep. FPL-15. 36 pp.

GRIFFIN, D.M. 1977. Water potential and wood-decay fungi. Annu. Rev. Phytopathol. *15*:319–329.

HADDOW, W.R. 1938. The disease caused by *Trametes pini* (Thore) Fries in white pine (*Pinus strobus* L.). R. Can. Inst. Trans. *47*:21–80.

HART, J.H., and D.M. SHRIMPTON. 1979. Role of stilbenes in resistance of wood to decay. Phytopathology *69*:1138–1143.

HARTIG, R. 1874. Important diseases of forest trees. Phytopathological Classics 12. English transl. by W. Merrill, D.H. Lambert, and W. Liese, 1975. American Phytopathological Society, St. Paul, Minn. 120 pp.

HEPTING, G.H. 1971. Diseases of forest and shade trees of the United States. USDA For. Serv. Agric. Handb. 386. 658 pp.

HIGHLEY, T.L. 1975. Properties of cellulases of two brown-rot fungi and two white-rot fungi. Wood and Fiber *6*:275–281.

HIGHLEY, T.L. 1976. Hemicellulases of white and brown rot fungi in relation to host preferences. Mater. Org. *11*:33–46.

HIGHLEY, T.L. 1977. Degradation of cellulose by culture filtrates of *Poria placenta*. Mater. Org. *12*:161–174.

HIGHLEY, T.L., and T.K. KIRK. 1979. Mechanisms of wood decay and unique features of heartrots. Phytopathology *69*:1151–1157.

HIGHLEY, T.L., and L.L. MURMANIS. 1984. Ultrastructural aspects of cellulose decomposition by white-rot fungi. Holzforschung *38*:73–78.

HIGHLEY, T.L., S.S. BAR-LEV, T.K. KIRK, and M.J. LARSEN. 1983. Influence of O_2 and CO_2 on wood decay by heartrot and saprot fungi. Phytopathology *73*:630–633.

HIGHLEY, T.L., L. MURMANIS, and J.G. PALMER. 1983. Electron microscopy of cellulose decomposition by brown-rot fungi. Holzforschung *37*:271–277.

KING, K.W., and M.I. VESSAL. 1969. Enzymes of the cellulase complex. *In* Cellulases and their applications. Adv. Chem. Ser. 95. American Chemical Society, Washington, D.C., pp. 7–25.

KIRK, T.K. 1975. Lignin-degrading enzyme system. Biotechnol. Bioeng. Symp. *5*:139–150.

KIRK, T.K. 1975. Chemistry of lignin degradation by wood destroying fungi. *In* Biological transformation of wood by microorganisms, ed. W. Liese. Springer-Verlag, New York, pp. 153–164.

KIRK, T.K. 1988. Lignin degradation by *Phanerochaete chrysosporium*. ISI Atlas of Science: Biochemistry *1*:71–76.

KIRK, T.K., and J.M. HARKIN. 1973. Lignin biodegradation and the bioconversion of wood. Am. Inst. Chem. Eng. Symp. Ser. *69*:124–126.

KIRK, T.K., and T.L. HIGHLEY. 1973. Quantitative changes in structural components of conifer woods during decay by white and brown rot fungi. Phytopathology *63*:1338–1342.

KIRK, T.K., and R.L. FARRELL. 1987. Enzymatic "combustion:" the microbial degradation of lignin. Annu. Rev. Microbiol. *41*:465–505.

KOENIGS, J.W. 1974. Hydrogen peroxide and iron: a proposed system for decomposition of wood by brown-rot Basidiomycetes. Wood and Fiber *6*:66–77.

LANE, P.H., and T.C. SCHEFFER. 1960. Water sprays protect hardwood logs from stain and decay. For. Prod. J. *10*:277–282.

LARSEN, M.J., M.F. JURGENSEN, A.E. HARVEY, and J.C. WARD. 1978. Dinitrogen fixation associated with sporophores of *Fomitopsis pinicola, Fomes fomentarius*, and *Echinodontium tinctorium*. Mycologia *70*:1217–1222.

LIESE, W. 1970. Ultrastructural aspects of woody tissue disintegration. Annu. Rev. Phytopathol. *8*:231–258.

LONSDALE, D. 1982. Available treatments for tree wounds: an assessment of their value. J. Arboric. *8*:99–107.

MANION, P.D., and D.W. FRENCH. 1968. Inoculation of living aspen trees with basidiospores of *Fomes igniarius* var. *populinus*. Phytopathology *58*:1302–1304.

MANION, P.D., and R.A. ZABEL. 1979. Stem decay perspectives: an introduction to tree defense and decay patterns. Phytopathology *69*:1136–1138.

McCRACKEN, F.I. 1978. Canker-rots in southern hardwoods. USDA For. Serv. Insect Dis. Leafl. 33. 4 pp.

MERRILL, W. 1970. Spore germination and host penetration by heartrotting Hymenomycetes. Annu. Rev. Phytopathol. *8*:281–300.

MERRILL, W., and E.B. COWLING. 1966. Role of nitrogen in wood deterioration: amount and distribution of nitrogen in fungi. Phytopathology *56*:1083–1090.

MERRILL, W., and A.L. SHIGO. 1979. An expanded concept of tree decay. Phytopathology *69*:1158–1160.

NOBLES, M.K. 1965. Identification of cultures of wood inhabiting Hymenomycetes. Can. J. Bot. *43*:1097–1139.

NYLAND, R.D. 1987. Selection system and its application to uneven-aged northern hardwoods. Managing North. Hardwoods Proc. Silvicultural Symp., June 23–25, 1986, Syracuse, N.Y. Soc. Am. For. Pub. 87-03.

NYLAND, R.D., and W.J. GABRIEL. 1971. Logging damage to partially cut hardwood stands in New York State. State Univ. Coll. For. Syracuse, N.Y. AFRI Res. Rep. 5. 38 pp.

OVERHOLTS, L.O. 1973. The Polyporaceae of the United States, Alaska, and Canada. University of Michigan Press, Ann Arbor, Mich. 466 pp.

PARTRIDGE, A.D., and D.L. MILLER. 1974. Major wood decays in the inland northwest. Univ. Idaho, Idaho, Res. Found. Nat. Res. Ser. 3. 125 pp.

PEGLER, D.N. 1967. Polyporaceae. Part II, with a key to world genera. Bull. Br. Mycol. Soc. *1*:17–36.

PEGLER, D.W. 1973. Aphyllophorales IV. Poriod Families. *In* The fungi: an advanced treatise, ed. G.C. Ainsworth, F.K. Sparrow, and A.S. Sussman. Academic Press, Inc., New York, pp. 397–420.

ROTH, E.R., and B. SLEETH. 1939. Butt rot in unburned sprout oak stands. USDA Tech. Bull. 684. 42 pp.

SCHEFFER, T.C., and E.B. COWLING. 1966. Natural resistance of wood to microbial deterioration. Annu. Rev. Phytopathol. *4*:147–170.

SCHEFFER, T.C., and A.F. VERRALL. 1973. Principles for protecting wood buildings from decay. USDA For. Serv. Res. Pap. FPL 190. 56 pp.

SHAFFER, R.L. 1968. Keys to the genera of higher fungi, 2nd ed. Univ. Mich. Biol. Sta. Ann. Arbor.

SHAIN, L. 1979. Dynamic responses of differentiated sapwood to injury and infection. Phytopathology *69*:1143–1147.

SHIGO, A.L. 1967. Succession of organisms in discoloration and decay of wood. Int. Rev. For. Res. *2*:237–299.

SHIGO, A.L. 1984. Tree decay and pruning. J. Arboric. *8*:1–12.

SHIGO, A.L. 1984. Compartmentalization: a conceptual framework for understanding how trees grow and defend themselves. Annu. Rev. Phytopathol. *22*:189–214.

SHIGO, A.L., and W.E. HILLIS. 1973. Heartwood, discolored wood, and microorganisms in living trees. Annu. Rev. Phytopathol. *11*:197–222.

SHIGO, A.L., and E.vH. LARSON. 1969. A photo guide to the patterns of discoloration and decay in living northern hardwood trees. USDA For. Serv. Res. Pap. NE-127. 100 p.

SHIGO, A.L., and H.G. MARX. 1977. Compartmentalization of decay in trees. USDA For. Serv. Agric. Inf. Bull. 405. 73 pp.

SHIGO, A.L., and E.M. SHARON. 1970. Mapping columns of discolored and decayed tissues in sugar maple, *Acer saccharum*. Phytopathology *60*:232–237.

SHIGO, A.L., and A. SHIGO. 1974. Detection of discoloration and decay in living trees and utility poles. USDA For. Serv. Res. Pap. NE 294. 11 pp.

SHIGO, A.L., and C.L. WILSON. 1971. Are tree wound dressings beneficial? Arborist's News *36*:85–88.

SHORTLE, W.C. 1979. Mechanisms of compartmentalization of decay in living trees. Phytopathology *69*:1147–1151.

SHORTLE, W.C., and E.B. COWLING. 1978. Development of discoloration, decay, and microorganisms following wounding of sweetgum and yellow poplar trees. Phytopathology *68*:609–616.

SHORTLE, W.C., and E.B. COWLING. 1978. Interaction of live sapwood and fungi commonly found in discolored and decayed wood. Phytopathology *68*:617–623.

STALPERS, J.A. 1978. Identification of wood-inhabiting Aphyllophorales in pure culture. Studies in Mycology 16. Centraalbureau voor Schimmelculture, Baarn, The Netherlands. 248 pp.

TOOLE, E.R. 1965. Deterioration of hardwood logging slash in the south. USDA For. Serv. Tech. Bull. 1328. 27 pp.

WAGENER, W.W., and R.W. DAVIDSON. 1954. Heart rots in living trees. Bot. Rev. *20*:61–134.

ZABEL, R., and R. ST. GEORGE. 1960. Wood protection from fungi and insects during storage and use. Proc. 5th World For. Congr. *3*:1530–1540.

15

FUNGI AS AGENTS OF TREE DISEASES: WOOD STAIN

- *TYPES OF WOOD STAINS AND MODE OF ACTION*
- *WOOD STAIN "DISEASE" CYCLE*
- *SYMPTOMS OF WOOD STAIN*
- *EXAMPLES OF WOOD STAIN AND CONTROL*

Wood stains are generally problems of wood products, but blue stain fungi have a number of features in common with fungi that cause wilts. The Ascomycotina genus *Ceratocystis* contains both wilt and stain fungi. Insect associations for dissemination are common to both groups. Invasion of parenchyma cells of dead or dying trees is common to both groups.

Association of the blue stain fungi *Ceratocystis* spp. with *Dendroctonus frontalis* (southern pine beetle), *D. brevicomis* (western pine beetle), *D. ponderosae* (mountain pine beetle), and other bark beetles is so general there have been questions since 1928 as to which is really the cause of the death of pines. Stain fungi will cause death in the absence of beetles in inoculation experiments, but one finds neither blue stain fungus nor beetles alone causing the death of trees in the field; both are involved.

Unlike wood decay, wood stains usually do not produce degrading because of strength loss but because of the appearance of the wood. Therefore, we are mainly concerned about stains in wood used for decorative purposes rather than for structural purposes alone.

Wood stains can be caused by incipient stages of decay fungi, by stain fungi, or strictly by chemical reactions in the absence of microorganisms. Chemically induced stains will be discussed in this chapter because the degrading of lumber is similar in concern to that of fungal-induced stains.

Incipient Decay

Early stages of invasion by wood decay fungi, incipient decay, usually produce stain (Fig. 15-1). Some stains are caused by the decay fungus, as in the case of *Phellinus pini (Fomes pini),* red rot. Others are caused by nondecay fungi and bacteria associated with the decay fungi. Decay was discussed in Chapter 14 and will not be elaborated upon except to contrast and compare decay and blue stain.

Blue Stains

Stains caused by blue stain fungi (Figs. 15-2 to 15-4) will be contrasted with decay fungi to point out specific details. Recall that decay fungi were generally from the Basidiomycotina. Stain fungi are generally from the Ascomycotina and Fungi Imperfecti. The stain fungi invade parenchyma cells and utilize the cell contents.

Figure 15-1 Incipient decay in pine as evidenced by the irregular dark zone lines.

Figure 15-2 Blue stain in a pine log.

The decay fungi, in contrast, invade and utilize the cell-wall materials of structural cells. Stain is restricted to the sapwood because, in trees with both sapwood and heartwood, the heartwood does not contain many living storage cells. The stain fungi have similar moisture, oxygen, and temperature requirements to the decay fungi, so the wood decay concepts of prevention relating to controlling environmental conditions apply also to biologically induced stain.

Figure 15-3 Pine lumber cut from blue-stained logs.

Types of Wood Stains and Mode of Action

Figure 15-4 Blue stain resulting from invasion of pine lumber after it was cut.

Bark beetles are often associated with blue stain fungi. They serve as vectors and provide wounds for infection by the fungi.

Chemical Stains

Chemical stains vary, depending upon the species of tree from which the wood is cut. Chemical stains may occur during the drying process because of the release of enzymes from the living cells during the cutting (Figs. 15-5 and 15-6). Maple and birch lumber will sometimes develop oxidation stains. They may result from chemical oxidation and reactions of phenols and sugars in the wood of white pine. Other chemical stains of oak occur because of chemical reactions between iron and tannins in wood to produce a dark iron tannate stain.

Molds

Many fungi can grow on the surface and discolor wood. These are particularly important in humid climates, where fungi grow on almost any surface (Fig. 15-7).

Weathering

Ultraviolet light, leaching of pigments by water, and mechanical abrasion of exposed wood produces a gray weathering stain to wood surfaces. Sometimes this is

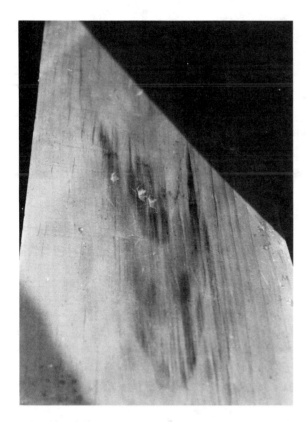

Figure 15-5 Chocolate brown or coffee stain of pine resulting from reaction of sugars and phenols during kiln drying.

desirable and other times it is undesirable. A bright red-brown redwood fence or patio will gray and need to be stained to renew the original appearance. On the other hand, old weathered barn boards are a premium decorative wood for internal paneling. It all depends on your point of view.

Figure 15-6 Sticker stain of sugar maple is seen as a slight color change going across the board at the tip of the pointer. This deep penetrating, yet subtle color change is associated with changes in chemical reaction in the area under the stickers used to separate the boards during air drying of the rough lumber.

Types of Wood Stains and Mode of Action

Figure 15-7 Mold fungi growing on the surface of this sugar maple board during drying may affect the clearness of the finished board.

Ambrosia Stains

Wood-boring beetles, in contrast to bark beetles, may feed on or have requirements for specific fungi. Some ambrosia beetles have developed specialized organs called mycangia for storing and carrying specific fungi. They inoculate the wood with their fungi and tend the crop like farmers, weeding and fertilizing as necessary. Localized stain associated with small ambrosia beetle bore holes is characteristic of this type of stain.

WOOD STAIN "DISEASE" CYCLE

Infection of wood by blue stain fungi can occur because of bark beetle activity in the trees or stored logs, or by contamination of boards by saws or handling equipment, by airborne or rain-splashed spores in the stacked lumber, or by contact with infected stickers used in the loose stacking of boards.

Invasion and utilization of stored materials of sapwood rays and parenchyma cells results in bluish or grayish radial streaking. The fungus hyphae generally penetrate from one cell to another through pits but occasionally produce bore holes.

As with wood decay, moisture is critical for stain fungi development. The minimum moisture necessary is fiber saturation. Ponding or sprinkling of logs reduces the oxygen and therefore inhibits the blue stain fungi.

Sporulation by conidia and/or ascospores may occur generally over the surface of moist lumber or may occur specifically in beetle galleries of infected logs and trees.

The asexual and sexual fruiting structures of *Ceratocystis* spp. and other stain fungi are ideally developed for insect dissemination of spores. Conidiophores are aggregated into stalks called coremia, with sticky masses of spores produced at the tip. Perithecia with long necks do not shoot off ascospores but accumulate mature spores in a sticky matrix at the top of the stalk. Both of these are ideally suited for projection into insect galleries to contaminate the insect vectors.

SYMPTOMS OF WOOD STAIN

Recognition of stain is sometimes complicated by natural color variations in the wood, so it is necessary to understand wood anatomy and normal color variation.

Radiating distinct blue to gray streaks in sapwood, when observed in cross section, may appear as interrupted lines of general blue to gray discoloration on the flat faces of boards. The color is due to pigmented hyphae and reaction products of the fungus invading the wood.

The chemical stains range from very dark pigment in the case of iron tannate stain of oak to very subtle coloring in the case of sticker stain of sugar maple. Sticker stain is recognized as a slight light-reflectance difference in the wood surface in the immediate area where it made contact with stickers. After surfacing and finishing, the band of different-colored wood is evident only when light reflects from the board at an oblique angle, so such stained boards sometimes are used in the manufacture of medium-quality furniture.

EXAMPLES OF WOOD STAIN AND CONTROL

Blue stain is prevented by minimizing the time between when the tree is cut and the board is dried below fiber saturation. Water ponding or sprinklers can be used to reduce stain during storage of logs. Rapid kiln-drying prevents stain during drying. If wood is air-dried, application of a fungicide dip to the boards as they come off the green chain is suggested.

Intercontinental shipment of green lumber in ships presents a serious stain problem. There is no need to dry the wood, because weight is no problem for the ships. Therefore, treatment of the lumber with tetrachlorophenol has been used. Good treatment and packaging to protect the lumber from leaching of preservative prevents serious deterioration even with up to 2 years' storage. Poor treatment, or allowing the preservative to leach out by exposure to rain, results in serious losses by stain, mold, and decay within 6 to 9 months. Present-day environmental concerns with the use of chlorinated phenols has caused a major problem for the lumber industry. The search for new chemicals to replace the clorinated phenols is currently in progress.

Chemical stains are individually specific problems. Treatment of green lumber with sodium azide prevents oxidation staining of hardwood lumber. A chocolate brown stain of white pine lumber occurs because of chemical reactions of sugars and phenols of the wood. If one uses moderate kiln-drying schedules, the stain can be prevented. Temperatures below 150°F and humidity below 65% are recommended.

Another common chemical stain occurs when iron comes in contact with moist oak lumber. A reaction between the iron and tannins of the wood produces a dark ink-like stain. Control is to avoid contact between wet oak lumber and iron materials.

REFERENCES

BATRA, L.R. 1963. Ecology of ambrosia fungi and their dissemination by beetles. Trans. Kans. Acad. Sci. *66:*213–236.

CAMPBELL, R.N. 1959. Fungus sap-stains of hardwoods. South. Lumberman *199:*115–120.

CZERJESI, A.J. and J.W. ROFF. 1975. Toxicity tests of some chemicals against certain wood-staining fungi. Int. Biodetect. Bull. *11:*90–96.

McGREGOR, M.D., and D.M. COLE. 1985. Integrating management strategies for the mountain pine beetle with multiple-resource management of lodgepole pine forests. U.S. For. Serv. Gen. Tech. Rept. INT-174. 68 pp.

MILLER, J.M., and F.P. KEEN. 1960. Biology and control of western pine beetle. USDA For. Serv. Misc. Publ. 800. 381 pp.

ROFF, J.W., A.J. CZERJESI, and G.W. SWANN. 1974. Prevention of sap stain and mold in packaged lumber. Dep. Environ. Can. For. Serv. Publ. 1325. 43 pp.

THATCHER, R.C., J.L. SEARCY, J.E. COSTER, and G.D. HERTEL. 1981. The southern pine beetle. U.S. For. Serv. Sci. and Ed. Adm. Tech. Bul. 1631. 267 pp.

ZABEL, R.A. 1953. Lumber stains and their control in northern white pine. J. For. Prod. Res. Soc. *3:*36–38.

ZABEL, R.A., and C.H. FOSTER. 1949. Effectiveness of stain control compounds on white pine seasoned in New York. N.Y. State Univ. Coll. For. Tech. Publ. 71.

16

FUNGI AS AGENTS OF TREE DISEASES: ROOT DISEASES

- *TYPES OF ROOT DISEASE*
- *MODE OF ACTION OF ROOT DISEASE FUNGI*
- *ROOT DISEASE CYCLE*
- *SYMPTOMS OF ROOT DISEASE*
- *METHODS FOR DIAGNOSIS OF ROOT DISEASE*
- *EXAMPLES OF ROOT DISEASES AND CONTROL*
- *FOREST PRACTICES AND ROOT DISEASES*

When an apparently healthy, mature tree or group of mature trees die for no apparent reason, we are often at a loss for an explanation. The difficulty in identifying the cause sometimes occurs because we are often looking at only half the tree. This preoccupation with the aboveground half of the tree is logistically imposed on us because of the difficulty in observing, collecting, and testing the root systems of trees. Although we have a great deal of information on a few root rot problems, we have just barely scratched the surface.

To give some perspective of the impact of root diseases, I will comment on reports from opposite sides of the continent. It was reported that 1% of the national forests of the northern Rocky mountains are occupied with readily discernible root disease centers. About one-third of the annual tree mortality of the area was associated with root disease.

From Virginia it was reported that 85% of the loblolly pine trees are infected with the root rot fungus *Heterobasidion annosum*. Most of the infected trees show no aboveground symptoms and therefore are only diagnosed as infected through excavation of the roots. It was further reported that the infected population (most of which showed no aboveground symptoms) experienced a 20% growth reduction over a period of 5 years when compared to noninfected individuals in the population.

These are just two of many examples that could be used to illustrate the point that there are two levels of impact of root disease. One very significant level is the readily observable and quantifiable mortality associated with infection centers. The second level of impact is growth reduction associated with root diseases. Growth reductions are much more difficult to identify and quantify. The second illustration also gives some indication of the prevalence of root disease problems in a population of trees.

Many people have become alarmed at reported growth reductions in, for example, southern pines of the United States (see Schefield and Cost, 1986). Those that do not understand the possible impact of root diseases and other factors on growth may suggest that air pollutants such as acid rain or ozone are the cause of this growth reduction. It is important to recognize that root diseases are an ever-present factor in forest productivity. It is also very important to recognize that harvesting and plantation establishment practices may set the stage for even greater impacts from root diseases.

TYPES OF ROOT DISEASE

Root diseases of seedlings in the nursery have some properties in common with those discussed in this chapter, but the tree nursery is a unique enough cultural system to warrant a specific discussion in Chapter 21.

At one time we separated root diseases into structural root rots caused by Basidiomycete decay fungi and feeder root rots caused by Oomycetes fungi. The separation is probably very artificial since both groups of organisms can interact with both structural and feeder roots. The differences among root diseases are better understood from perspective of the mode of pathogenesis of the causal organisms. Three types of root diseases are described under this system.

One type of root fungus interaction is described as a *pathogen-dominant tissue-nonspecific disease* where the pathogen is a very unspecialized root parasite that rapidly macerates root tissues. This destruction of the root system results in nutrient deficiency in plants and ultimately death. Pathogenic fungi of the subdivision Mastigomycotina, class Oomycetes, in the genera *Phytophthora* and *Pythium,* fit into this group. Some major nursery pathogens in the subdivision Fungi Imperfecti, such as *Cylindrocladium* and *Rhizoctonia,* are also in this group.

A second group is characterized as *host-dominant tissue-nonspecific disease* systems. In this group the fungi may exist with the root for an extended period of time leading to changes in root morphology. The pathogens may penetrate and parasitize the cambium of the root after a period of inoculum build up on the surface. These fungi can also function as saprobic decayers of dead stumps and trees or may function as heart rotters of the roots and lower bole of living trees. These pathogens may cause growth reductions in seemingly "healthy" trees and may also contribute to the death of trees. If the inoculum potential is very high, these pathogens may cause rapid death of vigorously growing trees. *Armillaria* spp., *Heterobasidion annosum,* and *Phellinus werii* are examples of Basidiomycete decay fungi that are best described by this group.

A third group of root diseases is characterized as *host-dominant tissue-specific diseases.* In this group are the root-infecting vascular parasites such as *Verticillium dahliae* and *Leptographium, (Verticicladiella) wageneri.* These diseases could be grouped with either the wilt diseases (Chapter 13) or the root diseases. To emphasize the point, Verticillium wilt was discussed with the wilt diseases and Leptographium black stain root disease will be included with the root diseases.

Some of the fungi of the host-dominant tissue-specific diseases colonize cortex tissue prior to invasion of the vascular tissues of the stele. In this cortical mode, these fungi develop a "quasisymbiotic" relationship with the host that is in some respect similar to mycorrhizae (see Chapter 8).

The root disease classification system just described illustrates a continuum of relationships from pathogen dominant nonspecific tissue maceration to reciprocal parasitism of specific tissues similar to mycorrhizae.

MODE OF ACTION OF ROOT DISEASE FUNGI

The three types of root diseases described above represent three modes of action. The pathogen-dominant tissue-nonspecific pathogens macerate succulent tissues. These nutrient-rich tissues could be utilized by many organisms. The key features of the pathogens are their ability to locate and take advantage of unprotected succulent tissues, their ability to interrupt or tolerate the host defense reactions, and their ability to rapidly colonize wounded or poorly protected tissues.

The host-dominant tissue-nonspecific pathogens exist primarily as saprobic decayers or as heart rot fungi of the lower stem and structural roots of trees. These fungi can also rapidly colonize the cambium of the roots and lower stem of susceptible hosts. As decayers they utilize cellulase and lignin peroxidase enzymes to derive nutrients by degrading cell wall material. They are constrained in the living host by the compartmentalization factors discussed in the wood decay chapter (Chapter 14). The containment of the decay fungi is highly influenced by the nutrient status and vigor of the host. Utilizing the base of nutrients derived from decay activities, the fungi can often successfully colonize the surface of an adjacent root. The activities on the root surface causes death of the cambium below and therefore a new focus for infection and host compartmentalization defense reactions to take place.

The mode of action of *L. wageneri,* an example of a host-dominant tissue-specific pathogen, is not well understood. In the living tree, the pathogen is restricted to the tracheids. Other tissues, including the cambium, are invaded after tree death. Activities in the infected tracheids result in phenolic and resin accumulations to produce a black staining. The stained tissues occur in arcs parallel to the annual rings. The mode of action of the early infection has not been described.

ROOT DISEASE CYCLE

The disease cycles of Basidiomycete fungi follow the same basic disease cycle as those of heart rot decay fungi. Infection of wounds by wind-disseminated basidiospores is followed by invasion and decay of the wood. The host response attempts to compartmentalize the infected tissue. At some point, sporulation completes the cycle.

The dissemination and infection phase of the disease cycle of *L. wageneri* may occur through the activities of root-feeding beetles and weevils. The sticky spores of the fungus are often produced in the insect galleries. The pathogen can also spread by contacts between infected and healthy roots.

Secondary spread from tree to tree may occur where the roots of trees contact each other. *Armillaria* spp. have another mechanism for secondary spread. Rhizomorphs (aggregates of fungal hyphae) that look like shoestrings grow many meters out into the soil. If the rhizomorph makes contact with a susceptible root, it may initiate infection.

The rhizomorphs are the main means for the spread for *Armillaria*. In a western ponderosa pine stand, the fungus has spread for hundreds of years by rhizomorphs alone. This conclusion was based on laboratory tests for incompatibility factors between a number of isolates of the fungus from the same area. If two isolates freely merge when grown together in a petri dish, this indicates a lack of incompatibility factors and a probable common vegetative origin for the isolates.

The disease cycle of a typical Oomycete root disease pathogen involves infection of wounds by motile zoospores. The zoospore, being motile, can move in water toward specific infection points. Gradients in concentrations of chemical diffusing from susceptible roots provide stimuli for zoospores to follow. Germination, infection, and invasion of the weakened root proceed in sequence to quickly utilize the substrate.

Soil moisture and temperature conditions are often ideal for fungi and other microorganisms. Therefore, great numbers of microorganisms compete for any available substrate. Evolutionary adaptations for successful competitive survival in soil are many. Rapid colonization of a substrate to exclude other competitors is one. Production of antibiotics that inhibit competitors is another. The ability to recognize chemical stimuli for triggering of spore formation, spore germination, or movement toward a point source is a third. The ability to produce a dormant stage to successfully survive periods during which a nutrient supply is lacking is a fourth adaptation of successful soil competitors.

The Oomycete fungi reproduces asexually by means of zoospores in sporangia or may go dormant by producing a sexual oospore. The oospore forms from the fusion of two modified hyphal structures, a small antheridium and a large oogonium. Transfer of nuclei from the antheridium to the oogonium brings compatible genomes together. Fusion of nuclei restores the diploid condition. Germination of the oospore to form a sporangium and zoospores reinitiates the active growth phase.

SYMPTOMS OF ROOT DISEASE

Initially, root rots look like nutrient deficiencies. With reduced water and mineral uptake, the foliage becomes smaller and yellowed. Growth is reduced, so that foliage may appear tufted at ends of branches.

Root pathogens spread among neighboring trees through root contacts or rhizomorphs. Therefore, infection centers, characterized by clusters of dead or dying trees surrounded by chlorotic and thin crowned trees, may develop. Excavation of root systems should reveal discolored xylem tissue and/or absence of feeder roots.

METHODS FOR DIAGNOSIS OF ROOT DISEASE

The aboveground symptoms, given above, provide some indication of root disease, but they also suggest other problems, related to nutrients, lack of mycorrhizae, nematodes, air pollution, and so on. The appearance of decayed roots is useful for diagnosis. These details are provided below, together with specific treatment for various root diseases.

The presence of characteristic fruit bodies or fungus structures of the causal organism are good diagnostic characters. Isolation and identification of the organism in culture is additional diagnostic evidence sometimes used when other diagnostic characters are lacking.

The fruiting structures of some root pathogens are microscopic and are therefore not useful as diagnostic criteria in the field. Isolation of microorganisms from roots on culture media seldom recovers the plant pathogen, because of the wide array of competitive saprobes which rapidly colonize the media. Therefore, a very useful technique for the diagnosis of root pathogens is to use a bait to selectively trap the pathogen and then culture from the bait. A number of baits have been used. For *Phytophthora cinnamomi* root rot, baits like apple, germinated lupine seed, or eucalyptus cotyledons have been used. A sample of soil is placed in a hole bored in the apple or a sample of soil mixed in water is used to float the germinating lupine seeds or eucalyptus cotyledons. After a few days, the fungi are isolated on culture media from browned tissues. Selective media for direct isolation from soil are also available.

EXAMPLES OF ROOT DISEASES AND CONTROL

Phytophthora cinnamomi

Phytophthora root rot is an example of a pathogen-dominant tissue-nonspecific disease. The life cycle of *P. cinnamomi* involves asexual (vegetative) production of motile zoospores in elongated sac-like structures call sporangia. These spores utilize flagella to swim to the infection site on the root, where they germinate, develop mycelium for infection and invasion of the host, and may again form more asexual sporangia. Under some conditions the fungus will form asexual resting spores called chlamydospores that will germinate to form sporangia once the environmental conditions are again favorable. The chlamydospores are about the size of sporangia. They are round and thick-walled to protect the fungus from desiccation. The sexual stage of the fungus involves the fusion of an oogonium from one line by an antheridium from another line. Prior to fusion, meosis takes place within the oogonium and the antheridium of this diploid fungus. The fusion results in a diploid oospore that will eventually germinate to form diploid hyphae. The oospores can also function as dormant resting spores.

Problems with *P. cinnamomi* usually occur on sites with excess soil water. In southwestern Australia the site relationship was a bit more complicated. In this area *P. cinnamomi* was a problem in jarrah *(Eucalyptus marginata)* forests (Fig. 16-1) on upland slopes. These slopes are very dry and unfavorable for *P. cinnamomi* during the summer. The understanding of the problem involved piecing together a number of factors.

One of the features of a jarrah dieback problem was the understory species

Figure 16-1 Jarrah dieback in a Southwestern Australian forest caused by *Phytophthora cinnamomi*. (Photograph compliments of Dr. Joanna Tippett)

Banksia grandis. This species is very susceptible to invasion by *P. cinnamomi.* Its deep spreading roots provide channels for movement of the pathogen.

Another feature of the problem is the concreted lateritic layer (hardpan) that occurs at depths of 10 to 100 cm. Root channels produced in the lateritic layer by Banksia provide an ideal site for the fungus to survive during dry periods. During wet periods the spores of the fungus are rapidly spread through the stand by water flowing across this lateritic layer. The fungus is therefore not uniformly distributed but rather is concentrated in and around roots that pass through channels in the hardpan. It was also shown that the fungus can survive in necrotic lesions of living roots without causing death of the trees.

The primary question was what combination of factors caused the death of jarrah on some sites and not on others. The concreted lateritic layer was one of the important factors. To survive drought periods on the soils with the hardpan layer, jarrah roots would utilize channels through the layer produced by the Banksia. These channels also represent the sites where *P. cinnamomi* is concentrated. During drought periods when the roots above the hardpan layer cannot survive, fungus invasion and death of roots that penetrate the hardpan result in the rapid death of the jarrah trees.

Clearly, the management of the problem will involve the recognition of sites with a hardpan as high-hazard areas. There is little prospect of control of the disease in these areas. On better-drained sites there is reasonable prospects for reducing the disease by reducing the density of Banksia in the understory and the promotion of more resistant legumes as understory species. The disease seems to expand rapidly following disturbances that change the drainage patterns and stress trees, and therefore minimizing soil disturbance is also recommended. In some areas, where *P. cinnamomi* is not yet established, it is important to avoid introduction of the pathogen on equipment from contaminated areas.

The role of site factors in *P. cinnamomi* root disease in the southeastern United States is also evident. A problem on shortleaf pine known as littleleaf disease occurs on sites with a hardpan layer (see Chapter 18 for more information on this problem).

Phytophthora cinnamomi is recognized as contributing to problems of tropical and subtropical forests throughout the world. Its wide distribution and presence on both affected and unaffected sites produces some confusion on the role of this fungus in various diseases. It is important to recognize the major interaction of site factors with this disease. It is also important to recognize the role of distribution and density of susceptible species in the development of this disease.

Armillaria spp.

Armillaria root rot (Fig. 16–2) is the first of three host-dominant tissue-nonspecific root diseases that will be presented in detail. All three of these produce a fibrous stringy white rot decay of wood. Armillaria root rot differs from the other two in that a white mycelial fan (Fig. 16–3) may develop beneath the bark where the fungus is actively parasitizing the living host. In addition, extensive resin accumulation around the base of the tree is common in conifers. Finally, the presence of distinc-

Examples of Root Diseases and Control **299**

Figure 16–2 Regeneration in this Oregon forest is being killed by *Armillaria* root disease.

tive rhizomorphs (Fig. 16–4) on the roots and beneath the bark of dead trees is a diagnostic feature of Armillaria root rot.

Until very recently, it had been assumed that one fungus *Armillaria mellea* was associated with a wide array of saprobic and parasitic situations on both conifers and hardwoods. There have been differing points of view on the pathogenicity of the

Figure 16–3 Light-brown-colored *Armillaria* mushrooms are common in the fall in northeastern conifer as well as in hardwood forests. (Photograph compliments of Dr. Savel Silverborg.)

Figure 16-4 White mycelial fan exposed by chopping some of the bark away is typical of *Armillaria* infection. Extensive resin production, not readily visible in the photograph, is also a symptom of infection in conifers.

fungus. We now recognize that there are many species of *Armillaria* with differing regional distribution, host ranges, and pathogenicity capabilities.

The different species of *Armillaria* may produce diseases with a range of symptoms. In mature trees there may be no crown symptoms since the fungus may function as a heart rot pathogen of the roots and butt section of the tree. In other situations, the infection and girdling of roots leads to growth reductions, yellowing of foliage, and thinning of the crown.

Unfortunately, it is not a simple task to determine which individual species or group of species are occurring in a given area. Specialists are still cataloging the geographic range, the abundance, and the roles of the various species. The species are identified on the basis of the appearance of the rhizomorphs and fruit bodies. They are also identified on the basis of interactions of cultures of unknown collections with previously classified cultures. Until much of this identification activity is completed it is probably appropriate to refer to the fungus as *Armillaria* spp.

The disease cycles of *Armillaria* spp. differ by species and region. The gilled mushrooms (Fig. 16-5) of these fungi may be common in the fall of the year in eastern forests but are rare in the coastal western forests. In the dryer interior western forests, fruiting is related to moisture conditions. The mushrooms occur in clumps, are tan to light brown in color, produce white spores, and have a ring of veil tissue that persists on the stipe.

Basidiospores from the mushrooms presumably infect through lower stem and root wounds. Freshly cut stumps may occasionally provide an infection site for the fungus, but the evidence for this is quite limited and circumstantial.

Infections through germination of basidiospores are not very common, but once the event does occur the fungus may become established and remain in the forest stand for hundreds of years or even through generations of plants. The fungus expands its base of influence through vegetative spread to healthy roots that make contact with infected roots. *Armillaria* spp. also produce collections of hyphae

Figure 16-5 Black shoestring-like rhizomorphs of *Armillaria* sp.

surrounded by a black protective layer that are capable of growing out into the soil for a number of meters. If these specialized hyphae, called rhizomorphs (Fig. 16-4), make contact with another root, a new infection lesion may be formed.

Invasion of the host from wounds in living trees is very slow. Host wound responses may check the invasion process. In this case the fungus may remain quiescent until the tree is wounded or stressed again. In the west harvesting seems to reactivate quiescent lesions.

There are both host-species-level differences in susceptibility to infection and individual plant differences. The individual plant differences are related to the sugars, alcohols, and nitrogen compounds of the roots. These compounds vary in response to an array of physical and biological stresses on the plant. Some of the changes stimulate growth of the fungus or may allow it to grow in the presence of phenolic inhibitors produced by the plant. Rapid invasion of susceptible tissue occurs as a mycelial fan in the cambium region beneath the bark (Fig. 16-3).

The various speices of *Armillaria* affect individual trees and regional forests differently. The range includes virulent pathogens of both conifers and hardwoods, pathogens of only conifers, pathogens of only stressed trees, and non pathogens. In northeastern forests there may be at least eight species, while in the west at least six species are reported.

These different fungi produce different diseases. In the moist coastal forests of the west, *Armillaria* is a butt rotter in old and dying trees. In plantations and natural stands less than 25 years of age, the infection centers are small, the impact rarely exceeds 5% of the forest, and the assumption is that the disease contributes to "natural" thinning.

The drier inland regions of the west have a more aggressive pathogen of pine, true firs, and Douglas fir. The disease expands into large infection centers of several hectares. Losses from the disease can be very serious, particularly following selective logging activities.

In the northeast the species of *Armillaria* are sometimes considered secondary pathogens. On hardwoods that are weakened by insect defoliation, drought, or many other stressing agents, the fungus can expand rapidly from large numbers of previously quiescent infections. The problems do not seem to develop in infection centers but rather emerge from a widely distributed inoculum source that is already in place.

Another type of *Armillaria* situation may occur in aspen stands regenerated through root suckering following harvesting. Infection in the old stumps can be transferred to the regeneration.

A third type of *Armillaria* situation occurs when conifers are planted on recently cleared hardwood forests. The large base of inoculum in the residue of the former forest is sufficient to allow infection and subsequent death of vigorously growing seedling to sapling-sized trees. Christmas tree plantings that are just about harvest size may be rapidly killed by *Armillaria* expanding from the former hardwood stumps.

There are a number of options for "control" of Armillaria root rots. As more information on individual species of *Armillaria* becomes available, there will be refinements in the control options. As a general rule it is important to evaluate the present and future impact critically before initiating control activities. The value of the crop may not warrant control. In other instances *Armillaria* may be functioning as a thinning agent or as a saprobic decayer.

One approach to control is to postpone planting a few years to allow time for the stumps to decompose sufficiently to be invaded by other fungi. A dramatic activity may involve stump removal with explosives or with large equipment. Another approach would consider replanting with species that are resistant to the *Armillaria* species present in the area. To utilize this option effectively, we will need to develop simple methods for recognition of different species and good information on the range of pathogenicity of the individual species. In some situations the key to reducing impacts of *Armillaria* will be tied to maintaining vigorous growth of the trees through appropriate thinnings. In other situations it will be necessary to use various management options to prevent buildup of defoliating insect populations that seriously stress trees.

Heterobasidion annosum

Heterobasidion annosum (Fomes annosus) is potentially an important root rot pathogen of conifers wherever they are grown. In the natural stand, the fungus is a typical root and butt rot pathogen of mature trees, much like *Phaeolus schweinitzii*. The root systems are decayed such that the tree is liable to windthrow.

In plantations of closely spaced pines, roots are often grafted to each other to produce one large root system with many stems. Thinning of the plantation results in a large number of cut stumps, which are really many wounds to the large multistemmed tree. Wounds are the normal point of infection for *H. annosum,* so the freshly exposed stumps provide ideal infection sites. Development of the fungus in the stump and associated root system causes death and decomposition of the infected tissue. Expansion of the area of invaded root system occurs through root grafts and root contacts. The root system supplying a number of stems is consumed, so that these stems die or are windthrown. The fungus continues to expand through the root system of the multiple-stemmed tree for a number of years. For some reason, the infection centers eventually stop expanding.

The amount of timber lost is related to the number of cut stumps where infections take place and the time after wounding. Losses in pine plantations are also related to soil conditions and the amount of inoculum present. Plantations on light sandy soil are more severely affected by *H. annosum*. The severity of the annosus problem is increased if inoculum is readily available from infection by previous cutting or adjacent stands.

The characteristic symptom of an annosus infection is an opening in the stand, with trees ranging from dead and windthrown in the center to chlorotic, thin-crowned trees on the periphery (Fig. 16-6). As the fungus slowly consumes more and more of the root system, the stems show chlorosis and the poor growth that is characteristic of nutrient deficiencies. Examination of the decay of affected trees should reveal a stringy white rot, characteristic of decay by *H. annosum* (Fig. 16-7). Fruit bodies of the fungus are brown on the top with a white pore surface underneath. They are formed on the stems under the duff or often on the underside of the stump and roots of fallen trees. (Fig. 16-8).

It is important to be able to recognize the presence of *H. annosum,* because if you fail to recognize the disease and take corrective measures, the future of the present stand, as well as future stands you may expect to grow on the site, may be in

Figure 16-6 *Heterobasidion annosum* infection center in mixed red and Scots pine in New York.

Figure 16-7 Fruiting body of *Heterobasidion annosum* on the root system of a fallen tree. Note also the stringy root rot characteristic of annosus decay.

Figure 16-8 *Heterobasidion annosum* fruit bodies observed by scratching back the duff from the base of a symptomatic tree. (Photograph compliments of Dr. Savel Silverborg.)

Examples of Root Diseases and Control

serious jeopardy. Although the general concept seems reasonable, it has been shown that infected pine stumps, in the southeastern United States, decay extremely rapidly and therefore do not represent a major hazard to survival of the next-generation stand.

If *H. annosum* is present in the vicinity of a plantation, it is advisable to treat stumps with urea, borax, or sodium nitrate at the time the stems are cut. Which treatment to use is a matter of local preference. The main concern is treatment immediately after cutting. No treatment provides an absolute guarantee that infection will not take place, so one should thin as few times as is reasonably acceptable for proper development of the stand.

Another possible treatment of the future is to inoculate with competitive microorganisms. *Peniophora gigantea* is a decay fungus that is effective as a stump treatment. The material is not generally available now, but it may be an alternative to chemical treatments in the future.

There are no good recommendations as to what to do for a seriously infected stand. Therefore, it is very important for forest managers to recognize the presence and seriousness of this problem and to prevent initial stand infection if they are to avoid destroying the productivity of conifer lands.

Phellinus weirii

The root rot caused by *Phellinus weirii* is known as laminated root rot because the decayed wood often seems to delaminate along the annual rings. The delamination is associated with differential decay of the spring and summer wood.

Phellinus weirii lacks the wide geographic distribution and host range of the former root rots but makes up for these by disrupting the potential future production of several of the most important timber species of North America: Douglas-fir, true firs, and hemlock in the Pacific northwest (Figs. 16-9 to 16-11).

Figure 16-9 *Phellinus weirii*-killed Douglas-fir in Oregon.

Figure 16-10 *Phellinus weirii* initially established in the old Douglas-fir stump in Oregon is responsible for mortality in these 10- to 15-in. (26- to 38-cm) Douglas-fir trees 50 years later.

Figure 16-11 Fallen tree on Vancouver Island, showing extensive stringy yellow brown rot of the root system typical of *Phellinus weirii*.

Examples of Root Diseases and Control

Like the former two pathogens *P. weirii* is a normal root rot fungus of mature trees. It becomes a serious threat to second-growth Douglas-fir about 50 years after the stand is established. Rapidly growing trees 10 to 20 in. in diameter (25 to 51 cm) are killed and fall like matchsticks, producing a very serious problem in maintaining full stocking for maximum production from the site.

The infection centers often appear to center on a large stump left from the previously cut stand. Aerial photographs of some infection centers appear ring-like. The inner portion of the ring or the center of the infection is invaded by hardwoods and brush species. The distinct, outer rim of the ring is the killing front.

The serious destruction of future stands by this disease has only recently been considered by foresters and pathologists of the Pacific northwest. Much about this disease is yet to be uncovered. No good prevention or control measure is known. If one can recognize the infection centers before cutting, these should not be allowed to regenerate to susceptible species. Pines are suggested for replanting in known infection centers.

Leptographium wageneri

Black-stain root disease of western conifers is an example of a host-dominant tissue-specific disease system. The recognition and research on this disease are relatively recent. Some of the concepts are therefore still in the formative stages.

Affected trees initially develop thin chlorotic crowns (Fig. 16-12) typical of many disorders. A progression of symptoms from crown thinning to death occurs on trees in groups. Clustering of affected individuals is a typical feature of root diseases. Black-stain root disease is positively diagnosed by the black staining of the cambium (Fig. 16-13) and arc-like bands of staining in the xylem. These arcs of stain are very different from the wedges of stain that are associated with blue stain fungal activity of the ray cells (see Chapter 15).

One of the first problems with this disorder was how to classify the disease. The pathogen, *Leptographium wageneri*, originally identified as *Verticicladiella wageneri* is classified in the Fungi Imperfecti in a group that generally has an Ascomycete perfect stage in the genus *Ceratocystis*. The perfect stage of *L. wageneri* has been reported to be *Ceratocystis wageneri*, but observations of this stage of the fungus are rare.

Fungi of this type are known to invade vessels and cause wilt diseases, such as Dutch elm disease. Others of the group invade parenchyma cells and cause blue stain. *L. wageneri* invades tracheids of the roots and lower stems of conifers. Invasion of clusters of tracheids in bands parallel to the annual rings causes dark resins and phenols to accumulate in the tissues. These, plus the dark fungus hyphae, form the black stain that can be seen under the bark of the affected plant and as arcs of dark-staining tissue within the stem.

Leptographium wageneri is therefore not a typical wilt pathogen, nor is it a typical stain pathogen, but it does have some of the characteristics of the fungi that cause these disorders. One of the features of this group of fungi is their association with bark beetles and other cambium-feeding insects. There is good evidence that *L. wageneri* is vectored by insects. One root-feeding weevil, *Hylastes nigrinus*, appears

Figure 16–12 Thin crowned ponderosa pine in northern California caused by black stain root disease. (Photograph compliments of Dr. Fields Cobb)

to be a major factor in dissemination of the fungus and in initiating new infection centers. The oviposition and feeding wounds are possible sites for infection. The insect activity increases following precommercial thinnings. These stands then become severely affected by black-stain root disease (Fig. 16-12).

Leptographium wageneri is not a typical root rot pathogen like *Armillaria* spp. It is very tissue specific in its invasion activities and does not decay wood significantly. It has some infection capabilities of typical root pathogens. The fungus can spread from one root to another through root contacts. It can also infect roots up to 6 cm away through growth of hyphae into the soil. This contact and hyphal spread result in clusters of affected trees.

The hosts of *L. wageneri* include ponderosa, lodgepole, Jeffery, and pinyon pines. It is also a problem of Douglas-fir, but true firs are apparently resistant. White pines can be infected if they are growing in association with susceptible species. There appear to be at least three host specialized races of the fungus that account for slightly different disease syndromes in different regions.

One of the factors that is important in the development of the disease is precommercial thinnings. Another factor on undisturbed sites is the moisture status. In the Sierra Nevada of northern California, the disease is associated with gullies and small creek drainages. Further south in pinyon pine stands, mortality is severe following years of heavy well-distributed rains during the summer. In Colorado the disease is most common on moist cool sites.

Examples of Root Diseases and Control

Figure 16-13 Dark streaks in the outer sapwood of the roots of a ponderosa pine are diagnostic symptoms of black stain root disease. (Photograph compliments of Dr. Fields Cobb)

Infection centers up to 10 ha or larger occur in ponderosa pine. The size of the infection center and spread rate of the pathogen is affected by the stand density. Spread rates of 3 m per year can occur under favorable conditions.

Management options to reduce the impacts are being developed. One possible option is to shift to a resistant species. Where there is concern, it may be appropriate to avoid or reduce precommercial thinnings. With more research effort and experience, there may also develop local cultural practices to reduce the impacts of this disease.

Leptographium procerum

Procerum root disease caused by *Leptographium procerum* is another more recently recognized root problem of eastern pines. The pathogen has a wide distribution with reports from Finland, Yugoslavia, New Zealand, and Canada. In eastern United States ornamental plantings, Christmas trees and seed orchards of eastern white, Scots, Austrian, red, ponderosa, Virginia, shortleaf, slash, sand, lodgepole, and jack pines are affected. Delayed bud break, reduced growth, chlorosis, wilting, and uniform browning of needles occurs on trees infected with *L. procerum*. Basal cankers, black staining of colonized tissue, and/or resinosis occurs at the base of the affected trees. Although the means of dissemination and infection are not fully understood, it is assumed that root and root crown feeding weevils and bark beetles are involved.

Procerum root disease is sometimes associated with pines on poorly drained soils. Root disturbances caused by roads or subsoiling in seed orchards can also be a problem. But, unfortunately, there is insufficient data available at this time to clearly interpret the relationship of procerum root disease and various site factors.

FOREST PRACTICES AND ROOT DISEASES

Root diseases are important problems primarily because of modern forest practices which emphasize tightly spaced pure stands or conversion to pure stands. Another aspect of modern forest practice is to grow the same species over and over on the same site. A third feature of modern forest practice is the emphasis on growing the most valuable species with little regard for important differences in site requirements. A fourth feature is the use of selective harvesting techniques that produce large numbers of infection sites and stress situations that favor root pathogens. Stop and think of how many of the major timber species grow as climax types in pure stands. Many of our commercial conifer species are pioneer types that would normally be replaced by other species if natural succession were allowed to progress. Root disease fungi are just doing what they have evolved to do. They are a significant part of successional pressures that break up one stand to allow colonization of the site by another. They are also a significant selection pressure against species and individuals that are poorly adapted to a site. Until we consider more natural forest management systems we will have to contend with the impacts of these pathogens.

REFERENCES

ARTMAN, J.D., and W.J. STAMBAUGH. 1970. A practical approach to the application of *Peniophora gigantea* for control of *Fomes annosus*. Plant Dis. Rep *54:*799–802.

BRADFORD, B., S.A. ALEXANDER, and J.M. SKELLY. 1978. Determination of growth loss of *Pinus taeda* L. caused by *Heterobasidion annosus* Fr. Bref. Eur. J. For. Pathol. *8:*129–134.

BRADFORD, B., J.M. SKELLY, and S.A. ALEXANDER. 1978. Incidence and severity of annosus root rot in loblolly pine plantations in Virginia. Eur. J. For. Pathol. *8:*135–145.

CHILDS, T.W. 1970. Laminated root rot of Douglas fir in western Oregon and Washington. USDA For. Serv. Res. Pap. PNW-102. 27 pp.

FILIP, G.M., and C.L. SCHMITT. 1979. Susceptibility of native conifers to laminated root rot east of the Cascade Range in Oregon and Washington. For. Sci. 25:261–265.

FROELICH, R.C., E.G. KUHLMAN, C.S. HODGES, M.J. WEISS, and J.D. NICHOLS. 1977. *Fomes annosus* root rot in the south, guidelines for prevention. USDA For. Serv. Southeast. Area State Private For. 17 pp.

GIBBS, J.N. 1967. The role of host vigour in susceptibility of pines to *Fomes annosus*. Ann. Bot. *31:*803–815.

GROSS, H.L. 1970. Root diseases of forest trees in Ontario. Can. For. Serv. Ont. Reg. For. Res. Lab., Sault Ste. Marie, Inf. Rep. O-X-137. 16 pp.

HARRINGTON, T.C., and F.W. COBB. 1988. Leptographium root diseases on conifers. The American Phytopathological Society, St. Paul, Minn. 149 pp.

HODGES, C.S. 1969. Modes of infection and spread of *Fomes annosus*. Annu. Rev. Phytopathol. *7:*247–266.

HODGES, C.W. 1974. Cost of treating stumps to prevent infection by *Fomes annosus*. J. For. *72:*402–404.

HODGES, C.W. 1974. Symptomatology and spread of *Fomes annosus* in southern pine plantations. USDA For. Serv. Res. Pap. SE-114. 10 pp.

HOUSTON, D.R., and H.G. ENO. 1969. Use of soil fumigants to control spread of *Fomes annosus*. USDA For. Serv. Res. Pap. NE-123. 23 pp.

HUNT, R.S., and D.J. MORRISON. 1986. Black-stain root disease on lodgepole pine in British Columbia. Can. J. For. Res. *16:*996–999.

HUNTLY, J.H., J.D. CAFLEY, and E. JORGENSEN. 1961. Armillaria root rot in Ontario. For. Chron. *37:*228–236.

JAMES, R.L., C.A. STEWART, and R.E. WILLIAMS. 1984. Estimating root disease losses in northern Rocky Mountain national forests. Can. J. For. Res. *14:*652–655.

KENERLEY, C.M., K. PAPKE, and R.I. BRUCK. 1984. Effect of flooding on development of Phytophthora root rot in Frasier fir seedlings. Phytopathology *74:*401–404.

KOMMEDAHL, T., and C.E. WINDELS. 1979. Fungi: pathogen or host dominance in disease. *In* Ecology of root pathogens, ed. S.V. Krupa and Y.R. Dommergues. Elsevier Science Publishing Co., Inc., New York, pp. 1–103.

KUHLMAN, E.G. 1986. Impact of annosus root rot minimal 22 years after planting pines on root rot infested sites. South. J. Appl. For. *10:*96–98.

LEAPHART, C.D. 1963. Armillaria root rot. USDA For. Serv. For. Pest Leafl. 78. 8 pp.

NEWHOOK, F.J., and F.D. PODGER. 1972. The role of *Phytophthora cinnamomi* in Australia and New Zealand forests. Annu. Rev. Phytopathol. *10:*299–326.

OTROSINA, W.J., and R.F. SCHARPF. 1989. Research and management of Annosus root disease *(Heterobasidion annosum)* in western North America. U.S. For. Serv. Gen. Tech. Rept. PSW-116. 177 pp.

PAWSEY, R.G., and M.A. RAHMAN. 1976. Chemical control of infection by honey fungus, *Armillaria mellea:* a review. J. Arboric. *2:*161–169.

PODGER, F.D. 1975. The role of *Phytophthora cinnamomi* in dieback disease of Australia eucalyp forests. *In* Biology and control of soil-borne plant pathogens, ed. G.W. Bruehl. The American Phytopathological Society, St. Paul, Minn., pp. 27–36.

REDFERN, D.B. 1975. The influence of food base on rhizomorph growth and pathogenicity of *Armillaria mellea* isolates. *In* Biology and control of soil-borne plant pathogens, ed. G.W. Bruehl. The American Phytopathological Society, St. Paul, Minn., pp. 69–73.

RISHBETH, J. 1975. Stump inoculation: a biological control of *Fomes annosus*. *In* Biology and control of soil-borne plant pathogens, ed. G. W. Bruehl. The American Phytopathological Society, St. Paul, Minn., pp. 158–162.

RISHBETH, J. 1985. Infection cycle of *Armillaria* and host response. Eur. J. For. Path *15:*332–341.

Ross, E.W. 1973. *Fomes annosus* in the southeastern United States: relation of environmental and biotic factors to stump colonization and losses in the residual stand. USDA For. Serv. Tech. Bull. 1459. 26 pp.

Schefield, R.M., and N.D. Cost. 1986. Behind the decline. J. For. *85:*29–33.

Shaw, C.G., III, and L.F. Roth. 1976. Persistence and distribution of a clone of *Armillaria mellea* in a ponderosa pine forest. Phytopathology *66:*1210–1213.

Shea, S.R., B.L. Shearer, J. Tippett, and P.M. Deegan. 1984. A new perspective of jarrah dieback. For. Focus *31:*3–11.

Sinclair, W.A. 1964. Root- and butt-rot of conifers caused by *Fomes annosus,* with special reference to inoculum dispersal and control of the disease in New York. Cornell Univ. Agric. Exp. Stn. Mem. 391. 54 pp.

Trappe, J.M. 1972. Regulation of soil organisms by red alder: potential biological system for control of *Poria weirii. In* Managing young forests in the Douglas fir region, ed. A.B. Berg. Ore. State Univ. School For., Corvallis, Ore., pp. 35–51.

Wargo, P.M., and C.G. Shaw III. 1985. Armillaria root rot: the puzzle is being solved. Plant Dis. *69:*826–832.

Weste, G., and G.C. Marks. 1987. The biology of *Phytophthora cinnamomi* in Australasian forests. Annu. Rev. Phytopathol. *25:*207–229.

Whitney, R.D. 1988. The hidden enemy. Root rot technology transfer. Can. For. Serv., Sault Ste. Marie, Ontario. 35 pp.

Wilcox, H.E. 1983. Fungal parasitism of woody plant roots from mycorrhizal relationships to plant disease. Annu. Rev. Phytopathol. *21:*221–242.

Williams, R.E., and M.E. Marsden. 1982. Modeling probability of root disease center occurrence in northern Idaho forests. Can. J. For. Res. *12:*876–882.

VanBroembsen, S.L., and F.J. Kruger. 1985. *Phytophthora cinnamomi* associated with mortality of native vegetation in South Africa. Plant Dis. *69:*715–717.

17

PARASITIC FLOWERING PLANTS AS AGENTS OF TREE DISEASES

- *TYPES OF PARASITIC FLOWERING PLANTS*
- *MODE OF ACTION OF PARASITIC FLOWERING PLANTS*
- *DISEASE CYCLE OF PARASITIC FLOWERING PLANTS*
- *SYMPTOMS OF DISEASES CAUSED BY PARASITIC FLOWERING PLANTS*
- *METHODS FOR RECOGNITION OF DISEASES CAUSED BY PARASITIC FLOWERING PLANTS*
- *EXAMPLES OF PARASITIC FLOWERING PLANTS AND CONTROL*

Coevolution of closely associated flowering plants has produced a number of obligate associations. Evolution of the parasitism of one plant on another has generally developed into a balanced relationship wherein the host population is not seriously affected by the parasite. But the imposition of human influence on natural processes of "undisturbed" areas, cultivation of trees in plantations, and management of natural forests for fiber upsets the balances to produce a number of problems.

TYPES OF PARASITIC FLOWERING PLANTS

Parasitic flowering plants are usually Dicotyledons. The exception is one parasitic conifer in New Caledonia. It is assumed that evolution toward parasitism has occurred at least eight times in unrelated groups of Dicotyledons.

The absence of one common ancestor for all parasitic plants has produced a number of different types of specific associations. Nevertheless, a certain amount of common structure and function is found in most parasites. The following groups of plants have some representative parasites. The individual parasitic representatives within each group may have more in common with nonparasitic members of the same group than with other parasites of other groups.

1. The order Santalales has a number of parasitic flowering plants in the families Loranthaceae (mistletoe), Viscaceae (dwarf mistletoe), Santalaceae (sandalwood), Olacaceae, and Myzodendraceae.

2. The families Scrophulariaceae and Orobanchaceae (figwarts and broomrapes) have 26 parasitic genera. In the Northeast, *Melampyrum* (cow-wheat), *Castilleja* (Indian paintbrush), *Euphrasia* (eyebright), *Gerardia* (downy false foxglove) and *Pedicularis* (lousewort) are in this group. *Striga* (witchweed), an introduced plant, and *Senna* (seymeria) are two southern parasitic plants in this group.

3. The families Rafflesiaceae and Hydnoraceae are mostly tropical parasitic flowering plants. *Rafflesia* produces the largest known flower (1 meter in diameter) from a subterranean root parasitic system.

4. The family Balanophoraceae is made up of small groups of tropical fungus-like plants. Tuberous rhizomes characterize these root parasites.

5. The family Cuscutaceae (dodders) resembles nonparasitic plants in the family Convolvulaceae more than parasitic plants, but these have evolved modifications typical of plant parasites. There are about 158 species of this widely distributed group of annual plants.

6. The genus *Cassytha* (scrub dodder) in the otherwise autrotrophic family Lauraceae is similar to dodder. In contrast to dodder, it is perennial and regionally confined to coastal areas.

7. The family Lennaceae is a small family of xerophytic parasitic plants found only in America.

8. The family Krameriaceae is a small group of xerophytic parasitic plants found mostly in Mexico.

Some of the plant parasites are totally dependent upon the host for preformed photosynthates (holoparasitic), while others are only partially dependent upon the host for photosynthates, because they still maintain some photosynthetic leaf surface (hemiparasites). Often the leaf is much reduced, so that photosynthetic capacity is minimal. Parasitic plants may be perennial or annual. Some, like sandalwood, are trees. Some parasitic plants are associated with roots. Others occur as stem parasites. A major portion of the life cycle of a parasite may be spent below the ground (subterranean) or within the host (endophytic). Systemic infection of meristems of hosts sometimes results in major expansion and development of the endophytic system.

MODE OF ACTION OF PARASITIC FLOWERING PLANTS

The mode of action of all parasitic plants revolves around the haustorium or parasitic nutrient-uptake organ. These organs of parasitic plants have no obvious morphological link to such organs as the roots and stems of autotrophic plants.

The haustorium functions like a root but has no root cap. Root hairs are absent or very reduced. Phloem is absent or highly reduced. The organ is composed mostly of vessel members and a few tracheids.

Haustoria may be compact organs that penetrate localized regions of host root or shoot tissues, or they may be much branched and ramify generally throughout host tissues. In some dwarf mistletoes, the haustorium or endophytic system grows just behind the apical meristem. In this way, the haustoria develop systemically in the most recent tissues.

It is not fully understood what causes haustoria to penetrate plants or what the haustoria remove from the plant. Haustorial cells align with host phloem or xylem elements, with emphasis on xylem vessels, and produce a continuity bridge between the two plants through pits or through resorbed host vessel walls.

Through the haustoria, elements are transferred from one plant to another. One might assume that carbohydrates and other complex organic materials are absorbed from phloem elements, but organic materials such as sugars, nitrogenous compounds, enzymes, and minerals, as well as water, can also be translocated from xylem vessels.

The continuity of the parasite with the host plant is sufficient to allow transfer of viruses and mycoplasmas between plants. Dodders are used by researchers to establish host transfer. A mycoplasma disease, sandal spike, is acquired by the sandal plant from a number of host trees.

There is generally a low degree of host specificity for flowering plant parasites, although many are restricted to a limited host range by ecological circumstances. Mechanisms of resistance other than escape are very poorly understood.

DISEASE CYCLE OF PARASITIC FLOWERING PLANTS

Seeds of parasites may be small, numerous, and wind-disseminated or large, few, and disseminated by birds. Dwarf mistletoe seeds have a unique forcible discharge mechanism for propelling the seeds up to 30 ft (9 m), farther if helped by the wind.

Germinating seeds form a radical that may be chemotactically attracted toward the host plant. Upon contact with the host, touch (thigmotropic) or chemical (chemotrophic) stimuli induce a cushion or holdfast to form a fine, intrusive organ to enzymatically and mechanically penetrate the host.

A primary haustorium or endophytic system develops from this penetration peg of the radical tip. Secondary haustoria may develop in some parasites from branching of the radical.

The endophytic existence for the parasite may be lengthy with a brief period of shoot growth for flowering, or the parasite may produce annual or perennial shoots with photosynthetic capacity.

SYMPTOMS OF DISEASES CAUSED BY PARASITIC FLOWERING PLANTS

Effects on parasitized hosts range from excessive hypertrophy and witches' brooms to almost no effect. Reduced reproductive capacity, chlorosis, and reduced growth are common symptoms. In some instances death occurs.

METHODS FOR RECOGNITION OF DISEASES CAUSED BY PARASITIC FLOWERING PLANTS

Recognition of diseases caused by parasitic plants always involves observation and identification of the parasite. Specific symptoms such as witches' brooming, branch swelling, or remnants of shoots of the parasite may be used for diagnostic purposes if one is rather familiar with a specific disease. Caution should be exercised in the diagnosis of parasitic plant diseases in the absence of an identifiable parasite because fungi, viruses, mycoplasmas, and other parasites, as well as environmental and genetic factors, will also induce these types of symptoms.

EXAMPLES OF PARASITIC FLOWERING PLANTS AND CONTROL

Although Indian paintbrush and many other flowering plants of the northeast are parasites of other plants, little is known of their effects on the parasitized hosts.

Witchweed

The root parasite witchweed (*Striga asiatica)* is an introduced plant from the Old World Tropics which represents a serious threat to corn, wheat, and other cultivated grasses and sedges of the south. The seeds of this parasite are difficult to separate from the grain, so that it is sometimes distributed with contaminated seed lots. Control is through use of herbicides and crop rotation. Because crabgrass is a host for witchweed, control may have to extend to areas surrounding fields.

Senna Seymeria

Seymeria cassioides is a newly recognized parasitic threat to slash pine and other southern pines. It weakens and sometimes kills pine seedlings if it becomes established in an area. It develops rapidly on reseeded or replanted sites where a few residual pines of a previous crop represent foci for infection centers. Control is by herbicide treatment of the succulent annual.

Dodders

Dodders *(Cuscuta* spp.) are of cosmopolitan occurrence and often result in large economic losses in field crops, particularly in the tropics (Figs. 17–1 and 17–2). Dodder seeds resemble seeds of many commercial crops, particularly legumes and flax. Before rigid methods of seed control were introduced, dodder seeds were often common in commercial seed lots; consequently, dodder species have been introduced into areas where they were not native. Dodders are seldom of importance in forest pathology. They have caused some problems in nurseries, particularly on black locust, green ash, and poplar. Morphologically, dodders are highly reduced plants having reduced leaves, a coiled habit, little vascular development, and a reduced chlorophyll content.

The radical from the germinating seed penetrates the ground sufficiently to support the seed and coiling shoot as it searches for a suitable host. Growth up to 35 cm over a 7-week period is possible before finding a host. Once contact is made with

Figure 17–1 Coiling shoots of dodder on understory forest vegetation.

Parasitic Flowering Plants as Agents of Tree Diseases   Chap. 17

Figure 17-2 Close-up of coiling shoot and haustoria penetrating red oak seedlings in the understory forest vegetation.

a host, the shoot coils around the host and produces penetration haustoria. The contact with the ground deteriorates and up to half a mile (0.8 km) of coiling shoot may develop.

Dodders have little or no ability for photosynthesis and thus depend on the host as a carbohydrate source. Their effects on the host often result in hypertrophies causing witches' brooms and galls. Preventing the introduction of dodder seeds into nurseries is an effective method of control. Sprays are available, but these are often injurious to the host plants. Soil fumigation, where economical, may be desirable.

Leafy Mistletoe

The true or leafy mistletoes have a long history and a considerable folklore. There are several species found in North America. The genus *Phoradendron* is the most significant to forest pathology (Figs. 17-3 to 17-5). In the arid and semiarid southwest, about a dozen species of this genus occur on a wide variety of hosts. They are most frequently associated with hardwoods, but several species occur on junipers, cypress, incense cedar, and true firs.

The leafy mistletoes are limited to warmer climates and do not occur above the 40 to 45° latitude (New Jersey-Oregon). Northward extension is probably limited by temperature. These plants are morphologically well-developed with moderate-size

Examples of Parasitic Flowering Plants and Control   **319**

Figure 17–3 *Phoradendron villosum* on emory oak.

Figure 17–4 Close-up of leafy mistletoe.

Parasitic Flowering Plants as Agents of Tree Diseases   Chap. 17

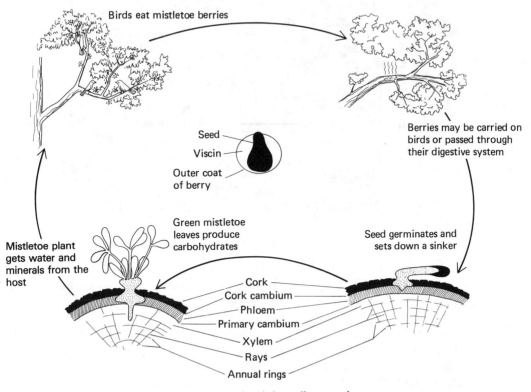

Cause: *Phoradendron* spp.
Hosts: Hardwoods, juniper, cypress, and incense cedar in southern U.S.

Birds eat mistletoe berries

Berries may be carried on birds or passed through their digestive system

Seed
Viscin
Outer coat of berry

Mistletoe plant gets water and minerals from the host

Green mistletoe leaves produce carbohydrates

Seed germinates and sets down a sinker

Cork
Cork cambium
Phloem
Primary cambium
Xylem
Rays
Annual rings

Figure 17–5 Leafy mistletoe disease cycle.

leaves and stems. Birds, feeding on mistletoe berries, disseminate the seeds, which pass through the birds undigested. Mistletoes support much of their carbohydrate requirements by photosynthesis. They use the host xylem as a water source and may therefore become a problem when water availability to the host is limited. Infected portions of the host often exhibit galls and brooms, and the uninfected host tissues may become deformed with reduced growth. The death of the host is rarely due to the mistletoe alone. Economic losses, in terms of volume and usable product, may occur.

Control is by pruning, fire, and sprays where economical, such as for high-value crops such as fruit crops. In Texas and Oklahoma the fruiting mistletoe may be gathered and sold as a Christmas ornament.

Dwarf Mistletoe

Dwarf mistletoes *(Arceuthobium* spp.) are the most serious parasitic higher plants in North America. About 45 species, usually host-specific, are important. These organisms cause large amounts of economic losses and may be the most serious problem in western conifer forests (Figs. 17–6 to 17–9). They are a major problem on ponderosa

Figure 17-6 Brown shoots of *Arceuthobium gillii* on Chihuahua pine in Arizona.

pine, lodgepole pine, Douglas-fir, true firs, western larch, and other western conifers. In the east, black and red spruce may be severely affected. Losses may result from mortality of older trees, growth reduction, poor quality of product, and an overall reduction in host vigor.

The portion of the tree invaded by the parasite becomes swollen, and witches' brooms sometimes occur. Physiological stimulation of infected portions result in reduced growth in other portions of the host. The disease occurs as patches of infected trees within a stand. Local spread of the disease occurs as a result of the

Figure 17-7 Douglas-fir branch systematically infected with *Arceuthobium douglasii* in Arizona.

Figure 17-8 Yellow orange shoots of *Arceuthobium vaginatum* on ponderosa pine in Arizona.

Figure 17-9 Close-up of *Arceuthobium vaginatum* on the main stem of a ponderosa pine in Arizona.

Examples of Parasitic Flowering Plants and Control

forcible ejection of viscous seed from the host (Fig, 17-10). Seeds shot from the host usually have a range of about 20 to 30 ft (6 to 10 m), depending on the height at which they were dispersed and on the wind. The seeds adhere to objects they strike, because of their sticky viscous coat (Fig. 17-11). When the viscous coating dries, it acts as a glue enabling the seed to hold fast to the host until germination and penetration occur. Seed dispersal usually occurs in the fall or late summer, but germination is usually delayed until spring. Once penetration of the host occurs, the parasite develops an extensive haustorial absorptive system, resulting in the production of "sinkers," which penetrate into the xylem of the host. They are subsequently embedded by additional growth of xylem tracheids. Vessel elements of the sinkers connect with tracheids of the host xylem, and the parasite derives most of its carbohydrate and probably all of its water and minerals from the host. The mistletoe life cycle, from infection to seed production, takes about 4 to 5 years (Fig. 17-12). Long-range dispersal, by external transport on birds, serves to introduce the parasite into new stands.

Control is by elimination of infected hosts by clear-cutting. Pruning and selective removal of large trees with crown infections may be practiced in high-value lightly to moderately infected stands.

Recognition that fire control is associated with the expanding importance of mistletoe has contributed somewhat to a second look at the role of fires in timber management. In some instances, the best management practice would be a major burn. Understanding the dynamics of seed dispersal and the disease cycle has developed a method for learning to live with minimal losses due to mistletoe. The growth impact is minimal unless the trees are highly infected. Highly infected trees generally develop from understory plants below an infected overstory.

An important consideration for learning to live with dwarf mistletoe was the development of a rating system to catalog individual trees into various dwarf

Figure 17-10 Close-up of *Arceuthobium* berries.

Figure 17-11 Seed of *Arceuthobium* sticking to pine needles.

mistletoe infection classes. The procedure involves visually separating the crown of the tree into upper, middle, and lower thirds. Each third is rated as to the intensity of infection using a scale of 0, 1 or 2, with 0 indicating no infection and 2 indicating heavy infection. The sum of the scores for the upper, lower, and middle third of the crown is the mistletoe rating for the tree.

The key to the system is its ability to characterize trees into current and future risk groups. There is minimal effect on growth of trees with a 0 to 3 rating and minimal potential for increase of the pathogen as long as the trees continue rapid height growth. If the upper third of the tree crowns are free of heavy infection by mistletoe, the forest manager can expect minimal impact from the mistletoe in the short run. The future impact on the present and next generation stand will be affected by how this level of infection is allowed to progress.

In contrast, a stand with many trees rated between 4 and 6 will sustain a significant growth impact due to dwarf mistletoe and should be scheduled for immediate harvest. A stand with a few trees rated between 4 and 6 can be maintained or improved by a thinning. In this case the rating system is used to select which trees to remove.

The system can also be used to facilitate management decisions for individual trees and stands of trees in recreation sites. The integration of the rating system into management decisions provides a means for continued amenity or economic production from mistletoe-infected stands.

Extensive evaluation of dwarf mistletoe development in western forests has led to an integration of timber yield and loss figures into computer-based management option systems. The programs provide resource managers with projected yields for various management options. If a particular stand has a high incidence of upper-crown infection, the yields over the short and long run can be predicted for various levels of cutting. Lower crown infection of varying intensity results in different

Examples of Parasitic Flowering Plants and Control

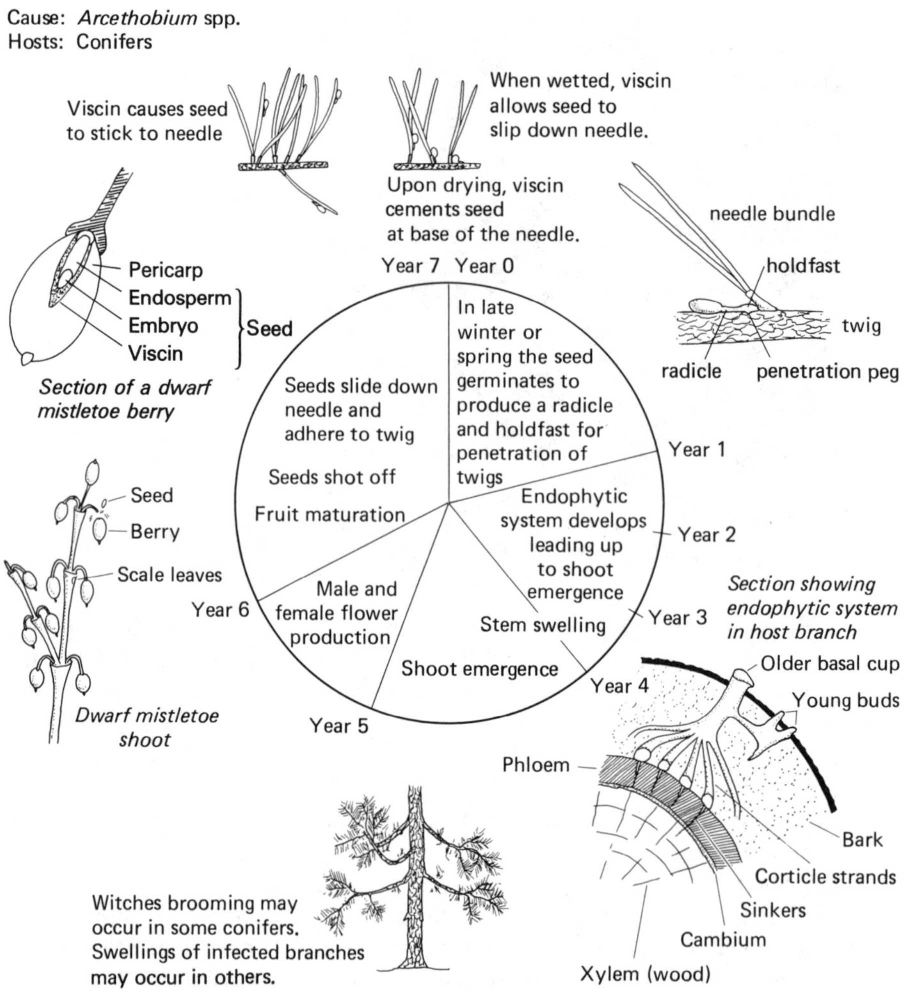

Cause: *Arcethobium* spp.
Hosts: Conifers

Viscin causes seed to stick to needle

When wetted, viscin allows seed to slip down needle.

Upon drying, viscin cements seed at base of the needle.

needle bundle

holdfast

twig

radicle   penetration peg

Pericarp
Endosperm
Embryo   } Seed
Viscin

Section of a dwarf mistletoe berry

Year 7   Year 0

In late winter or spring the seed germinates to produce a radicle and holdfast for penetration of twigs

Seeds slide down needle and adhere to twig

Seeds shot off

Fruit maturation

Year 1

Endophytic system develops leading up to shoot emergence

Year 2

Section showing endophytic system in host branch

Older basal cup

Young buds

Phloem

Male and female flower production

Year 6

Stem swelling

Year 3

Shoot emergence

Year 4

Year 5

Seed
Berry
Scale leaves

Dwarf mistletoe shoot

Bark

Corticle strands

Sinkers

Cambium

Xylem (wood)

Witches brooming may occur in some conifers. Swellings of infected branches may occur in others.

Figure 17-12 Dwarf mistletoe disease cycle.

impacts on yields, depending upon thinning and pruning practices. Yields from two-storied stands can also be predicted, based on the intensity of the overstory infection and various cutting and thinning options.

It is important to recognize the relationship of forest management activities to dwarf mistletoe impact. For example, unevenaged management is more conducive to the spread and intensification of dwarf mistletoe than even aged management. These consequences need to be factored properly into management decisions.

The integration of growth and disease-loss relationships, in relation to management decisions, represents one of the most advanced on-line contributions that forest pathologists have made to forest management. By recognizing the impact of various management practices on disease, it develops a philosophy of minimizing losses by learning to live with a disease agent. This philosophy should be a universal doctrine of both pathologists and forest resource managers.

REFERENCES

CALDER, M., and P. BERNHARDT, eds. 1983. The biology of mistletoes. Academic Press, Inc., Sydney, Australia. 348 pp.

GILL, L.S., and F.G. HAWKSWORTH. 1961. The mistletoes: a literature review. USDA For. Serv. Tech. Bull. 1242. 87 pp.

GRAHAM, D.P. 1967. A training aid on dwarf mistletoe and its control. Issued by USDA For. Serv. Pac. Northwest Reg. Insect Disease Control Br., 49 pp.

HAWKSWORTH, F.G. 1973. Dwarf mistletoe and its role in lodgepole pine ecosystems. *In* Management of lodgepole pine ecosystems, symposium, ed. D.M. Baumgartner. Wash. State Univ., Pullman, Wash., pp. 342–358.

HAWKSWORTH, F.G. 1977. The 6-class dwarf mistletoe rating system. USDA For. Serv. Gen. Tech. Rep. RM-48. 7 pp.

HAWKSWORTH, F.G., T.E. HINDS, D.W. JOHNSON, and T.D. LANDIS. 1977. Silvicultural control of dwarf mistletoe in young lodgepole pine stands. USDA For. Serv. Tech. Rep. R2-10. 11 pp.

HAWKSWORTH, F.G., and D.W. JOHNSON. 1989. Biology and management of dwarf mistletoe in lodgepole pine in the Rocky Mountains. USDA For. Serv. Gen. Tech. Rep. RM-169. 38 pp.

HAWKSWORTH, F.G., and D. WIENS. 1972. Biology and classification of dwarf mistletoes *Arceuthobium).* USDA For. Serv. Agric. Handb. 401. 234 pp.

KUIJT, J. 1969. The biology of parasitic flowering plants. Univ. Calif. Press, Berkeley, Calif. 246 pp.

MALCOLM, W.M. 1966. Root parasitism of *Castilleja coccinea.* Ecology *47:*179–186.

MANN, W.F., H.E. GRELEN, and B.C. WILLIAMSON. 1969. *Seymeria cassioides,* a parasitic weed on slash pine. For. Sci. *15:*318–319.

SCHARPF, R.F., and J.R. PARMETER, JR., eds. 1978. Proceedings of the symposium on dwarf mistletoe control through forest management. USDA For. Serv. Gen. Tech. Rep. PSW–31. 190 pp.

THOMSON, A.J., R.I. ALFARO, W.J. BLOOMBERG, and R.B. SMITH. 1985. Impact of dwarf mistletoe on growth of western hemlock trees having different patterns of suppression and release. Can J. For. Res. *15:*665–668.

18

DECLINE DISEASES
OF COMPLEX BIOTIC
AND ABIOTIC ORIGIN

- *THE DECLINE SYNDROME*
- *COMMON DENOMINATORS OF DECLINES*
- *SYMPTOMS OF DECLINES*
- *DIAGNOSIS OF DECLINE DISEASES*
- *EXAMPLES OF DECLINES*
- *ECOLOGICAL ROLE OF DECLINES*

Preceding chapters have dealt with single-causal-factor abiotic and biotic diseases. This chapter develops a third category, the decline diseases, which are caused by the interaction of a number of interchangeable, specifically ordered abiotic and biotic factors to produce a gradual general deterioration, often ending in the death of trees.

As you progress through this chapter, it will become clearer what is meant by the definition. You will see that a number of different stress inducing factors presumably produce somewhat similar responses in trees, and therefore I suggest that these factors can be somewhat interchangeable from one place to another or one time to another. In addition, the various stress inducing factors are clustered into three specifically ordered groups. A decline disease will involve at least one factor from each group.

The topic of decline has been popularized in recent years because of the perceived association of acid rain and deterioration in forest health, forest decline (see Chapter 4). This notoriety has had both positive and negative benefits. On the positive side, there has been some very much needed additional thinking and evolution of the basic concepts. On the negative side, there have been some misconceptions that have been widely accepted as fact. It has also become evident that different disciplines have evolved their own terms and concepts. I have attempted to note and integrate aspects of the historical and current thinking into the present chapter, but the reader should recognize that we are working with a conceptual base that is still in the formative and testing stages.

The significance of decline disease and forest decline has become somewhat confusing. Like other diseases, decline diseases are usually normal, appropriate factors contributing to the death of individuals in an ecosystem. Diseased and dying individuals or of whole forests (forest decline) may become a problem when expectations of people are considered. The difficulty comes in accepting that death of individual trees or even large numbers of trees is not necessarily abnormal. Death may be required for proper development of the next generation.

Popularized misconception generally assumes that forest declines lead to the death of ecosystems, just like declines diseases lead to death of trees. Ecosystems have certainly been destroyed through activities of people, but I am not certain that the tree decline model developed in this chapter is the best way to interpret ecosystem destruction. At the very minimum, "natural" forest decline producing normal cycling of populations may be misinterpreted as pollutant-induced "unnatural" forest decline, leading presumably to ecosystem destruction. Little real progress will be made on unnatural forest declines until there is a better understanding of natural forest declines.

The concept of declines as a distinct category of disease is not totally accepted by forest pathologists. Some would assume that the category decline is a collection of diseases with incompletely understood etiology. Declines are often poorly understood, and therefore some of the diseases we call declines may eventually be shown, by additional investigation, to have specific single causal agents.

Others would suggest that diseases of complex etiology are the rule rather than the exception. Terms such as *unthrifty, debilitated, retarded growth, decline,* and *dieback* are used to characterize these complex disorders. Since Koch's postulates (see Chapter 1) cannot be used to define a specific single causal agent (pathogen), the interacting biotic and abiotic agents are better referred to as determinants or associated factors.

Still others would question the use of a common word *decline* for diseases with specific etiological connotation. Decline can be interpreted as a gradual failure in health regardless of the cause or causes. Trees can decline in health from simple injury or single biotic disease agents. There is also a natural decline in annual increment as trees age. Since there is a confusion in terms, it may be appropriate to use the designation *decline disease* for diseases of complex abiotic and biotic origin, the topic of this chapter.

Hepting (1963) suggested the involvement of significant climatic changes in decline diseases. He presented evidence of a climatic warming trend during the middle part of this century. It is very difficult to verify the role of climatic changes in disease because it is difficult to even document climatic change.

Plant development or deterioration is intimately tied to weather conditions.

Weather is just a short-term reflection of climate. Most declines are associated with some type of local or regional climatic aberration. Therefore, it is reasonable to accept the role of continental climatic changes in declines.

Most diseases are diagnosed on the basis of abnormal physiological functions of specific plant parts, associated with invasion by specific pathogens. With decline diseases the physiological disruption is somewhat general and nonspecific. The general deterioration in vigor can be caused by an array of stress-inducing factors. Decline diseases may be better understood from an ecological point of view.

THE DECLINE SYNDROME

Details of specific declines will be covered later, but it is useful for development of the concept at this point to refer to a specific decline complex. Attempts to understand maple decline as a single standard pathogen-induced disease or a cause-effect relationship has identified some involvement of soil moisture, soil aeration, nematodes, Armillaria root rot, sugar maple borer, high temperatures, Verticillium wilt, canker fungi, insect defoliators, root excavation, salt, injuries, and air pollution (Fig. 18-1). Although there are many possible factors, none can be shown to produce maple decline individually.

Sinclair (1965) was the first person to suggest that maple decline and other similar problems may develop from the interaction of three or more sets of factors (Fig. 18-2). The first were long-term slowly changing factors such as soil, site, and climate. These *predisposing* factors alter the trees' ability to withstand or respond to injury-inducing agents. A second group of factors were called *inciting* factors. Their action is short in duration. They may be physiological or biological in nature. These generally produce dieback of small branches. Examples of incitants are insect defoliators, late spring frost, drought, and salt spray. A third group of factors, called *contributing,* were suggested to include a collection of environmental factors as well as biotic agents such as canker and decay fungi and bark and wood boring insects.

The foregoing concept developed from research efforts to identify the causes of three important problems in the northeast: ash dieback, maple decline, and oak decline. In the first edition of *Tree Disease Concepts,* the decline concept was expanded, modified, and developed into a teaching tool based on additional information on the original three problems plus other diseases which I considered to have similar etiology. The outcome of this was the Manion decline spiral, modified and updated slightly in Fig. 18-3.

Age of the trees was one factor of every decline that seldom received much recognition. It was and still is my impression that decline diseases occur only in reproductively mature populations of trees. Therefore, I would suggest that the interaction of age with environmental and site variables predisposes trees to decline.

Another addition to the original concept of predisposition is the role of genetic potential in decline diseases. The impact of decline diseases are generally most evident on the dominant and codominant trees of a stand.

A third addition to the concept is to suggest that viruses as genetic determi-

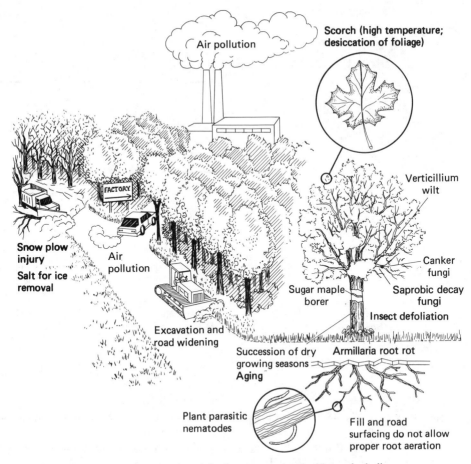

Air pollution

Scorch (high temperature; desiccation of foliage)

FACTORY

Snow plow injury

Salt for ice removal

Air pollution

Verticillium wilt

Canker fungi

Sugar maple borer

Saprobic decay fungi

Insect defoliation

Excavation and road widening

Succession of dry growing seasons

Aging

Armillaria root rot

Plant parasitic nematodes

Fill and road surfacing do not allow proper root aeration

Figure 18–1 Summary of the factors associated with maple decline.

nants may adjust the genetic potential of a tree and therefore contribute to the predisposition.

Each of these will be discussed in more detail later. For now it is important to recognize that the interaction of age, genetic potential, and viruses with environmental and site variables may predispose or predetermine which trees or stands are going to decline.

Any tree exposed to short-term biotic or abiotic events, inciting factors, will often respond with twig dieback. Vigorously growing plants usually recover quickly if the incitant-caused stress is mild enough. Therefore, dieback does not always lead to decline. Those plants previously modified by predisposing factors do not recover quickly.

In the debilitated state caused by predisposing and inciting factors, they are additionally vulnerable to other stressing agents, called contributing factors. Secondary insects such as bark beetles, opportunistic canker and root rot fungi, and possibly other organisms that specialize in the colonization of weakened plants, invade and produce additional dieback, leading progressively toward death. These

The Decline Syndrome

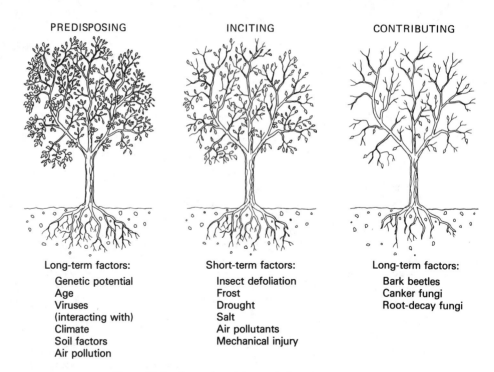

PREDISPOSING	INCITING	CONTRIBUTING
Long-term factors:	Short-term factors:	Long-term factors:
Genetic potential Age Viruses (interacting with) Climate Soil factors Air pollution	Insect defoliation Frost Drought Salt Air pollutants Mechanical injury	Bark beetles Canker fungi Root-decay fungi

Figure 18–2 Categories of factors influencing declines.

organisms are persistent and are often blamed for the condition of the host. They are better understood as indicators of weakened hosts. Eventually, the plant dies or is rendered useless as an ornamental or forest tree.

I like to think of the relationships between the various determinants or factors as a dynamic spiral obstacle course (Fig. 18–3). The factors are like barriers in the course. During the life of any tree, it will continuously and repeatedly encounter an array of stress factors. The predisposing factors nudge the tree inward to the second level of the spiral. The inciting factors accentuate the inward spiral of predisposed trees. The contributing factors of the inner spiral compete for a niche on the weakened and dying tree.

Others describe the relationships with different terms and models. Wallace (1978) visualized the plant at the center of a collection of outward-radiating lines. At various distances along the lines are the determinants. Those determinants that are close to the plant are more important than those far from the plant. Cross-links between determinants also occur in the Wallace model.

Houston (1981) suggested a linear model involving transition from healthy trees through dieback to decline and eventually death. Environmental stress on healthy plants often initiates a dieback of buds, twigs, and branches. With abatement of stress, trees often recover but additional stress by the same agent or a different agent on the altered trees initiates further dieback. Sufficiently stressed trees are subjected to the effects of organisms of secondary action, resulting in decline and eventually death.

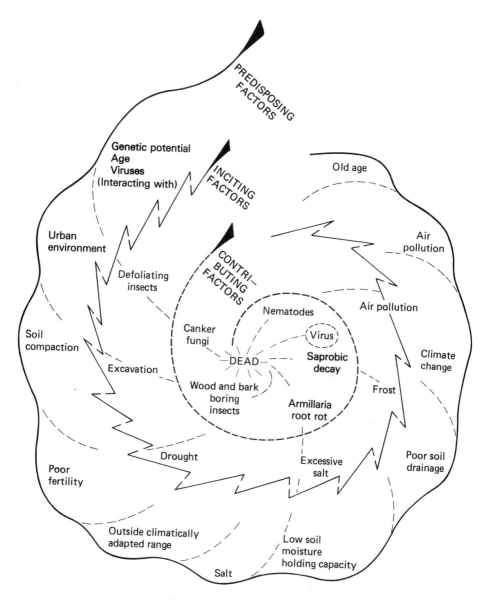

Figure 18-3 Decline disease spiral.

From a different perspective, Mueller-Dombois (1983) suggests an ecological synchronous cohort senescence model. In this model a cohort (population of common origin) of plants develops over time to a level of advanced maturation (senescence) based on the genetic potential and ecological parameters of the site. Environmental stress factors such as drought and hurricanes simultaneously stress the previously synchronized senescing cohort. Dieback is initiated and is further promoted by secondary pathogens and secondary insects. The key to this model is that "normal" senescence factors first set the stage for these disorders.

The Decline Syndrome **333**

These brief descriptions of different decline models are presented to show that a number of people have recognized the existence plant problems caused by multiple factors. These people have recently independently developed models and concepts to deal with these problems.

Returning to the decline spiral model, I have listed below the common denominators and symptoms of what I consider to be decline diseases. Many but not all of these features are common to the other interpretations of decline.

COMMON DENOMINATORS OF DECLINES

1. Many interchangeable factors are involved in the decline syndrome. This implies that more than one factor can produce a given effect and, therefore, if punch number one does not produce an effect, punch number two or three may be involved.

2. At least three factors are involved in a decline, one each from the categories of predisposing, inciting, and contributing factors.

3. Opportunistic fungus pathogens and insects are often involved as contributing factors and may be given more credit for the condition than they really deserve.

4. Climatic or site factors are almost always major predisposing or inciting factors in the decline syndrome. Examination of site or climatic history of a decline syndrome will often show a weak, but nevertheless real correlation with site or climate. The reason an absolute correlation does not occur is because of the importance of sequential timing of the inciting and contributing factors.

5. Feeder roots and mycorrhizae (see Chapter 9) degenerate prior to onset of symptoms in the aboveground portion of the trees. Quantification of root conditions is a difficult task, particularly with large trees. A balance normally develops between uptake capacity and demands upon the root system by the crown of the tree. Degeneration and regeneration of feeder roots and mycorrhizae is a continual process necessary to keep pace with parasitic microorganisms in the soil and seasonal growth demands of the crown. Imbalance in the regenerative capacity of roots or excessive demands of the top will accentuate degeneration of roots.

6. Declining trees usually have a serious depletion of storage-reserve carbohydrates. Storage reserves are used to start spring growth or regenerate fungus- or insect-defoliated leaves. Excessive demands on the reserves deplete the storage reserves and limit the tree's ability to respond.

7. *Armillaria* root rot is a common feature of many declines. The variation in pathogenicity of various *Armillaria* species (grouped, in the past, under *A. mellea)* was discussed in Chapter 16. Some species of *Armillaria* function as primary pathogens while others require prestressed hosts. The impact of *Armillaria* varies depending on the species, the amount and location of the infection on the roots, and the level of previous plant stress.

8. Viruses are another common denominator of at least three hardwood declines and may be a part of the Norway spruce decline in Europe. Because of the

difficulty in working with viruses with trees, they have only been investigated in a limited number of declines. Viruses were discussed more fully in Chapter 6.

We do not understand the role of viruses in the decline of trees. They can be detected in apparently healthy as well as declining trees. One concept of viral action suggests that they predispose the plant to increased impact of other inciting and contributing factors. Another suggests that they are present in most trees and function as contributing factors once the tree is under stress. To illustrate this point, I suggest that viruses are like the nonproductive members of a household, such as dogs, cats, and other pets. In times of plenty (ideal environment) the consuming aspects of pets (viruses) is not noticeable in a smooth-running, balanced household. But in times of stress (drought), the consumption of food by pets may reduce the amount available for the productive members of the household (photosynthetic leaves, transporting phloem, storage parenchyma, absorbing roots). The effects of viruses may therefore be accentuated by conditions that reduce the vigor of the already infected host.

9. Another item that is common to declines is the age of the host. Declines generally occur in physiologically mature trees at or after the age when juvenile recuperative vigor is lost. In the forest, the impacted trees are predominantly in the dominant and codominant crown classes. Declines are therefore a problem of well-established upper crown trees that theoretically represent the "best" genotypes.

Fast growth and large size may provide a competitive advantage for some genotypes of trees as long as the environmental conditions remain within acceptable limits. The slow-growing, less competitive genotypes are weeded from the population. This natural thinning is assumed to be an example of natural selection leading toward survival of the fittest, the foundations of evolution.

But the environment is not constant. Unusual drought, temperature, or wind events may occur only a few times during the life span of a population of trees. The biggest trees of a mature population are highly dependent on very large amounts of water and minerals from the site. These trees, severely impacted by the deficiencies imposed by the unusual events, respond by allowing branch tips to die back. The reduced photosynthetic surface lessens the demand for moisture but also results in less stem growth. Since stem growth maintains the transport system for tying the roots and crown together, the growth response is felt over a number of years. Young trees and understory trees with smaller demands quickly reestablish the balance of crown, roots, and stem transport. Older and larger trees with greater demands respond more slowly and may even shunt energy into seed production for a new generation rather than repair and maintenance of the current generation. These are the trees that decline.

10. Some authors have speculated that air pollution is a common denominator of a number of declines. This is the central issue of extensive research today. The possible involvement of air pollutants is difficult to test or verify properly (see Chapter 4). Unfortunately, possible involvement of air pollution in tree declines is a political and social issue that sometimes leads to speculation and hypotheses on air pollutants that capitalize on emotion rather than scientific judgment based on reasonable diagnostic procedures.

1. Reduced growth is often the first noticeable symptom. Shoot growth and diameter growth are reduced.

2. Reduced growth produces shorter internodes and therefore often a tufted appearance of foliage near the ends of twigs.

3. Roots and mycorrhizae often degenerate prior to the appearance of above-ground symptoms, but these symptoms are usually not recognized until after symptoms are seen in twigs and foliage.

4. Chemical analysis of the roots shows a reduction in stored food reserves.

5. Premature fall coloration in late summer or early fall is common in hardwoods.

6. Yellowing and undersized leaves, similar to symptoms caused by mineral deficiencies, are observed.

7. Twigs and branches often die during the winter, and facultatively parasitic fungi often invade the dying twigs and aggravate the dieback.

8. Portions of the crown die, giving an irregular or asymmetric form to the crown.

9. In hardwoods, sprouts develop from dormant buds on the main stem and large branches, producing clusters of foliage.

10. Root rot decay fungi, such as *Armillaria* spp., parasitize the root system and promote the decline syndrome. Rhizomorphs and mycelial fan signs of *Armillaria* are common.

11. The symptoms and signs listed above persist and intensify over a number of years.

12. Individual trees showing these symptoms are generally randomly dispersed in the population. Clustering of trees with similar symptoms would suggest that the problem is caused by a simple biotic disease or selective injury-inducing agents, not by a decline disease syndrome.

DIAGNOSIS OF DECLINE DISEASES

Diagnosis involves extensive observations on a population of trees. One cannot readily diagnose a decline disease with a quick visit or superficial inspection of a population of trees. In the beginning it is appropriate to consider the possibility that the disease of concern is caused by a single pathogen or injury inducing agent. Detailed examination of symptoms, signs, weather, and site variables may suggest a simple explanation. After exhaustive consideration of simple explanations, the possibility of a decline disease syndrome may begin to emerge.

It has been my observation that people who are not trained in plant pathology dismiss the simple explanations much too rapidly. On the other hand, extensively trained plant pathologists are often very reluctant to abandon the search for a simple explanation. These points of view often lead to different interpretations of the same problem.

A reasonable diagnosis of a decline disease will require extensive use of correlation statistics. Since the foundations of statistics are based on representative random samples, the need for appropriately representative field data is critical. The field sites must sample the geographical and symptom expression range of the problem. Individual observations need to describe and quantify accurately the symptoms, signs, site, and environmental variables.

If the problem is a decline disease, simple correlation analysis of symptoms in relation to signs, weather, or site variables will often show weak relationships of the variables. A strong correlation would suggest that the problem is not a decline disease syndrome but is probably caused by a single agent or stimulus.

Multiple regression analysis of representative well-defined field data is an excellent way of identifying an array of possible factors in the decline syndrome. By logical selection and grouping of variables, one can generate one or more hypotheses on the identity of the interacting causal factors. Keep in mind that there may be more than one decline syndrome operating. In other words, there may be different factors affecting the trees from one location to another. The verification for the statistically derived hypotheses for the decline syndromes will need to be tested with additional field samples.

Obviously, the diagnosis of a decline is much more complex than diagnosis of most other diseases. In the past it was acceptable to describe declines based on subjective evaluation of obvious factors. This has lead to serious disagreement and limited progress. Declines have also been defined on the basis of a few symptoms or on selected hypotheses of poorly defined interactions among factors. This has lead to misconceptions and has masked the true identity of the associated factors. It is therefore appropriate to recognize that the causes of declines are best defined only after well-executed detailed field surveys and rigorous statistical analysis.

EXAMPLES OF DECLINES

A few general comments are necessary to explain my interpretation of the factors affecting declines. One characteristic that is common to all declines is lack of agreement among various researchers on the cause and importance of the specific factors implicated in declines. This controversial nature of declines is probably even more characteristic than the common denominators listed above. Each investigation attempts to find the "cause." With at least three factors involved and an inter-changeability of the individual factors with others, often on a regional basis, it is not unlikely that controversies over interpretation of "cause" arise. No one has satisfactorily demonstrated pathogenicity or cause relationships using Koch's rules for proof of pathogenicity for any of the declines. I will present my interpretation of the main factors associated with specific declines, but it must be remembered that these do not exhaust the various factors involved with each of the problems, nor do they necessarily represent the opinions of the original researchers in each case.

Birch Dieback

Birch dieback of yellow and paper birches emerged as forest problems between 1930 and 1950 in New Brunswick and Nova Scotia, Canada, and the northeastern United States (Figs. 18-4 and 18-5). Birch dieback is still a potential forest problem throughout eastern Canada and the northeastern United States. Another type of birch dieback is associated with ornamental birches.

Predisposing factors. Early observations indicated that the degree of damage and mortality was directly related to the age and size class. There was also some indication that the problems were most severe on wet sites. More recently, apple mosaic virus has been identified from declining birch trees. Predisposing factors of ornamental birch are also not well established but it can be suggested that premature tree aging on exposed planting sites is a factor.

Inciting factors. Stand opening by logging, as well as increased soil temperatures, cause an increase in air temperature. Birch rootlet mortality increased from 6% to 60% as a result of a 2°C increase in soil temperatures above the normal summer temperature. Leaf miner, leaf skeletonizing, and other foliage insects, as well as late spring or early fall frosts, incite birch dieback in localized situations.

Figure 18-4 Birch dieback in New York associated with stand opening by logging.

Figure 18–5 Birch dieback in eastern Canada.

Contributing factors. Bronze birch borer is a destructive wood-boring insect of weakened birches. *Armillaria* sp. contributes to intensification of birch dieback by invading the root systems of weakened trees.

Ash Dieback

Ash dieback was first observed in the late 1930s but developed into a serious concern for foresters in the 1950s in the northeastern United States (fig. 18–6). The disease is still commonly observed in New York.

Predisposing factors. Trees growing in exposed hedgerows were very commonly diseased. Although not explicitly considered important at the time, it now appears from the data available that heavier soils were common to ash dieback sites. Recent evidence indicates that three viruses (tobacco ringspot, tomato ringspot, and tobacco mosaic virus) may be commonly associated with the ash population in New York. Mycoplasma-like organisms (MLOs) are also generally present in declining ash of all ages. The MLO agent is capable of causing many of the specific symptoms of ash dieback and therefore it is appropriate to recognize that current research is suggesting that ash dieback may be caused by MLO. The name *ash yellows* is gaining prominence for this disease. The possible single-agent etiology of this disease of various-aged ash suggests that ash dieback may no longer fit within the characteristics of the diseases of this chapter. In contrast, the absence of MLO from some declining ash complicates the interpretation. The individual and combined roles of factors in ash decline are still under investigation and therefore it may be appropriate to reserve judgment for the present.

Inciting factors. Many people have suggested a relationship of drought periods to ash dieback. Unfortunately, the data analysis to support this relationship

Examples of Declines

Figure 18-6 Ash dieback in a New York woodlot.

is lacking. There is no general agreement as to what constitutes drought, and therefore it is difficult to identify clearly when drought really occurs. To confuse things further, there are often considerable differences in rainfall across a region as well as daily and seasonal distribution differences. These considerations make it difficult to interpret drought properly without resorting to detailed analysis of very local weather conditions.

Contributing factors. Two fungi, *Cytophoma pruinosa* and *Fusiococcum* spp., were commonly isolated from cankers on declining ash. The fungi, when inoculated in ash tress, caused cankers only in weakened trees.

Maple Decline

The beginnings of the maple decline problem can be traced to 1913, but serious interest was generated in the problem during the 1950s and early 1960s in the northeastern United States and eastern Canada (Figs. 18–7 and 18–8). There are at least three different maple declines. Maples growing along roadsides, in sugarbushes, and in the forest are affected by different conditions and therefore have different decline syndromes.

Figure 18-7 Maple decline of roadside sugar maple in New York.

Figure 18-8 Decline of a streetside Norway maple in Syracuse, New York.

Predisposing factors.   The roadside and urban conditions of soil compaction, impeded drainage, poor soil aeration, salt, heat, and interacting with tree age are all predisposing factors. In the sugarbush age and overtapping may be important predisposing factors. In other situations the grazing of cattle or use of heavy equipment for removal of sap has compacted the soil. The selective removal of non-productive trees may in some instances open the stand extensively, thereby modifying the microclimate. Open stands are subjected to more rapid and more extreme fluctuations in temperature and moisture than is a dense forest. Variation in soil nutrients due to parent material, site, climate, and vegetation may be a part of some maple declines. Without recognizing the importance of natural variation, some have assumed that air pollution in the form of acid deposition is the cause of changes in soil nutrients. More objective analysis will be required to substantiate these assumptions.

Maple decline is less common in the natural forest than in the disturbed urban or sugarbush forest. Unfortunately, there have been few detailed investigations of maple decline in the natural forest and therefore our understanding of the decline syndrome is less clear. Tree age and soil drainage conditions seem to be involved as predisposing factors in some instances. Shallow soils over subsurface hardpans or ledges may cause problems with root development and function during either wet or dry periods. The roots of aging trees are less able to adjust and recover to keep pace with the changing moisture conditions.

Inciting factors.   The drought years of the 1950s and 1960s were a factor in maple decline. The excess rain periods of the 1970s may also be a part of some maple declines. In some instances logging induces climatic changes for the remaining trees. Death of the American elms in urban environments due to Dutch elm disease produced climatic effects on the associated maples similar to those caused by logging in natural stands. Late winter conditions characterized by a warm period followed by a deep-freezing period may be associated with some maple declines. Extreme cold periods in winters with little snow cover may also be involved. Defoliating insects such as the forest tent caterpillar, the saddled prominent, the loopers, and many others are locally important inciting factors in the forest, sugarbush, or urban environment. Root damage from excavation for road improvements and other construction are common inciting factors in the urban and roadside environment. Root damage due to heavy equipment is also a problem in the forest and sugarbush.

Contributing factors.   Armillaria root rot accentuates the degeneration of weakened maples in the forest and sugarbush. In the urban environment the vascular wilt fungus *Verticillium* may be involved as a contributing factor, even though pathologists assume that it is a pathogen capable of causing disease by itself (see Chapter 13). In the sugarbush and managed forest the sapstreak fungus, *Ceratocystis coerulescens,* may function like *Verticillium* in the urban environment. The sugar maple borer, opportunistic canker fungi, and saprobic decay fungi may be contributing factors in some maple decline situations.

Pole Blight of Western White Pine

Western white pine of the Inland Empire of Idaho and Montana was severely affected by pole blight during the period from 1916 through the middle of the century. Pole-sized trees were affected, thereby giving rise to the name.

Predisposing factors. Shallow soils and soils with low moisture storage capacity were common to pole blight areas. Sites with less than 5 in. (13 cm) of storage capacity in the upper 3 ft (90 cm) of soil were subjected to pole blight.

Inciting factors. By measuring annual growth rings of trees in the Inland Empire, it was determined that the period 1916–1940 experienced the most severe drought the area had sustained in 180 years.

Contributing factors. Armillaria root rot and *Leptographium* sp. were associated commonly with roots. *Ophiostoma (Europhium) trinacriforme* was associated with elongate cankers on the stems of declining trees. None of these were capable of inducing pole blight but may contribute to the disease.

Littleleaf Disease

During the early part of this century, abandoned cotton-growing areas in the Piedmont region of the southeastern United States regenerated naturally or were planted to shortleaf and loblolly pines. Decline of pole-sized trees (approximately 20 years of age) was first reported in the mid-1930s and by the 1940s was recognized as a serious threat to shortleaf pine (Fig. 18–9). Littleleaf continues to be a problem today. From initial symptoms to tree death takes from 6 to 15 years.

Predisposing factors. Sheet erosion during cotton cropping eliminated most of the top soil of the Piedmont. Fertility was depleted. Drainage was impeded commonly by an impervious hard-packed lower horizon. Periodic moisture stress followed by moisture excess were characteristics of the sites. Poor soil aeration resulted from excessive moisture.

Inciting factors. Although emphasis has never been placed on the inciting factors for the disease, it is apparent from the literature that the disease does not progress at the same rate from year to year. The period 1946–1951 was marked by a rapid increase in the disease and, therefore, weather conditions over that period may be considered as potential inciting agents.

Contributing factors. *Phytophthora cinnamomi,* which parasitizes and kills feeder roots of pines, is a contributing factor (see Chapter 16 for more discussion of *P. cinnamomi).* The detection of *P. cinnamomi* on sites where littleleaf was a problem led researchers to attribute the littleleaf syndrome primarily to root

Figure 18-9 Littleleaf disease of shortleaf pine in Georgia as seen by yellowed, sparce tufted foliage on the right. (Photograph compliments of Dr. Savel Silverborg.)

damage by this fungus. An alternative interpretation, however, is that littleleaf is incited by deficiencies in water and/or oxygen, and that *P. cinnamomi* merely contributes to root damage.

The spread pattern for littleleaf is not typical of expanding foci expected for disease caused by root pathogens. *P. cinnamomi* is as ubiquitous to littleleaf as *Armillaria* spp. are to other declines. *Armillaria* is also found in littleleaf sites. Diseased as well as nondiseased stands are infested by *P. cinnamomi*. Therefore, the distinction of "causal agent" for *P. cinnamomi* is inappropriate.

Fungi in the *Pythium* complex, *P. irregulare* and *debaryanum,* are also found in sites with or without littleleaf. Although these are shown to be pathogenic on shortleaf and loblolly pines, there is no reason to assume that they are any different from *P. cinnamomi.*

Oak Declines

Oak declines have been described at various times, but one syndrome in the northeast has received more recent emphasis because of the involvement of the gypsy moth. In the southeast there have developed a number of localized oak decline syndromes associated with site and weather conditions. There is also a live oak decline problem in the south that is associated with the oak wilt fungus, *Ceratocystis fagacearum* (see Chapter 13). The live oak decline is a good example situation where additional research has demonstrated involvement of a specific pathogen. The issue is still not completely resolved because there is some live oak decline occurring in the absence of *C. fagacearum.*

Predisposing factors.   Soil drainage problems are important predisposing factors in oak declines of both the north and the south. Poorly drained areas may also correspond to low frost pockets. In the northeast, the thin stony soils of the ridge tops may predispose trees on some sites. Tree age is also a predisposing factor.

Inciting factors.   Drought and frost are common inciting factors. Defoliating insects, such as the gypsy moth in the northeast, are major inciting factors. Foliar pathogens such as anthracnose are considered to be a factor in the south.

Contribuing factors.   In the northeast, root damage by *Armillaria* and trunk girdling by two-lined chestnut borer larvae are the most prominent contributing factors. In the south the fungus *Hypoxylon atropunctatum* is involved along with a number of root decay fungi, including *Ganoderma* sp., *Inonotus (Polyporus) dryadeus, Clitocybe tabescens, Scytinostroma (Cotricium) galactinum,* and other decay fungi.

Spruce Declines

Some perceive that there exists a red spruce decline situation in the mountainous areas of the eastern United States and a Norway spruce decline situation in Europe, with special reference to Germany. The etiologies of both of these have been so tied into the political rhetoric of acid rain and air pollution that it is very difficult to determine the degree and extent of the problems. It is also very difficult to properly characterize the roles and possible interactions of the identified and suspected causal agents. Huge amounts of money have been expended and numerous competitive hypotheses have been, and continue to be, projected into the scientific literature and news media as the proper answer.

It is important to recognize that the activities on these two problems have been going on since the mid-1970s and that some ideas have changed over this period. For example, extensive and accelerating decline of Norway spruce caused by air pollutants was supposed to be threatening the very survival of West German forests. One should recognize that the German forests are not natural forests but plantations that have been managed for a number of generations (hundreds of years).

The highly emotional term *Waldsterben* (forest death) was used to describe the situation in Germany. Initial symptoms were characterized as crown thinning of older and larger trees. An array of other equally nonspecific symptoms have been used to characterize the problem. The forest death concept was an effective political tool, but any unbiased look at the situation had to seriously question what was really happening. Recent official surveys have demonstrated little change and in some instances improvement in the health of the forest.

If one discards the initial forest death or Waldsterben scenario, there still exists a very specific problem characterized by chlorosis of older needles of high-elevation spruce plantations that can usually be related to magnesium deficiencies in the soil. The term *newartige Waldschäden* (novel forest damage) has been used for this condition. The chemical composition of the parent material for these soils is generally rather low in magnesium and calcium. The possible involvement of acid deposition in accelerating this condition and the effect of the chlorosis on the long-term health of the forests are some of the topics being investigated. The possible involvement of ozone is also under investigation, but for now it is difficult to characterize novel forest damage as a decline.

The topic of red spruce decline in the mountains of the eastern United States was spawned from the politically explosive topic of Norway spruce decline in Germany. Huge amounts of money have been spent to investigate numerous competitive hypotheses. Death of the outer crown branches is a common symptom, but in contrast to the plantation grown Norway spruce situation, there is extensive mortality in this natural high-elevation forest. Synchronized growth reductions have also been suggested based on studies of increment cores.

Some involvement of air pollutants is usually hypothesized in most investigations, but the scientific evidence for direct effects of air pollutants seems to be rather elusive. Natural factors such as wind, ice, drought, and winter injury are clearly involved. Insects, fungi, and dwarf mistletoe are also problems in some situations. The evidence to date suggests that red spruce "decline" is better characterized as a collection of individual site-specific problems rather than a widespread decline with a common complex of interacting factors.

ECOLOGICAL ROLE OF DECLINES

A final consideration is the ecological role of declines. Trees are continually expanding their range to the limit of their site tolerance. Those on the periphery of their site tolerance are most affected by declines. It is interesting that declines occur well within the natural geographic range of the species and not on the periphery of the geographic range. Within the geographic range, variations between specific sites are the conditions that seem to affect declines.

Are we really observing natural ecological succession in action? If declines are nature's way of moving successions along or preventing expansion of site tolerance for tree species, it is rather futile to expect that control is possible. We may need to learn to recognize declines as a type of natural balance phenomenon and adjust our expectations accordingly.

REFERENCES

BAUCE, E.S. 1989. Sugar maple, *Acer saccharum* Marsh., decline associated with past disturbances. Ph.D. thesis, State University of New York, College of Environmental Science and Forestry, Syracuse, N.Y. 141 pp.

BRANDT, C.J. 1987. Forest decline in central Europe and North America. Pp. 175–206 *In* Acid precipitation formation and impact on terrestrial ecosystems. Verein Deutscher Ingenieure Kommission Reinhaltung der Luft, Deusseldorf.

CAMPBELL, W.A. 1961. Littleleaf disease of shortleaf pine: present status and future needs. *In* Recent advances in botany. Univ. Toronto Press, Toronto, pp. 1529–1532.

CAMPBELL, W.H., O.L. COPELAND, JR., and G.H. HEPTING. 1953. Managing shortleaf pine in littleleaf disease areas. USDA For. Serv. Southeast. For. Exp. Stn. Pap. 25. 12 pp.

CASTELLO, J.D., S.B. SILVERBORG, and P.D. MANION. 1985. Intensification of ash decline in New York state from 1962-1980. Plant Dis. *69:*243-246.

GREGORY, R.A., M.W. WILLIAMS, JR., B.L. WONG, and G.J. HAWLEY. 1986. Proposed scenario for dieback and decline of *Acer saccarum* in Northeastern USA and Southern Canada. IAWA Bull. n.s. *7:*357-369.

HANSBROUGH, J.R., V.S. JENSEN, H.J. MACALONEY, and R.W. NASH. 1950. Excessive birch mortality in the northeast. USDA For. Serv. Tree Pest Leafl. 52. 4 pp.

HEPTING, G.H. 1963. Climate and forest disease. Annu. Rev. Phytopathol. *1:*31-50.

HODGES, C.S., K.T. ADEE, J.D. STEIN, H.B. WOOD, and R.D. DOTY. 1986. Decline of ohia *(Metrosideros polymorpha)* in Hawaii: a review. USDA For. Serv. Gen. Tech. Rep. PSW-86. 22 pp.

HOUSTON, D.R. 1973. Dieback and declines: diseases initiated by stress, including defoliation. Int. Shade Tree Conf. Proc. *49:*73-76.

HOUSTON, D.R. 1981. Stress triggered tree diseases the diebacks and declines. USDA For. Serv. NE-INF-41-81. 36 pp.

HOUSTON, D.R. 1987. Forest tree declines of past and present: current understanding. Can. J. Plant Pathol. *9:*349-360.

KANDLER, O. 1990. Epidemiological evaluation of the development of "Waldsterben" in Germany. Plant Disease *74:*4-12.

LACHANCE, D. 1988. Sugar maple decline, acid rain, pest interactions in the northeast. Proc. Soc. Am. For. 1988 Conv., Oct. 16-19. Rochester, N.Y., pp. 102-105.

LEAPHART, C.D. 1958. Pole blight: how it may influence western white pine management in light of current knowledge. J. For. *56:*746-751.

LEAPHART, C.D., and A.R. STAGE. 1971. Climate: a factor in the origin of the pole blight disease of *Pinus monticola* Dougl. Ecology *52:*229-239.

MANION, P.D. 1985. Factors contributing to the decline of forests: a conceptual overview. Proc. symposium: effects of air pollutants on forest ecosystems. Acid Rain Foundation, Inc., Minneapolis, Minn., pp. 63-73.

MANION, P.D. 1986. Decline as a phenomenon in forests: pathological and ecological considerations. Proc. NATO Adv. Res. Workshop, Effects of acidic deposition on forests, wetlands, and agricultural ecosystems. Springer-Verlag, Berlin.

MANION, P.D. 1988. Hardwood forest declines: concepts and management. Proc. Soc. Am. For. 1988 Conv., Oct. 16-19. Rochester, N.Y., pp. 127-130.

McKRACKEN, F.I. 1985. Oak decline and mortality in the south. Proc. 3rd Symp. Southeast. Hardwoods, Atlanta, Ga., pp. 77-81.

McILVEEN, W.D., S.T. RUTHERFORD, and S.N. LINZON. 1986. A historical perspective of sugar maple decline within Ontario and outside of Ontario. Ontario Ministry of the Environment, Toronto, Ontario, Canada. 40 pp.

MUELLER-DOMBOIS, D. 1983. Canopy dieback and successional processes in Pacific forests. Pac. Sci. *37:*317-325.

MUELLER-DOMBOIS, D., J.E. CANFIELD, R.A. HOLT, and G.P. BUELOW. 1983. Tree-group death in North American and Hawaiian forest: a pathological problem or a new problem for vegetation ecology. Phytocoenologia *11:*117-137.

OAK, S.W., and F.H. TAINTER. 1988. How to identify and control littleleaf disease. USDA For. Serv. Prot. Rep. R8-PR 12. 14 pp.

OTROSINA, W.J., and D.H. MARX. 1975. Populations of *Phytophthora cinnamomi* and *Pythium* spp. under shortleaf and loblolly pines in littleleaf disease sites. Phytopathology *65:*1224–1229.

PARKER, A.K. 1957. The nature of the association of *Europhium trinacriforme* with pole blight lesions. Can. J. Bot. *35:*845–856.

PARKER, J. 1970. Effects of defoliation and drought on root food reserves in sugar maple seedlings. USDA For. Serv. Res. Pap. NE–169.

PITELKA, L.F., and D.J. RAYNAL. 1989. Forest decline and acid deposition. Ecology *70:*2–10.

POMERLEAU, R. 1953. The relationship between environmental conditions and the dying of birches and other hardwoods. *In* Report on the symposium on birch dieback. For. Biol. Div. Can. Dep. Agric. Sci. Serv. 182 pp.

REDMOND, D.R. 1955. Studies in forest pathology XV. Rootlets, mycorrhiza, and soil temperature in relation to birch dieback. Can. J. Bot. *33:*595–627.

REDMOND, D.R. 1957. The future of birch from the viewpoint of diseases and insects. For. Chron. *33:*25–30.

ROSS, E.W. 1966. Ash dieback etiological and developmental studies. N.Y. State Univ. Coll. For. Syracuse Tech. Publ. 88. 80 pp.

ROTH, E.R. 1954. Spread and intensification of littleleaf disease of pine. J. For. *52:*592–596.

SINCLAIR, W.A. 1965. Comparisons of recent declines of white ash, oak, and sugar maple in northeastern woodlands. Cornell Plant. *20:*62–67.

SINCLAIR, W.A. 1967. Decline of hardwoods: possible causes. Proc. Int. Shade Tree Conf. *42:*17–32.

SINCLAIR, W.A., and G.W. HUDLER. 1988. Tree declines: four concepts of causality. J. Arboric. *14:*29–35.

SPAULDING, P., and H.J. MACALONEY. 1931. A study of organic factors concerned in the decadence of birch on cut-over lands in northern New England. J. For. *29:*1134–1149.

STALEY, J.M. 1965. Decline and mortality of red and scarlet oaks. For. Sci. *11:*2–17.

STONE, E.L., R.R. MORROW, and D.S. WELCH. 1954. A malady of red pine on poorly drained sites. J. For. *52:*104–114.

WALLACE, H.R. 1978. The diagnosis of plant diseases of complex etiology. Annu. Rev. Phytopath. *16:*379–402.

WESTING, A.H. 1966. Sugar maple decline: an evaluation. Econ. Bot. *20:*196–212.

ZAK, B. 1957. Littleleaf of pine. USDA For. Serv. For. Pest Leafl. 20. 4 pp.

ZAK, B. 1961. Aeration and other soil factors affecting southern pines as related to littleleaf disease. USDA For. Serv. Tech. Bull. 1248. 30 pp.

19

PLANT DISEASE EPIDEMICS

- *CALCULATION OF DISEASE INCREASE RATES*
- *USES OF EPIDEMIOLOGICAL INFORMATION*
- *SAMPLE PROBLEMS OF DISEASE EPIDEMICS*
- *SOLUTION AND DISCUSSION OF PROBLEMS*

Up to this point, we have concerned ourselves primarily with types of organisms and nonliving agents that cause disease. A number of specific diseases have been emphasized to demonstrate the etiology (disease cycle) of disease-causing agents and control procedures. All of this has a very narrow application unless one develops a perspective regarding how a disease-causing agent produces disease in a population of plants. Control of plant disease in a single plant is like trying to fight brush fires by stepping on each burning ember. Control is best applied to and evaluated on a population of plants.

Sound understanding of disease epidemics and the influence of control procedures on the development of epidemics would make presentation of our case to the general public a lot easier. Very few control measures completely stop a disease. If a disease can be slowed to the extent that replanting or regeneration compensates for the amount lost, the control measure is successful.

We often oversell the effectiveness of a control measure and then later disappoint the public because the disease is not eliminated. For example, in Syracuse, New York, the sanitation control measure for Dutch elm disease was accepted as the cure for DED. When it did not eliminate the problem, city officials became disillusioned with the value of controlling the disease and in a tight fiscal year cut the funds. Would it not have been better to demonstrate the effectiveness of the control procedure in increasing the life expectancy of the population of elms?

CALCULATION OF DISEASE INCREASE RATES

A disease epidemic is an increase in the amount of disease with time. Some diseases increase rapidly, others more slowly. Disease increase can sometimes be equated to money invested at compound interest. The basic formula for compound interest of money or disease is

$$r_1 = \frac{1}{t_2 - t_1} \log_e \frac{x_2}{x_1}$$

The rate of interest or increase (r), the length of time ($t_2 - t_1$), the amount of initial principal or initial disease (x_1), and the final amount of principal or disease (x_2) are the four terms of the formula. Obviously, this formula can be used to determine r, t_2, or x_2, depending upon which values you substitute for in the formula.

The basic formula applies to compound-interest diseases only when fewer than 5% of the population of plants are diseased. Beyond 5% diseased, actual disease increase is less than compound interest would predict. Why? The compound-interest formula is based on an infinite supply of money or additional infectible plants. As more and more plants become diseased, the infectible population becomes smaller and more widely spaced. Differences in the spatial distribution of infectible plants change the ratio of successful infection to total inoculum production. As the population of healthy susceptible plants becomes smaller, the percent of inoculum infecting new hosts is reduced almost to zero.

To compensate for a changing population of infectible plants we apply a correction factor $x/(1 - x)$ instead of using x. The amount of disease (x) is expressed as a decimal proportion. If we insert $x/(1 - x)$ for the amount of disease terms and arrange the log fraction into a subtraction, our working formula for disease becomes:

$$r_1 = \frac{1}{t_2 - t_1} \left(\log_e \frac{x_2}{1 - x_2} - \log_e \frac{x_1}{1 - x_1} \right)$$

$\log_e x/(1 - x)$ values from 0.001 to 0.999 can be calculated and are provided as a table of logit values (Table 19-2).

With a logit table available, it is a simple matter to look up the logit values and plug these into the formula

$$r_1 = \frac{1}{t_2 - t_1} (\text{logit of } x_2 - \text{logit of } x_1)$$

Another way that a disease can increase is like money invested at simple interest. With simple interest, one does not collect interest on previous interest dividends. The formula

$$QR = \frac{1}{t_2 - t_1} \left(\log_e \frac{1}{1 - x_2} - \log_e \frac{1}{1 - x_1} \right)$$

represents the simple-interest disease. QR expresses the rate of increase for simple-interest diseases. Note the correction factor $1/(1 - x)$ for simple-interest diseases. A physiological disease may follow a simple interest increase because the loss of an individual plant does not contribute inoculum to infect others. Verticillium wilt follows a simple-interest increase. It is the same as always removing the interest dividends in a savings account. Any dividends (in disease terms, disease losses) do not add to the principal and therefore do not influence the size of the next dividend.

If one plots the accumulative percent disease against time, either a sigmoid growth curve or a straight-line plot is obtained (Fig. 19-1). A compound-interest disease will generate a typical sigmoid growth-curve plot. The compound-interest disease-increase curve has a lag phase, a logarithmic phase, and a leveling-off phase.

The simple-interest disease, in contrast, plots as a straight line. There is no log phase or compound-interest phase to a simple-interest disease plot. But, as with compound interest, there is a leveling-off phase due to a limited amount of infectible host population.

It is difficult to work with the curved lines of the compound interest plot, so we generally plot the logit of the cumulative amount of disease (Fig. 19-2). This technique produces a linear plot which can be compared to other linear plots. One can readily see differences with straight lines that are difficult to detect in the original curved-line plot.

In the real world, one does not necessarily know if a disease increases by compound or simple interest. The first test is to plot the cumulative disease increase. As already described, the two plots are rather distinctive.

A compound-interest disease is usually caused by a pathogen that reproduces on the infected host to produce additional inoculum for secondary spread. The secondary spread from infected plants results in foci of infection (pockets of infection).

The simple-interest disease does not produce inoculum on infected hosts for secondary disease spread. In the absence of secondary spread, the infection is usually random. Randomness or nonrandomness of infected individuals is one of the best indicators of simple-interest or compound-interest diseases, respectively.

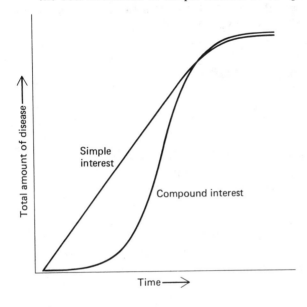

Figure 19-1 Disease increase plots for simple- and compound-interest diseases.

Calculation of Disease Increase Rates

351

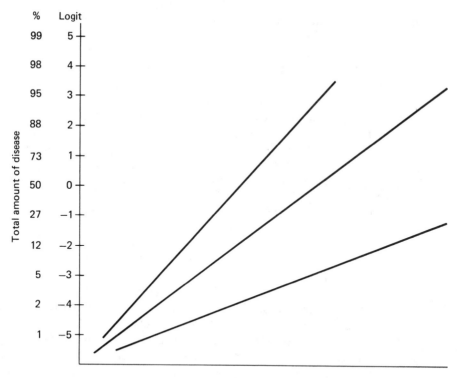

Figure 19-2 Compound-interest plotted on a logit scale for percent disease.

Why would it be important to recognize simple- from compound-interest diseases? If one looks at the plots of compound- and simple-interest diseases, it is apparent that predicting future losses for a compound-interest disease using a simple-interest formula will greatly underestimate the actual loss. Similarly, if one uses a compound-interest disease formula to predict future losses of a simple-interest disease, one greatly exaggerates the actual losses.

USES OF EPIDEMIOLOGICAL INFORMATION

Table 19-1 presents rate of increases (r values) calculated for a number of diseases. Why does Dutch elm disease or oak wilt have different r values in different locations?

Differences may occur because of differing factors such as climate, density of hosts, and control practice. Specific differences in disease increase rates may give clues as to how effective certain control practices or how important climatic and site factors are on disease increases.

Note that Hypoxylon canker is assigned a simple-interest rate of increase (QR). The field data better fit a simple-interest model. This may occur because of the long period between infection and production of inoculum. It may also occur if the amount of disease is more controlled by periodic environmental events than the density of inoculum.

TABLE 19-1 RATE OF DISEASE INCREASE PER UNIT PER YEAR FOR SEVERAL TREE DISEASES[a]

Disease	r_1	Location
White pine blister rust	0.50	British Columbia
	0.67	Minnesota
	0.50	Maine
Dutch elm disease	1.37	Illinois
	0.50	Connecticut
	0.25	Quebec
Chestnut blight	1.42	Pennsylvania
	1.10	Virginia
	0.83	Connecticut
Oak wilt	0.10	Pennsylvania
	0.31	West Virginia
	0.22	Illinois
	0.20	Tennessee
	0.36	Arkansas
Elm phloem necrosis (Elm yellows)	0.46	Indiana
Tympanis canker (red pine)	0.08–0.15	Connecticut
Polyporus tomentosus (spruce)	0.03–0.05	Saskatchewan
Cytospora canker (spruce)	0.14	Quebec
Hypoxylon canker (trembling aspen)	0.02 QR	New York

[a] r_1 values based on Merrill (1967); QR value based on Manion and Blume (1975).

It is important to recognize that compound-interest increase is not particularly normal for any natural population. Compound-interest disease increase is more characteristic of an artificial population such as an agricultural field or a native population of plants being invaded by an introduced pathogen. Natural populations of plants and pathogens evolve to some homeostatic state that prevents domination of one over the other. Environmental factors affecting inoculum production, host predisposition infection points, or other features of the disease cycle allow rapid short-term change but provide limits on long-term changes.

Utility of the mathematical models of disease increase can therefore go well beyond just predicting the disease situation at some time in the future. The shape of the disease progress curve can be used to recognize introduced pathogens situations, to recognize when artificial management activities have upset natural balance forces, to identify specific environmental triggers for disease increase, and for many other purposes.

SAMPLE PROBLEMS OF DISEASE EPIDEMICS

It is difficult to really understand this chapter without actually doing some calculations of disease increase rates and then using these to make management decisions. For this reason, three sample problems are included. The first two are hypothetical problems, but the third is based on the Dutch elm disease epidemic for Syracuse, New York.

A word of caution before starting the calculations. Be certain to express percent disease (x) as a cumulative percent disease, not just the annual percent loss. Table 19-2 gives the logit values that you will need to solve these problems. The solution to the problems, plus a discussion, are presented at the end of the chapter.

Sample Problem 1: Cottonwood Quick Butt Rot

Mr. J.I. Fink, city forester for Burning Stump, Arkansas, decided to demonstrate to the city fathers the seriousness of the current epidemic of cottonwood quick butt rot. He recorded the number of cars crushed by falling trees during the summers of 1962, 1963, and 1964. Out of a population of 1000 cars, 10 were hit in 1962, 11 were hit in 1963, and 12 were hit in 1964. What is the disease increase rate of the car destruction epidemic, and if there is a very large population of cottonwoods, how long will it be before 20% of the cars are destroyed?

TABLE 19-2. LOGIT VALUES [$\log_e (x/1 - x)$ FOR DECIMAL PROPORTIONS FROM 0.001 TO 0.999

x	0	1	2	3	4	5	6	7	8	9		
					Thousandths, for x (in left column)							
0.00		6.91	6.21	5.81	5.52	5.29	5.11	4.95	4.82	4.70	4.60	0.99
0.01	4.60	4.50	4.41	4.33	4.25	4.18	4.12	4.06	4.00	3.94	3.89	0.98
0.02	3.89	3.84	3.79	3.74	3.71	3.66	3.62	3.58	3.55	3.51	3.48	0.97
0.03	3.48	3.44	3.41	3.38	3.35	3.32	3.29	3.26	3.23	3.20	3.18	0.96
0.04	3.18	3.15	3.13	3.10	3.08	3.06	3.03	3.01	2.99	2.97	2.94	0.95
0.05	2.94	2.92	2.90	2.88	2.86	2.84	2.82	2.81	2.79	2.77	2.75	0.94
0.06	2.75	2.73	2.72	2.70	2.68	2.67	2.65	2.63	2.62	2.60	2.59	0.93
0.07	2.59	2.57	2.56	2.54	2.53	2.51	2.50	2.48	2.47	2.45	2.44	0.92
0.08	2.44	2.43	2.42	2.40	2.39	2.38	2.36	2.35	2.34	2.33	2.31	0.91
0.09	2.31	2.30	2.29	2.28	2.27	2.25	2.24	2.23	2.22	2.21	2.20	0.90
0.10	2.20	2.19	2.18	2.16	2.15	2.14	2.13	2.12	2.11	2.10	2.09	0.89
0.11	2.09	2.08	2.07	2.06	2.05	2.04	2.03	2.02	2.01	2.00	1.99	0.88
0.12	1.99	1.98	1.97	1.96	1.96	1.95	1.94	1.93	1.92	1.91	0.90	0.87
0.13	1.90	1.89	1.88	1.87	1.87	1.86	1.85	1.84	1.83	1.82	1.82	0.86
0.14	1.82	1.81	1.80	1.79	1.78	1.77	1.77	1.76	1.75	1.74	1.73	0.85
0.15	1.73	1.73	1.72	1.71	1.70	1.70	1.69	1.68	1.67	1.67	1.66	0.84
0.16	1.66	1.65	1.64	1.64	1.63	1.62	1.61	1.61	1.60	1.59	1.59	0.83
0.17	1.59	1.58	1.57	1.56	1.56	1.55	1.54	1.54	1.53	1.52	1.52	0.82
0.18	1.52	1.51	1.50	1.50	1.49	1.48	1.48	1.47	1.46	1.46	1.45	0.81
0.19	1.45	1.44	1.44	1.43	1.42	1.42	1.41	1.41	1.40	1.39	1.39	0.80
0.20	1.39	1.38	1.37	1.37	1.36	1.36	1.35	1.34	1.34	1.33	1.32	0.79
0.21	1.32	1.32	1.31	1.31	1.30	1.30	1.29	1.28	1.28	1.27	1.27	0.78
0.22	1.27	1.26	1.25	1.25	1.24	1.24	1.23	1.22	1.22	1.21	1.21	0.77
0.23	1.21	1.20	1.20	1.19	1.19	1.18	1.17	1.17	1.16	1.16	1.15	0.76
0.24	1.15	1.15	1.14	1.14	1.13	1.13	1.12	1.11	1.11	1.10	1.10	0.75
		9	8	7	6	5	4	3	2	1	0	x

Sample Problem 2: Red Christmas Trees

Rather than develop an advertising campaign to push red, sparsely foliated Christmas trees, Ms. L.V. Schrooge decided to invest some money in fungicide sprays. Being concerned about unnecessary expenditures, she ordered an evaluation of various fungicides and application timings. The normal course of the foliage epidemic has an r value of 0.4 per unit per day, during the month of June. No spread occurs after July 1 or before June 1.

Ecologically Blessed is a fungicide whose action is short-lived (1 day) and is effective only against established needle infection. Big Zap is a contact fungicide that kills anything and everything that moves or does not move except trees and is effective for a period of 30 days. It cannot penetrate the needles to affect the parasite once inside the plant.

Plot the normal course of the epidemic if a sampling determined that 1 in 1000

Thousandths, for x (in left column)

x	0	1	2	3	4	5	6	7	8	9		
0.25	1.10	1.09	1.09	1.08	1.08	1.07	1.07	1.06	1.06	1.05	1.05	0.74
0.26	1.05	1.04	1.03	1.03	1.03	1.02	1.02	1.01	1.00	1.00	0.99	0.73
0.27	0.99	0.99	0.98	0.97	0.97	0.97	0.96	0.96	0.95	0.95	0.94	0.72
0.28	0.94	0.94	0.93	0.93	0.92	0.92	0.91	0.91	0.91	0.90	0.90	0.71
0.29	0.90	0.89	0.89	0.88	0.88	0.87	0.87	0.86	0.86	0.85	0.85	0.70
0.30	0.85	0.84	0.84	0.83	0.83	0.82	0.82	0.81	0.81	0.80	0.80	0.69
0.31	0.80	0.80	0.79	0.79	0.78	0.78	0.77	0.77	0.76	0.76	0.75	0.68
0.32	0.75	0.75	0.74	0.74	0.74	0.73	0.73	0.72	0.72	0.71	0.71	0.67
0.33	0.71	0.70	0.70	0.69	0.69	0.69	0.68	0.68	0.67	0.67	0.66	0.66
0.34	0.66	0.66	0.65	0.76	0.65	0.64	0.64	0.63	0.63	0.62	0.62	0.65
0.35	0.62	0.61	0.61	0.61	0.60	0.60	0.59	0.59	0.58	0.58	0.58	0.64
0.36	0.58	0.57	0.57	0.56	0.56	0.55	0.55	0.55	0.54	0.54	0.53	0.63
0.37	0.53	0.53	0.52	0.52	0.52	0.51	0.51	0.50	0.50	0.49	0.49	0.62
0.38	0.40	0.49	0.48	0.48	0.47	0.47	0.46	0.46	0.46	0.45	0.45	0.61
0.39	0.45	0.44	0.44	0.43	0.43	0.43	0.42	0.42	0.41	0.41	0.41	0.60
0.40	0.41	0.40	0.40	0.39	0.39	0.39	0.38	0.38	0.37	0.37	0.36	0.59
0.41	0.36	0.36	0.36	0.35	0.35	0.34	0.34	0.34	0.33	0.33	0.32	0.58
0.42	0.32	0.32	0.31	0.31	0.31	0.30	0.30	0.29	0.29	0.29	0.28	0.57
0.43	0.28	0.28	0.27	0.27	0.27	0.26	0.26	0.25	0.25	0.25	0.24	0.56
0.44	0.24	0.24	0.23	0.23	0.22	0.22	0.22	0.21	0.21	0.20	0.20	0.55
0.45	0.20	0.20	0.19	0.19	0.18	0.18	0.18	0.17	0.17	0.16	0.16	0.54
0.46	0.16	0.16	0.15	0.15	0.14	0.14	0.14	0.13	0.13	0.12	0.12	0.53
0.47	0.12	0.12	0.11	0.11	0.10	0.10	0.10	0.09	0.09	0.08	0.08	0.52
0.48	0.08	0.08	0.07	0.07	0.06	0.06	0.06	0.05	0.05	0.04	0.04	0.51
0.49	0.04	0.04	0.03	0.03	0.02	0.02	0.02	0.01	0.01	0.00	0.00	0.50
		9	8	7	6	5	4	3	2	1	0	x

Thousandths, for x (in right column)

 For x less than 0.50 (on left), logit is negative. For x greater than 0.50 (on right), logit is positive.

trees produced viable inoculum on June 1. What percent of the trees are infected by July 1?

What will happen if Ecologically Blessed is used as a spray treatment on the trees on June 15? Assume that the fungicide is 90% effective; that is, 1 in 10 trees do not receive sufficient treatment to kill internal pathogens.

What will happen if Big Zap is applied on June 10 and June 20? New growth of foliage causes some foliage to be unprotected, thereby reducing the effectiveness of the fungicide. The spray is, therefore, 100% effective only for 5 days after application. The cost of the fungicide prohibits additional applications.

Graph each treatment. Which is best? What is the effect of Big Zap on the rate of increase from June 1 to July 1? Would it make any difference if the plantation contained 10,000 trees or 1,000,000 trees?

Sample Problem 3: Dutch Elm Disease in Syracuse, New York

The yearly losses of American elms to Dutch elm disease are given in Table 19-3. Working space is provided for you to calculate the rate of increase (r_1) values for the disease epidemic.

Calculate an average rate of increase (r_1) values for the three periods and

TABLE 19-3 YEARLY LOSSES OF AMERICAN ELMS TO DUTCH ELM DISEASE IN SYRACUSE, NEW YORK[a]

Year	Number of trees killed	Accumulative losses to DED	Accumulative percent loss	Logit value	Rate of increase
(Minimum sanitation)					
1951	1	_____	_____	____	_____
1952	7	_____	_____	____	_____
1953	19	_____	_____	____	_____
1954	181	_____	_____	____	_____
1955	657	_____	_____	____	_____
1956	864	_____	_____	____	_____
1957	1066	_____	_____	____	_____
(Maximum sanitation)					
1958	425	_____	_____	____	_____
1959	817	_____	_____	____	_____
1960	581	_____	_____	____	_____
1961	510	_____	_____	____	_____
1962	529	_____	_____	____	_____
1963	751	_____	_____	____	_____
1964	748	_____	_____	____	_____
(No sanitation)					
1965	2597	_____	_____	____	_____
1966	4033	_____	_____	____	_____
1967	5687	_____	_____	____	_____

[a]The original population contained 53,618 elms.

determine the projected date for loss of 90% of the trees, based on the value for each period.

Can you justify sanitation as a control for DED based on these figures? The sanitation program involved removal of dead and dying elms. The rate of increase during the sanitation period reflects root graft spread and infection by European elm bark beetles produced on elm material outside the control area. Do you suppose that a more effective control program is justified?

SOLUTION AND DISCUSSION OF PROBLEMS

Cottonwood Quick Butt Rot

If we assume that the number of cars is directly proportional to the number of trees that fall over due to the butt rot, the number of crushed cars is a good quantitative measure of the tree disease.

Stop and think about the disease cycle of a root rot. You should recall that the time interval between infection and inoculum production is long. Root rot diseases, because of the long time interval and the usual association with some environmental event, generally increase at a simple interest rate. Therefore, you should use the simple-interest formula. Some might be tempted to use the compound-interest formula because the numbers seem to be increasing each year. One has to be cautious in seeing trends based on slight differences and weigh these observations against other bits of information.

If you use a compound-interest formula, you will see that the disease increase rate keeps changing. If the disease increase rate does not generally remain constant, you are probably using the wrong formula.

In this case, the rate of increase is 0.01 and it would take 21 years, or until 1983, to destroy 20% of the cars.

Cottonwood quick butt rot is a simple example, but it demonstrates how a root rot disease would progress through a population of non-regenerating trees. A recreation site or a street setting, where large mature trees are not being replaced, would be drastically changed by the loss of 20% of the trees. The obvious answer is to balance losses with replacements.

Red Christmas Trees

The graphic plot of this disease epidemic, and the effects of control activities, are presented in Fig. 19-3. Note the utility of the logit scale of the vertical axis. This allows one to plot straight lines to more accurately predict future disease levels.

It should be evident that a disease epidemic with a 0.4 per unit per day rate of increase can totally infect the crop in one season. Neither fungicide is effective in reducing losses to a reasonable level using the present spray schedules. Many applications of fungicide would be needed to control the disease.

A more effective approach to control would be to look for resistant varieties or different growing conditions to reduce the rate of disease increase. If the disease

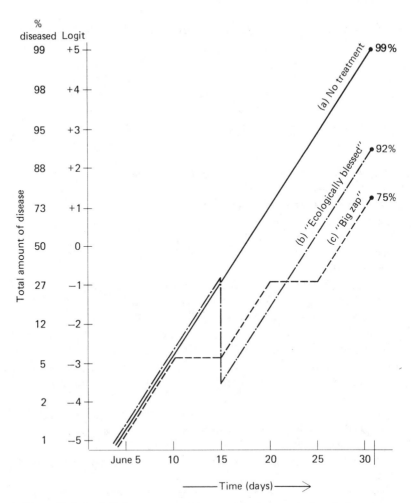

Figure 19-3 Red Christmas tree problem graphic plot on a logit scale for percent disease.

increase rate could be slowed somewhat, the application of fungicides could be economically justified.

Dutch Elm Disease in Syracuse, New York

An understanding of the Dutch elm disease cycle would predict that the disease should follow a compound-disease increase rate. But if one calculates the yearly increase rates using Table 19-3, or if the data are plotted graphically, it is obvious that the rate of increase values are not constant.

A closer look at the data shows three rates of increase. During the minimal-sanitation period, the rates change yearly, but during the 1954–1957 period, the rate averages 0.87. During the maximum-sanitation period, the rates are more constant and average 0.15. During the no-sanitation period, the rate of increase averages 0.43 or 0.88, depending on the base-level population that is used.

358   Plant Disease Epidemics   Chap. 19

The changing rate of increase values for the early period are due in part to the effects of multiple introductions of the disease into the area. Disease increase is made up of a combination of new introductions and secondary spread from established infection centers. When the numbers are small, just a few additional introductions can produce major changes.

A second aspect of the early epidemic is the period of training needed for people diagnosing the disease. If there are very few diseased trees, it is difficult to find them all. As there are more diseased trees, the crew will cover larger areas and find infections that occurred earlier but were not detected. Therefore, the yearly losses to disease may reflect both deflated and inflated numbers. Under these circumstances, the early figures are best evaluated as a group.

Disease control through sanitation was obviously very successful. The disease increase rate remained relatively stable and low. The rate of increase probably reflects the effects of root graft transmission of the disease and the effects of bark beetles from outside the control area moving into the area. With a disease increase rate of 0.15 in 1960, one could predict that 50% of the trees would be dead by 1975 and that 90% would be dead by 1990.

These projections may not look too promising for the future of elms, but they should be compared to the projected losses without control. If one projects from 1957 with a 0.87 rate of increase, 50% loss will occur by 1960, and 90% loss by 1963. The disease is not eliminated by the control activity, but the costs of removal and replacement are spread out over a longer period of time. The benefits of sanitation look even better if one considers that many of the trees, planted shortly after the Civil War, would not have survived until 1990 even in the absence of Dutch elm disease.

The no-sanitation period is marked by an increase in the disease increase rate. If one assumes that the population of elm trees and the amount of beetles were not seriously affected by the sanitation program, an average rate of increase value of 0.43 is calculated for the period 1964–1967.

It is probably better to assume that the removal of the trees during the sanitation period changed the base population of both the disease vector and the nondiseased trees. Therefore, a new epidemic, beginning in 1965, starts with an initial population of 46,462 trees rather than 53,618. The base diseased population is 2597 trees in 1965. Using these figures, an average increase rate of 0.88 is calculated for the 1965–1967 period.

Projections of 90% disease losses, using the 0.43 and 0.88 rates of increase, target 1974 and 1971, respectively. Although accurate yearly figures are not available for the period, it now appears that the 0.88 rate of increase more closely predicted the actual course for the epidemic.

REFERENCES

LEONARD, K.J., and W.E. FRY, eds. 1986. Plant disease epidemiology. Macmillan Publishing Company, New York. 372 pp.

MANION, P.D., and M. BLUME. 1975. Epidemiology of hypoxylon canker of aspen. Proc. Am. Phytopathol. Soc. *2*:101.

MERRILL, W. 1967a. Analyses of some epidemics of forest tree diseases. Phytopathology *57*:822.

MERRILL, W. 1967b. The oak wilt epidemics of Pennsylvania and West Virginia: analysis. Phytopathology *57*:1206–1210.

MERRILL, W. 1968. Effect of control programs on development of epidemics of Dutch elm disease. Phytopathology *58*:1060.

MILLER, H.C., S.B. SILVERBORG, and R.J. CAMPANA. 1969. Dutch elm disease: relation of spread and intensification to control by sanitation in Syracuse, New York. Plant Dis. Rep. *53*:551–555.

VAN DER PLANK, J.E. 1963. Plant diseases: epidemics and control. Academic Press, Inc., New York. 349 pp.

WEIDENSAUL, T.C., and F.A. WOOD. 1974. Analysis of a maple canker epidemic in Pennsylvania. Phytopathology *64*:1024–1027.

20

DISEASE CONTROL THROUGH GENETIC RESISTANCE

- *METHODS OF DISEASE CONTROL*
- *CONSIDERATIONS RELATIVE TO A RESISTANCE BREEDING PROGRAM WITH TREES*
- *METHODS OF INHERITANCE OF RESISTANCE*
- *EXAMPLES OF RESISTANCE BREEDING STRATEGIES*

Throughout this book, we have tried to develop an appreciation of the importance of tree diseases. It may be frustrating to you — at least it is to me — that once one appreciates the role of disease and even learns much about the biology of the interaction, not very much can be done. Most control practices are totally impractical on a large scale. We professional forest pathologists, therefore, spend most of our applied time developing theories of why something died, and putting down scattered problems with techniques that are too expensive and not completely effective.

METHODS OF DISEASE CONTROL

I like to make the analogy between problem solving in forest pathology and problem solving of international conflicts. Conflicts between nations can be resolved in three ways. One is war, which may solve the problem but also destroy the reason for solving it. We can apply direct control procedures to a disease in localized spots, but do we really accomplish anything significant? Direct control involves tremendous amounts of money. In the end, is it worth the cost of both dollars and environmental impact? This is a question currently being asked of all uses of pesticides.

A second way of resolving a conflict is isolation, just not communicating at all. This works and nations survive. But it is the potential progress of both inhibited by the lack of interaction? I equate quarantines with this type of activity. One has to continuously keep building the barriers and avoiding confrontation or a conflict (disease) will flare up.

The third type of solution involves getting together and resolving the differences. Both cultures will be affected by the interaction such that neither will be as it was before. The interaction process does not stop with one meeting but must continue indefinitely. Once embarked on such a course, one must maintain interaction to resolve minor misunderstandings that will develop. This type of approach I equate with breeding for disease resistance and silvicultural stand manipulations.

This chapter will emphasize disease resistance breeding. In Chapter 23 we look at the total picture of silvicultural stand manipulations in intensive forest management.

There have been times in history when each of the three approaches to control was indeed the most appropriate. With increasing population and changes in values, the third type of solution emerges as the most appropriate. We are presently in a stage where increasing pressure on our tree resources is forcing us to move in the direction that agriculture started to move in about the turn of the century. We must develop resistance or tolerance in our trees to the biotic and physical pressures of our environment. And we must learn to manage both the diseases and the trees through silvicultural practices.

Forest pathologists do not universally agree that disease-resistance breeding is a viable approach to disease control. The skeptics point to the many problems and possible pitfalls. They also say that it will take generations of pathologists to accomplish anything significant. As we shall see, there are possible problems, but they are not absolute barriers. We shall also see that the time period is not necessarily measured in generations of pathologists.

As noted in Chapter 11, we have already made substantial progress through genetic resistance with white pine blister rust in the west and fusiform rust in the south. Genetically improved western white pine seed that should produce a 65% rust-free population of trees is presently available. With southern pines the disease resistance programs are producing seed that will ultimately result in slash pine plantations with a 50% reduction in fusiform rust and loblolly pine plantations of 40% reduction in rust. These substantial gains have been made in less than one generation of pathologists.

The current biotechnology with capabilities for transferring genetic information among widely different organisms will provide a whole new dimension to disease control through genetic resistance (see Chapter 7). One should also recognize that breeding programs and biotechnology are not the only way to produce genetically superior populations of trees. Selection harvesting systems that do not harvest the best-quality trees until they have produced seed for the next generation should produce better form and less diseased populations of trees (see Chapter 14).

CONSIDERATIONS RELATIVE TO A RESISTANCE BREEDING PROGRAM WITH TREES

With that as an introduction, let me now develop the topic of breeding for disease resistance in trees with three general considerations.

First, there is no simple, inexpensive method of developing a disease-resistant plant, so whatever program is started must indeed warrant the cost. High value and widely used ornamentals, or intensively managed fiber or timber crop trees, are the only ones to consider.

Second, once committed to a breeding program, there is no ending point. Resistance breeding does not stop once a resistant variety is released. It is important to monitor the performance of the new variety as it is used in the field. Unknown or unpredicted problems may flare up and destroy everything you have accomplished.

Third, one should not develop a resistant line based on just one resistant plant. Genetic diversity of natural populations provides a certain amount of elasticity to the population for survival in a variable environment. A broad-based search for a number of resistant plants with varying types of resistance is an important first step.

METHODS OF INHERITANCE OF RESISTANCE

You may already be aware of the various types of inheritance from introductory biology or genetics. Resistance is just like any other character. It can be controlled by a single gene and will be inherited much like the Mendelian characters. It can be controlled by many interacting genes and be inherited much like height growth in human beings. It can also be controlled by cytoplasmic factors and will therefore demonstrate a greater contribution from the female parent. It is important to determine how the resistance is inherited, because the tree breeder utilizes different techniques to obtain genetic improvement with each of these methods of inheritance.

Many of our resistant varieties of agricultural crops are based on single-gene resistance. This type of resistance is ideal for incorporation into a hybrid population.

Hybrid plant varieties are usually produced through forced interfertilization of two independently inbred plant lines. One line may be bred for growth form or quality characters, the other may be for disease resistance or other characters. The hybrid, produced by crossing the two lines, is a very uniform population with both resistance and growth form.

If one of the parents carried a homozygous-dominant resistance gene, all the hybrid plants will carry at least one dominant gene for resistance. Although the hybrid population is then totally resistant, a second generation of plants produced by crossing the hybrids usually results in a 3:1 segregation for resistance.

Production of a resistant variety based on quantitatively inherited factors is somewhat more difficult. One difficulty arises from the combined effect of single-gene and quantitative inheritance. Quantitatively inherited genetic improvement is difficult to accurately determine if single genes for resistance are maintained in the population. The single genes may mask the quantitative genes for resistance. The plant breeder must utilize a number of test crosses to verify the absence of single genes.

Genetic improvement of a plant species using quantitative inheritance is usually less spectacular than with single-gene inheritance. Quantitative inheritance of resistance will produce varying improvements in the population, but never total resistance as seen with single-gene inheritance.

Cytoplasmic inheritance is utilized effectively in the production of hybrid varieties if seed is collected only from one plant line. The female parent contributes the largest volume of cytoplasm to the embryo and therefore the greatest probability of cytoplasmic-inherited characters.

Cytoplasmically inherited characteristics have not been utilized in tree breeding but are commonly used in hybrid corn breeding. Cytoplasmically inherited male sterility makes it possible to avoid self-fertilization. In the past, the tassels had to be cut to avoid self-fertilization.

A corn leaf blight epidemic in 1970 is discussed in Chapter 1. This disease, caused by *Helminthosporium maydis,* was extremely damaging to varieties that contained a specific cytoplasmically inherited male sterility factor. To control the disease, a different cytoplasmically inherited male sterility gene was utilized.

EXAMPLES OF RESISTANCE BREEDING STRATEGIES

Single-Gene Inheritance

If selection of resistant individuals and controlled crosses determines that a single gene controls resistance, it is best not to produce a long-term crop such as trees using only that one gene for resistance. Examples in agriculture have demonstrated that pathogens can also change by mutation or genetic recombination. It takes 10 to 15 years for the wheat stem rust fungus *(Puccinia graminis* var. *tritici)* to develop a large population of a new infective race capable of destroying the wheat crop. Therefore, every 10 to 15 years a new variety with different single genes controlling resistance must be available for release to farmers.

The corn leaf blight epidemic of 1970 developed because a single gene was maintained in the population. Genetic variation in the *Helminthosporium* fungus eventually resulted in a pathogen population capable of infecting the corn population. Rapid increase in this virulent pathogen population resulted in a serious disease epidemic.

One method of overcoming this difficulty is to produce a synthetic variety in which you incorporate a number of different genes for resistance in a population of plants. Borlaug (1966) describes such a method, which is summarized in Fig. 20-1.

The synthetic-variety approach incorporates single-gene resistance of single-cross, backcross, and double-cross programs into a mass-selection synthetic seed orchard. The plant breeder starts by locating disease-resistant and superior-growth-form parent trees. The first crosses attempt to get both resistance and form in the

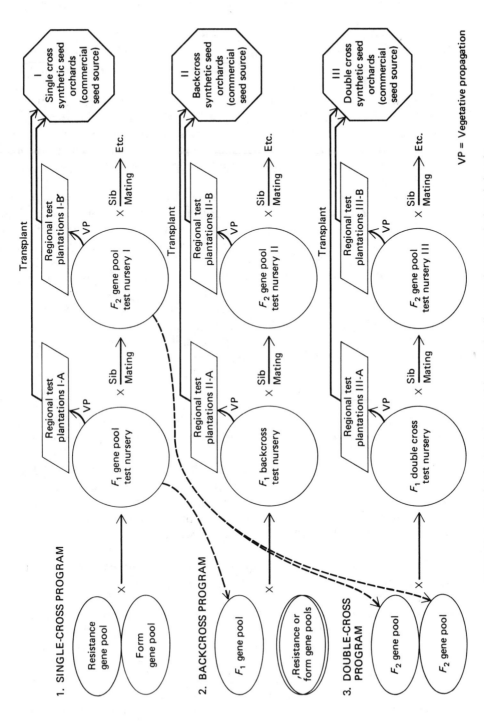

Figure 20–1 Plan for the development of fast-growing disease-resistant synthetic varieties of a cross-pollinated forest crop. (From Borlaug, 1966.)

same plants. Siblings of selected resistant, good-growth-form trees are mated to stabilize the characteristics.

The regional test plantations are set up to evaluate the resistance and growth form of the best plants over a range of environmental conditions. These plantations should be in a number of locations that represent the environmental variability over which the plant materials may eventually be used.

Those trees that look good in the regional test plantation are transplanted to a seed orchard. Open pollination among the various trees in the seed orchard should produce a seed lot with a thorough mixing of genes for growth form and resistance. The synthetic variety thus produced should be genetically improved, yet contain sufficient variability to buffer against rapid buildup of a new race of pathogens or against any other pathogens.

The backcross and double-cross programs are breeding techniques for combining different genetic traits into a new line. Backcrossing an F_1 hybrid to one of the parents is done to enhance the number of genetic characters of that parent in the progeny. If the F_1 hybrids of the single-cross program have a good resistance but a poor form, the breeder will backcross to the form parent. The progeny of the backcross should have more of the good form characteristics and yet retain some of the resistance.

The double-cross program might be used to incorporate more than one set of resistance or form genes in the progeny. The plan for the development of synthetic resistant varieties demonstrates some of the complexity in developing genetically improved varieties when using single genes for resistance.

A second approach to the use of single genes is to utilize a system where stabilizing, directional, or balancing selection pressure keeps the pathogen change in check. There is a general rule that says that the pathogen most fit to survive has the minimum number of virulence genes. In other words, if there is no need for virulence genes, they will be selected against and reduced to low numbers in the wild population.

The selection-pressure approach works well where a pathogen has two distinct and separate phases of growth and development. One example would be the *Fusarium* wilt of tomato. The fungus has a pathogenic existence during the growing season on tomatoes, and then during the winter it survives saprophytically by growing on dead organic matter. Any excess virulence genes developed on the live tomato are of a negative advantage and are selected against in its saprophytic existence. Therefore, a serious buildup of disease due to virulence in the pathogen does not occur. A single-gene-resistant tomato variety can survive indefinitely as an agricultural variety.

We have no good examples of stabilizing selection in diseases of trees, but one might speculate that stabilizing selection might operate if we could find single genes for resistance in aspen for Hypoxylon canker. The pathogen must survive and compete for 2 to 3 years as a saprobe on killed tissue prior to sporulation. The capacity for pathogenicity may be of selective disadvantage during this saprobic sporulation phase. Single genes for resistance, which appear to induce callusing of inoculation wounds, appear to be present in the aspen population, but these are only effective against the least-virulent isolates of the fungus. Stabilizing selection might also work for white pine blister rust since more virulent lines would potentially be at a selective disadvantage on the alternate host *Ribes*.

If one can recognize those disease situations where stabilizing selection pressure exists, genetic improvement should be "relatively easy" to develop, and the improved varieties should remain resistant to that one disease. The progeny of single-cross, backcross, or double-cross programs could be used directly as an improved variety.

If screening of F_1 hybrids shows that particular parent trees produce resistant progeny, the genetic improvement can be accomplished through development of a seed orchard. Branches from the resistant tree are grafted to root stock. These grafted plants, planted in seed orchards, will usually flower and produce seed while small trees. Therefore, genetic improvement can be accomplished within a relatively short period of time.

Tissue culture provides a means for rapidly producing large numbers of selected individuals. Pathogen screening tests on tissue cultured plantlets may reduce the time and space required for field tests. Superior individuals can be multiplied and rooted to eventually become part of a seed orchard or replicated field test plantation.

One must be cautious with single-gene resistance because the genetic diversity of the variety is narrowed and other pathogens, less affected by stabilizing selections, may become problems. Varieties may have to be shifted or other control activities initiated to maintain the forest productivity, so that periodic surveillance of the plantation is necessary.

Quantitative Inheritance

Most phenotypic characters, including disease resistance, are quantitatively inherited. The methods used to develop resistant varieties using quantitative resistance are somewhat more difficult than single genes, because crosses between resistant trees do not necessarily produce resistant seedlings. The progeny have variable levels of resistance, normally distributed, ranging from nonresistant to somewhat resistant. Many of the progeny are less resistant than the parents are.

The degree to which the seedlings resemble the parents varies. Therefore, one attempts to select parents with resistance factors that are readily inherited by a good proportion of the progeny. The degree of similarity between parents and progeny is the heritability. A heritability of 1.0 would occur if the progeny were identical to the parents. Most of the quantitative genetic improvement is done with heritabilities less than 0.6. With Hypoxylon canker of aspen, the heritability for reducing canker enlargement ranged from 0.01 to 0.25.

The potential for genetic improvement is not totally expressed in the heritability figure. If one selects parents with outstanding resistance characteristics, genetic improvement can be accomplished even if the heritability is low. The potential genetic improvement through breeding is the product of the heritability of the trait and the amount of deviation from the average of the selected parents.

The extra effort in utilizing quantitative selection is well worth the time spent since the resistant (tolerant, not immune) plants do not place a strong selection pressure on the pathogen. With limited use of backcrosses and sib matings, one avoids unconscious narrowing of the gene pool, so that a high degree of hetero-geneity can be maintained in the population. The original pathogen plus many unknown future pathogens will have considerable difficulty developing diseases of epidemic proportions. A truly mutual tolerance among pathogen, plant, and man is then developed through the use of quantitative resistance.

Genetic improvement through the use of quantitatively inherited resistance need not take generations of pathologists. In the Hypoxylon resistance program in aspens, the characterization of parents with good heritability took less than 8 years. If genetically improved aspens with resistance to Hypoxylon canker were considered important by industry, one could start a seed orchard, consisting of sucker-reproduced plants, that could be expected to start producing genetically improved seed within another 6 to 8 years.

REFERENCES

ANONYMOUS. 1972. Genetic vulnerability of major crops. National Academy of Sciences, Washington, D.C. 307 pp.

BELANGER, R.R., P.D. MANION, and D.H. GRIFFIN. 1989. *Hypoxylon mammatum* as-cospore infection of *Populus tremuloides* clones: effects of moisture stress in tissue culture. Phytopathology *79*:315–317.

BINGHAM, R.T., R.J. HOFF, and G.I. McDONALD. 1971. Disease resistance in forest trees. Annu. Rev. Phytopathol. *9*:433–452.

BORLAUG, N.E. 1966. Basic concepts which influence the choice of methods for use in breeding for resistance in cross pollinated and self-pollinated crop plants. *In* Breeding pest resistant trees, ed. H.D. Gerhold et al. Pergamon Press, Ltd., Oxford, pp. 327–348. 505 pp.

CARSON, S.D., and M.J. CARSON. 1989. Breeding for resistance in forest trees: a quantitative approach. Annu. Rev. Phytopathol. *27*:373–395.

FLOR, H.H. 1971. Current status of the gene-for-gene concept. Annu. Rev. Phytopathol. *9*:275–296.

HEYBROCK, H.M., B. STEPHAN, and K. VON WEISSENBERG, eds. 1982. Resistance to diseases and pests in forest trees. Proc. 3rd Int. Workshop Genetics Host-Parasite Interactions For., Sept. 1980, Wageningen, The Netherlands. Centre for Agricultural Publishing and Documentation, Wageningen, The Netherlands. 503 pp.

KINLOCH, B.B., and J.W. BYLER. 1981. Relative effectiveness and stability of different resistance mechanisms to white pine blister rust in sugar pine. Phytopathology *71*:386–391.

LEONARD, K.J., and R.J. CZOCHOR. 1980. Theory of genetic interactions among populations of plants and their pathogens. Annu. Rev. Phytopathol. *18*:237–258.

LEPPIK, E.E. 1970. Gene centers of plants as sources of disease resistance. Annu. Rev. Phytopathol. *8*:323–344.

METTER, L.E., and T.G. GREGG. 1969. Population genetics and evolution. Prentice-Hall, Inc., Englewood Cliffs, N.J., 212 pp.

SEDEROFF, R.R., and F.T. LEDIS. 1985. Increasing forest productivity and value through biotechnology. *in Weyerhaeuser Science Symposium, Forest Potentials Productivity and Value, V4.*Weyerhaeuser Co., Tacoma Washington. pp. 253–276.

VALENTINE, F.A., P.D. MANION, and K.E. MOORE. 1976. Genetic control of resistance to Hypoxylon infection and canker development in *Populus tremuloides.* Proc. 12th Lakes States For. Tree Improv. Conf. USDA For. Serv. Gen. Tech. Rep. NC-26, pp. 132–146.

VAN DER PLANK, J.E. 1964. Plant diseases: epidemics and control. Academic Press, Inc., New York. 349 pp.

WENZEL, G. 1985. Strategies in unconventional breeding for disease resistance. Annu. Rev. Phytopathol. *23*:149–172.

WILLIAMS, P.H. 1975. Genetics of resistance in plants. Genetics (supplement). *79*:409–419.

WOLFE, M.S. 1985. The current status and prospects of multiline cultivars and variety mixtures for disease resistance. Annu. Rev. Phytopathol. *23*:251–273.

ZOBEL, B., AND J. TALBERT. 1984. Applied forest tree improvement. John Wiley & Sons, New York.

21

DISEASES OF SEEDLINGS
IN THE NURSERY

- *TYPES OF SEEDLING DISEASES AND RELATED PROBLEMS*
- *DAMPING-OFF*
- *ROOT DISEASES OF NURSERY-GROWN SEEDLINGS*
- *FOLIAGE AND STEM DISEASES*

Management of a tree nursery is a specialized forestry activity with a limited number of participants. Although the vast majority of the readers of this book will never participate in nursery management, many will work with plant materials produced by someone else in the nursery. Some disease management problems associated with forest and urban plantings are directly affected by nursery management practices.

Many of the diseases of seedlings in the nursery are the same diseases we have discussed in Chapters 5, 7, 9, 10, 11, 12, and 16. Nursery problems are also associated with abiotic stress agents, as discussed in Chapters 2, 3, and 4.

Even though most of the basic disease concepts were discussed in earlier chapters, it seems appropriate to bring the threads of the topic together at this point. The resource manager who will never directly participate in nursery management decisions can therefore better understand the origin of some problems and better appreciate the complexity of the nursery manager's problems.

Intensive, repeated cultivation of tree seedlings in nurseries is an ideal environment for the buildup of diseases that do not generally represent a serious threat to the natural regeneration of trees. In the nursery, foliage disease, canker diseases, and nematode problems intensify rapidly. They may pose an immediate problem to the nursery operator and/or a long-term problem to the forest manager of the out-planted stock. Diseases such as fusiform rust, brown spot needle blight, and Scleroderris canker are accentuated as field problems because of initial infection in the nursery.

Damping-off is a problem unique to germinating seeds and is therefore a consideration of the nursery operator. Damping-off will be discussed as a specific problem of tree seedlings, but is not unique to trees. Many agricultural crop seeds are treated with fungicides to reduce damping-off.

Trees grown in nurseries have root disease problems as were discussed in Chapter 16. The damping-off and root diseases may have many similarities, including the same fungal pathogens. The nursery situation is ideal for the development of root-based problems.

Another consideration for the nursery operator is mycorrhizae. These were discussed earlier, but it is important to recognize here that chemicals used for the control of seedling diseases and weeds may have detrimental effects on mycorrhizal development.

Nursery practices are rapidly changing to accommodate to mechanization and maximum production. The practice of growing seedlings in synthetic soil in containers will generate a completely new set of problems. Containerized seedling production is often done in greenhouses with overhead sprinkler systems. These are ideal conditions for the development of foliage pathogens. The seedlings are fertilized heavily, thereby producing an environment unfavorable for mycorrhizae and favorable for some root pathogens. Steamed or fumigated potting mixtures are devoid of a natural microflora of fungi and bacteria. Pathogenic microorganisms can more rapidly colonize a medium in which competition and the antagonism of natural microflora are missing.

DAMPING-OFF

Anyone who has planted seeds and carefully watched for the plants to emerge recognizes that some seeds do not result in plants. Some seeds are not viable, but others are viable yet are killed before the shoot emerges from the soil. Some of the seedlings that emerge seem to collapse and die. These last two phenomena are called damping-off, a problem that may regularly produce losses of 15% or more (Figs. 21-1 to 21-4).

Preemergence damping-off occurs as a result of fungus attack of the radicle prior to seedling emergence from the ground. Postemergence damping-off occurs because of fungus attack at the base of the seedling stem after it emerges from the ground. Both are caused by fungi that invade the succulent stem tissues. Once cells

Figure 21-1 Western conifer nursery.

Figure 21-2 Many of the missing trees in these rows of Douglas-fir were probably lost to damping-off and root rot.

Figure 21–3 Damping-off of Austrian pine seedlings in a greenhouse demonstration.

Figure 21–4 Close-up of damping-off in Austrian pine.

Damping-Off

have stopped expanding and their walls become lignified, they become resistant to infection by damping-off fungi.

Oomycete fungi in the genera *Pythium* and *Phytophthora* are commonly involved in damping-off. Other fungi are *Rhizoctonia* spp., *Fusarium* spp., and *Sclerotium* spp. in the Fungi Imperfecti.

Cool, wet, highly organic, neutral-to-basic soils favor damping-off. Close spacing and continuous cropping also favor damping-off.

Most nursery operators expect a certain amount of damping-off, but keep it to a minimum by avoiding conditions that favor the problem. Soil fumigation with methyl bromide, chloropicrin, or Vorlex are common for weed and root rot control. The fumigation also reduces damping-off. Soil drench with H_2SO_4 to reduce the pH has been used specifically for damping-off. Some other damping-off fungi are carried on the seed coat so proper handling of seed collection will reduce problems later.

ROOT DISEASES OF NURSERY-GROWN SEEDLINGS

The damping-off problems grade into root disease problems such that a line of separation is not possible. One common root disease of nurseries is caused by *Fusarium oxysporum*. The fungus may be introduced to the nursery with contaminated seed. Population buildup occurs when debris of culled seedlings is incorporated into the soil for the next crop. Organic matter such as horse manure and some cover crops contribute to the population increase. A period of cool weather during the time that seeds are germinating sets the stage for infection and some damping-off and future root disease that may not become evident for a few months. *F. oxysporum* may also take advantage of hot, dry spells during the germination and early development of seedlings. The nursery operator needs to avoid any type of moisture or temperature stress during the germination period.

The pathogen forms chlamydospores that persist for long periods of time in the organic matter in the soil. Infection occurs when the roots of young seedlings make contact with organic matter containing the fungus. At the Saratoga nursery, New York, up to 92,000 propagules of *F. oxysporum* per gram of soil were detected in some areas. Fumigation before planting with methyl bromide is often necessary to reduce populations of the pathogen once the problem has become well established in the nursery. Management of organic matter, particularly infected organic matter, is an important consideration for the nursery manager.

Another root disease problem of forest tree nurseries is Cylindrocladium root rot caused by *Cylindrocladium scoparium*. This root rot occurs in older seedlings in transplant beds and has been responsible for the abandonment of at least one nursery in the Lake states.

Charcoal root disease of western forest nurseries is caused by *Macrophomina phaseoli (Sclerotium bataticola),* a fungus that causes root diseases of more than 300 species of plants. This pathogen forms sclerotia (small black resting structures) for survival in the soil in the absence of hosts. These sclerotia make cultural treatments other than thorough fumigation impractical for controlling this disease.

Control of root rot problems is achieved by fumigation of nursery beds prior to seeding. The use of sterilants such as methyl bromide and Vorlex provides a measure of weed and nematode control in addition to root rot control. The sterilants must be used before every crop, in some instances, to prevent major losses. Even with fumigation, damping-off and root rot cause mortality of seedlings.

FOLIAGE AND STEM DISEASES

The close spacing and moist conditions maintained in the nursery are ideal for the rapid spread of foliage and canker pathogens. If allowed to go unchecked, these organisms would seriously reduce or eliminate production.

The category of stem diseases of trees in the nursery would include stem rusts such as white pine blister rust, fusiform rust, western gall rust, and others that infect through foliage and young shoots. It would include Scleroderris canker and other cankers.

Crown gall, described in Chapter 7, is a major concern of hardwood nursery operations. They try to prevent introduction of the bacterium on transplant stock, because once an area is infested, it may need to be taken out of production and grown for 2 years with a resistant crop such as oats or cowpeas. Treatment of plants with *Agrobacterium radiobacter* strain 84 is also used to prevent crown gall (see Chapter 7).

Foliage problems are Lophodermium needle cast, brown spot needle blight, Dothistroma needle blight, Phomopsis blight, Sirococcus blight, and others on conifers. On hardwoods, rusts, anthracnose, and other foliage diseases are problems.

A common feature of the shoot blights and foliage problems of nurseries is that they develop rapidly during some years and then are totally absent in others. It is difficult to predict and properly manage the problem years. By the time the nursery operator identifies the cause of the problem and develops a suitable chemical treatment schedule, the problem often disappears. Fluctuations in climatic conditions and the availability of inoculum from surrounding areas or contaminated seeds interact to produce problem years.

Most stem and foliage diseases are problems of nursery seedlings as well as trees in the field. If the problems are controlled in the nursery, we might expect less of a problem in the field plantings. This is not always the case, particularly where the field planting is done in areas where high levels of natural inoculum occur. Under these conditions, the large artificially maintained population, in the absence of protective fungicides, is rapidly reduced to a much smaller population of naturally resistant individuals. For example, one might question the logic of producing large numbers of seedlings susceptible to fusiform rust, by use of fungicides in the nursery, when these are going to be subjected to natural selection for survival in an environment with high levels of rust. Would it not be better to allow the rust to do some selective thinning in the nursery prior to outplanting?

This is a difficult question to answer because of another possibility—that infected nursery stock may actually contribute to the distribution of highly infective populations of pathogens. Under these circumstances, it is very appropriate to control the pathogens in the nursery.

These and other questions will only be answered by very close ties between the nursery operator, who is producing the stock, and the forest manager, who attempts to maintain the trees for products and environmental benefits over a period of years.

REFERENCES

BLOOMBERG, W.J. 1979. A model of damping-off and root rot of Douglas-fir seedlings caused by *Fusarium oxysporum*. Phytopathology *69*:74–81.

BLOOMBERG, W.J. 1985. The epidemiology of forest nursery diseases. Annu. Rev. Phytopathol. *23*:83–96.

CORDELL, C.E., and T.H. FILER, JR. 1985. Integrated nursery pest management. *In* Southern pine nursery handbook, ed. C.W. Lantz. USDA Forest Service, Southern Region.

DORWORTH, C.E., H.L. GROSS, and D.T. MYREN. 1975. Diseases in Ontario forest tree nurseries, 1966 to 1974. Can. For. Serv. Great Lakes For. Res. Cent. Rep. O-X-230.

HARTLEY, C. 1921. Damping-off in forest nurseries. USDA Bull. 934. 99 pp.

HUANG, J.W., and E.G. KUHLMAN. 1990. Fungi associated with damping-off of slash pine seedlings in Georgia. Plant. Dis. *74*:27–30.

PETERSON, G.W., and R. S. SMITH, JR., tech. coords. 1975. Forest nursery diseases in the United States. USDA For. Serv. Agric. Handb. 470. 125 pp.

PLUMLEY, K.A. 1986. Fusarium as a cause of seedling mortality at the Saratoga tree nursery, Saratoga Springs, New York. M.S. thesis, SUNY College of Environmental Science and Forestry, Syracuse, N.Y. 57 pp.

ROWAN, S.J., T.H. FILER, and W.R. PHELPS. 1972. Nursery diseases of southern hardwoods. USDA For. Serv. For. Pest Leafl. 137. 7 pp.

SUTHERLAND, J.R., W. LOCKS, and F.H. FARRIS. 1981. Sirococcus blight: a seed-borne disease of container-grown spruce seedlings in coastal British Columbia forest nurseries. Can. J. Bot. *59*:559–562.

SUTHERLAND, J.R., T. MILLER, and R.S. QUINARD. 1987. Cone and seed diseases of North American conifers. North American Forestry Commission Publ. 1, Victoria, British Columbia, Canada. 77 pp.

TINUS, R.E., W.I. STEIN, and W.E. BALMER, eds. 1974. Proc. North Am. Containerized For. Tree Seedling Symp. Great Plains Agric. Counc. Publ. 68. 458 pp.

WALL, R.E. 1984. Effects of recently incorporated organic amendments on damping-off of conifer seedlings. Plant. Dis. *68*:60–61.

22

PATHOLOGICAL CONSIDERATIONS OF URBAN TREE MANAGEMENT

- *INITIAL CONSIDERATIONS*
- *GOALS OF TREE MANAGEMENT FOR URBAN ENVIRONMENTS*
- *ADMINISTRATION OF URBAN TREE MANAGEMENT*
- *ESTABLISHED TREE MANAGEMENT*
- *PLANNING AND MANAGEMENT OF NEW PLANTINGS*

Pathological considerations of trees growing in urban and heavily used recreation sites should be a fundamental part of management decisions. Proper planning and maintenance, for maximum benefit with minimum disruption and cost, cannot be accomplished in the absence of pathological concerns. Yet, in practice, much urban and recreation site management is accomplished without serious planning or thought of the pathological consequences. The pathologist is often called in to diagnose problems, prescribe band-aid or sugar-pill therapy, or write a proper epitaph, in situations where some initial involvement would have avoided or reduced the extent of the problem.

The pathology of urban trees is best understood in the context of the urban tree management plan. Therefore, I will present my concept of what is involved in urban tree management, even though some of my concepts go beyond the realm of pathology.

INITIAL CONSIDERATIONS

Before beginning with the topic of urban tree management, it is appropriate to first recognize both the negative and positive aspects of maintaining a tree population. It should be very obvious that, even though many of us consider trees as an important part of an acceptable environment, they really are not essential.

Why do municipalities appropriate funds to plant and maintain trees? They are expensive investments which may become expensive liabilities when they die and a nuisance to remove when reconstructing roads and buildings. They are also a serious source of property damage during wind or ice storms.

It is hard to determine exactly why trees are planted and maintained. Tradition plays a role. Shade from hot summer sun is also involved. Today, city planners consider such aspects as softening of harsh man-made structural lines and screening or framing of vistas as contributions of vegetation. Trees generally increase property values. A whole array of other benefits are ascribed to urban trees, but most of them are emotional rather than factually based benefits.

GOALS OF TREE MANAGEMENT FOR URBAN ENVIRONMENTS

The primary goal is to produce an aesthetically appealing community. A second goal is to minimize catastrophic and expensive losses of trees. A third goal is to maximize public support for what you are doing, and the fourth goal is to efficiently utilize a limited resource for the accomplishment of the first three goals.

Urban tree management will be broken down into administration, established tree management, and planning and management of new plantings.

ADMINISTRATION OF URBAN TREE MANAGEMENT

Personnel. The administrative personnel in charge of urban tree populations must be holistic enough to recognize the complexity of managing a long-term living resource. Neglect or mismanagement at one time will be felt many years later. One cannot properly manage a tree resource with elected or short-term appointed personnel who are concerned with immediate rather than long-term problems.

Inventory. An inventory of location, age, and condition is important for the planning and allocation of resources. Inventories should be periodically updated either by ground survey and/or aerial photographic reconnaissance. It is also important to inventory and categorize sites by soil conditions, including fertility, texture, and moisture-holding capacity. Air pollution may vary and should be recognized. Different uses, such as residential, recreational, industrial, and business, should be recognized because these require different management considerations.

Records. Readily accessible inventory and maintenance records should be maintained and utilized, to determine trends and to coordinate management.

ESTABLISHED TREE MANAGEMENT

Mature trees can represent an asset or a liability. They have demonstrated a tolerance for an array of stress-inducing factors associated with the urban environment. Barring a major disease or insect introduction, or a major street or sidewalk renovation, they can generally be expected to continue to survive as a population for a number of additional years.

It is important to recognize that populations of urban trees are not immortal. The "natural" mortality rate is a subject that should be considered in estimating disease impacts and projecting how long tree populations will survive. The mortality rate should also be a factor in developing a desirability or suitability ranking of trees for urban plantings.

Losses of elms due to Dutch elm disease in Syracuse, New York, for the period 1958–1964 were maintained at 0.84 to 1.64% per year by good sanitation practices. From 1967 to 1985 losses in sugar and Norway maples in Syracuse and Rochester, New York, averaged 3.5 and 1.9% per year, respectively. Surveys of two neighborhoods in Urbana, Illinois, in 1932 and 1982 found that 75% of the 1932 population of sugar maples were lost during the 50-year period. This is about 1.5% per year. These figures provide some examples of actual losses in established tree populations. They suggest that Dutch elm disease losses can be maintained within acceptable levels simply by removing dead and dying trees. They also show that elms can be managed in the urban population as effectively as maples. See Chapter 13 for more information on elm management in relation to Dutch elm disease.

It will be interesting to develop mortality rate figures from effective inventories and to use these figures for making decisions on various urban tree populations. I would expect that the figures for recent plantings will be quite different from those for established plantings. This will reflect some of the conditions associated with the planting activity but may also reflect aspects of nursery management and a general shift to exotics and cultivars rather than the use of locally adapted native plants. It will take time to acquire the quantitative data, but I would predict that an efficient urban forestry program should be based on native trees with a few introductions to add diversity.

The pathological and insect problems of native mature trees are rather straightforward and generally predictable. The high value of individual trees often warrants direct control. Most problems, including Dutch elm disease, can be handled with normal tree maintenance. Pruning of cankered, dead, or broken branches eliminates future problems in most instances, and reduces the hazard of decay infection.

Recent surveys of street trees in two upstate New York cities found up to one-third of the large trees with large broken or poorly pruned branch stubs in the upper crown. Good indications of decay, such as conks, punk knots and open decayed wounds, were very common. In some cities, the population of mature trees is obviously not being properly maintained. Such neglect will reduce the potential value derived from these trees, by increasing future problems and reducing the useful life expectancy.

Foliage problems can be kept to a minimum by raking and disposing of leaves. A certain amount of scattered incidence of endemic wilt disease (Verticillium wilt), root rot, decay, and normal old-age deterioration can be expected but will not represent a major expenditure for the established urban tree population.

A major problem of the larger trees is associated with pruning for utility wires. Death of large branches in the upper crown of Norway maples was found to be related to the degree of crown disruption for utility wires in a survey of Syracuse, New York, street trees. Severe pruning of some species of mature trees apparently leads to subsequent deterioration of the crown.

The most serious threat to a mature tree population is from decline. The trees are generally growing under predisposed or stress conditions, so that any inciting factor may start the decline syndrome rolling. Declines are discussed in Chapter 18. Therefore, the subject will not be elaborated further, except to emphasize that prevention of losses due to decline must be directed toward avoiding predisposing factors. In a practical sense, it is very difficult to avoid predisposing trees in the urban environment, so that early detection and amelioration of inciting factors is appropriate. If one can avoid massive increases in salt applied to roads, or avoid excavation near trees, or prevent major insect defoliation or any of the other inciting factors, declines can be prevented. If it is impossible to avoid an inciting factor, it is important to help the younger trees recover by pruning and fertilizing. The overmature trees, or trees on highly deficient sites, will not recover, so planning for removal and replacement should begin at the time the inciting factor occurs.

The tools for quantifying decline in urban maple populations have been established. At Syracuse we developed a 1 to 5 decline rating system from regression analysis of a number of variables.

$$\text{Decline class} = 0.84 + (0.10 \times \text{CS}) + (0.21 \times \text{CD}) + (0.22 \times \text{SLD})$$

Crown shape (CS) identifies how much of the crown is alive. Full-crown trees are scaled using odd numbers. Trees with the upper portion of the crown dead or missing are scaled using even numbers. A full-crown healthy tree is rated 1. If the upper portion of the crown is dead or has been pruned, the rating is 2. The topped trees' rating is changed to 4, 6, or 8 if one-quarter, one-half, or three-quarters of the lower crown are dead or missing, respectively. The full-crown tree (nontopped) is rated 3, 5, or 7 if one- to three-quarters of the side branches of the crown are dead or missing.

Crown density (CD) is rated 0 to 9 to indicate how much sunlight passes through the crown. A healthy sugar or Norway maple interrupts almost all the sunlight and is given a rating of 0. Loss of foliage in declining trees allows progressively more sunlight through the crown. A severely diseased tree may allow 90% of the sunlight to pass through the crown. This type of tree is scaled 9. The small dead limbs (SDL) are also rated to the 0 to 9 scale. The categories represent mortality of 0 to 90% of the small branches on the outer crown.

The decline rating system can be used to develop a predictive model for short-term mortality or survival probabilities. A population of maples in Syracuse and Rochester, New York, scaled as healthy, decline class 1, in 1976 had 9.5% total mortality by 1985. Those classed as 2, 3, 4, and 5 had 33.3, 57.7, 75.0, and 100% mortality, respectively, by 1985. These figures could be used to make judgments on which trees to cut and for planning purposes to distribute removal costs over appropriate periods of time.

Large, mature trees are often neglected during a sidewalk, curb, street, or utility line construction project. They are in the way during the work phase, and may actually require additional costs to work around. After the construction is completed, they die and need to be removed. Would it not have been better to cut these doomed trees before the project begins, and utilize the money to prune and rejuvenate the more vigorous trees or for the planting of new trees?

There is a point in the development of a tree when it changes from an asset to a liability. In Chapter 14 the concept of pathological rotation was introduced. Pathological rotation in forest trees is the point in time when added increment each year just compensates for the amount of wood lost due to decay fungi. Obviously, when added increment each year is consumed by decay fungi it is appropriate to cut, since the maximum value of the stand has been reached.

I would like to introduce a similar pathological rotation concept for urban trees. Let me call it negative asset rotation. The arborist's method of determining the estimated value of shade trees is based primarily on replacement cost. Obviously, the larger the tree, the larger the replacement cost. Adjustments are made in the replacement cost, depending on species, condition class, and groupings of trees.

Comparison of value increment over time with cost of maintenance or removal indicates that value increment eventually starts to level off while expenses and liabilities are rapidly increasing.

It would seem inappropriate to continue expenditures on maintenance in excess of value increment, and therefore when assets begin to drop (negative asset rotation), one should consider removal of the tree.

PLANNING AND MANAGEMENT OF NEW PLANTINGS

Specifications

One of the first considerations is tree species. The list to select from is very large, as indicated by the list of species planted in the communities of Syracuse, Rochester, and Poughkeepsie, New York, provided in Table 22-1. It is obvious that the species composition of the urban forest is shifting. All three of these cities once had large elm populations. The larger trees today are primarily maples. We see a major shift from native species to exotics in the present plantings.

Considerations for selection of species in actual practice include availability of stock, previous satisfaction, and design specification for shape or size. There seems to be a desire to try anything new. Disease problems are not usually considered, except to discriminate against American elm because of Dutch elm disease.

TABLE 22-1 STREET TREES OF SYRACUSE, ROCHESTER, AND POUGHKEEPSIE, NEW YORK

Older tree population	Younger tree population	
60% of the population	45% of the population	
Norway maple	*Malus* spp.	Honeylocust
	Linden	Sugar maple
25% of the population	Green ash	London plane
Silver maple		
Sugar maple	25% of the population	
	Norway maple	
10% of the population		
Red maple	20% of the population	
White ash	Red maple	*Prunus* spp.
Cottonwood	*Zelkva velcova*	Red oak
Basswood	Silver maple	Cork tree
Linden	Tulip tree	Pin oak
Boxelder	White birch	Ginkgo
Honeylocust	Golden rain tree	
London plane		
	10% of the population	
5% of the population	Hawthorn	Red bud
American elm	White ash	Willow
Horsechestnut	European mountain ash	Basswood
Catalpa	Hackberry	Horsechestnut
Hackberry	American beech	Tree of heaven
Green ash	Boxelder	Black locust
Sycamore	Kentucky coffeetree	Others
Others	Katsura	

Source: Based on Valentine et al. (1978).

Because the selection of suitable species is a key to minimum maintenance and maximum satisfaction, let me suggest a list of criteria for consideration before selecting which tree species to plant. One should recognize all the positive and negative attributes of each species under consideration and generate a species-suitability index considering the following criteria:

1. *Retail cost.* There is a great variety of production costs among tree species. Low-cost, easily obtainable trees are most suitable for most plantings.

2. *Site requirements.* A major asset for an ideal urban tree is tolerance of very poor sites. Tolerance of poor soil aeration makes swamp tree species preferable to upland tree species.

3. *Maintenance requirements.* Some trees require more pruning and shaping than others. Water sprouting along the main stem by sycamores, London plane, and American basswood need to be repeatedly removed and therefore are also a consideration.

Growing trees in the open causes lower branches to enlarge quickly. If not pruned early and regularly, Norway maples and other species will develop a low spreading crown which interferes with both sidewalk and street traffic. Some columnar varieties of maple avoid much of this problem. American elm requires minimal shaping during early growth.

4. *Susceptibility to breakage.* Ice and wind storms cause breakage of some tree

species more than others. An ice storm in the spring of 1976 in Syracuse caused much breakage of boxelder, Carolina poplar, and silver maple. Broad-spreading Norway maples were also seriously damaged. More upright trees were less damaged. Besides the immediate problem, breakage represents infection sites for decay fungi and future problems.

5. *Life expectancy*. Most people expect a tree to be long-lived. This assumption is correct for sugar maple and oak, but is not necessarily correct for many of the species presently being planted.

As a general rule, long-lived trees such as sugar maple are desirable in parks and recreation areas, but the shorter-lived (30 to 50 years) trees such as honey locust are more appropriate for street plantings. Long-lived trees grow more slowly. Short-lived trees have the advantage of growing quickly, supplying shade to those who planted and paid for them.

As a general rule, urban trees planted today have an average life expectancy of 20 years. If this is correct, we are not planting many shade trees for tomorrow. It might be appropriate to consider planting rapid-growing, long-lived trees such as elms and silver maples and then cutting them before they become too large. These species will develop into shade trees quickly.

6. *Major disease problems*. Susceptibility to a lethal disease-causing organism is a major disadvantage but should not preclude the use of a tree species. Reasonable and effective control measures can offset the negative aspects of a lethal disease.

Actually, a major lethal disease may be an asset for an urban tree. One of the more difficult problems in managing urban maples is deciding when a tree is aesthetically dead. Considerable controversy sometimes arises over cutting trees that a city forester considers functionally dead and the local residents do not. There is usually little doubt when a Dutch-elm-diseased American elm needs to be removed. A major lethal disease also prevents the development of a large overmature, rather hazardous, population of trees. Therefore, from a management standpoint, a major lethal disease should not prohibit the use of a particular species.

7. *Minor disease and insect pests*. All trees have diseases and insect pest problems. Consideration of the visual and psychological impact on people may intensify the negative aspects of a biologically minor problem. The impact of insects and minor diseases is poorly understood, particularly in cities, where the interaction of pests with stressed trees may change the overall effect on trees of insects and minor pathogens.

8. *Genetic diversity*. Domestication of plants, as a general rule, reduces the genetic diversity. Selection for uniformity of desirable traits also unconsciously selects for uniformity of other traits, some of which may be undesirable. Cloning or vegetative propagation is the extreme in uniformity. A desirable trait, uniform columnar growth form in the clone lombardy poplar, is counterbalanced by common susceptibility to environmental stress and weak canker pathogens. The potential for rapid buildup of destructive insects and pathogens is reduced if a population has a degree of genetic diversity. Trees of mixed seed origin or synthetic varieties should be used in preference to clones and narrow-genetic-base varieties. Further elaboration on the genetic considerations was presented in Chapter 20.

9. *Tree form and growth characteristics*. Urban trees are best known for their

form and desirable or undesirable growth characteristics. Although size should remain a major factor in species selection, form and other growth characteristics should not be a dominant consideration. How much does exotic form really mean to the general public? Indirect consequences of the other selection criteria are probably more important.

10. *Probability of the unexpected.* The final ingredient of a species suitability index should include a safety factor for the unexpected. There is a real advantage to using native varieties and species that have been widely planted for a number of years or generations. We know what to expect in the way of problems. The emphasis today on new exotics and cultivars should be weighed heavily against the probability of unexpected pests and pathogens.

If the positive and negative aspects of various trees using these 10 criteria are quantified, a ranking of suitable and unsuitable species is obtained. This type of list would be invaluable for planning purposes. At the present time, it is very difficult to quantify or rank species within the minor diseases and insect pest category because of insufficient data. Assessment of the impact of minor pathogens and insects will greatly aid in the eventual establishment of a species-suitability index.

Selection of suitable species is not enough to ensure maximum satisfaction. There are within-species differences for adaptation to variation in climate. This topic, described in Chapter 3, emphasized the use of local seed sources as best adapted to the climate of an area. With ornamentals, local seed sources may not be available, so a period of 3 to 5 years of screening in a regional or city nursery is suggested. A city can hardly afford not to grow the trees for a few years in its own nursery. Those not adapted to the climate can be easily removed from the population. It is also much easier and less costly to evaluate future insect and pathogen problems in a city nursery.

Maintenance

Maintenance often has a lower priority than planting and removal in the urban tree management budget. This is ironic, since good maintenance often reduces the number of removals and, consequently, the need for planting.

Periodic pruning, to shape trees or remove broken and dead branches, produces immediate visual improvements which the general public can quickly recognize, and also produces long-run benefits such as less decay, fewer cankers, and less future breakage.

Although fertilizing has been demonstrated to produce beneficial effects on tree growth, improved growth may not be what established urban trees need. It may be appropriate to stimulate the growth of young trees to help them recover from an environmental or pest problem. But little documented evidence is available to demonstrate the long-term benefits of urban tree fertilization, so the use of fertilizer should have a low priority in a tight-budget management program.

Pesticides to control insects and diseases are sometimes applied to alleviate entomophobia or pathophobia in people rather than to control problems on trees. They never solve the problem. The unending costs from year to year are rather discouraging.

Careful selection of pesticides and application only where and when absolutely needed should be a general rule in urban tree management.

Pest management is better accomplished through mixing of 5 to 10 species of trees selected on the basis of the suitability criteria presented. By using a mixture, one has a natural buffer against rapid disease increase and catastrophic changes in the population. By working with a reasonably small number of species, it is possible to optimize management activities.

Insect populations are less capable of increasing in a mixed population. In the future, they may be further held at reasonable levels by biological control agents such as parasites, predators, and pathogens. Another tool for insect manipulation may be pheromones, chemicals produced by insects as signals to other insects. An aggregation sex attractant is being experimented with for manipulation of the smaller European elm bark beetle in Dutch elm disease control.

Periodic inventory and assessment should be used to locate trees that are prone to disease or insect problems. These trees represent concentration centers of pest populations. Selective removal of these problem trees should be beneficial to the remaining population.

Rotation

It is appropriate to recognize the useful-life expectancy of trees at the time of planting. Contrary to popular belief, most of the trees planted today will not be around for two or three generations. The stress-inducing pressures of the urban environment greatly reduce the life expectancy of even long-lived trees. There is also a shift toward smaller, shorter-lived tree species.

The resource manager should recognize the useful-life expectancy of trees and anticipate when they should be removed. The negative asset rotation concept, as discussed earlier, is applicable.

By mixing short, medium, and long rotation species, we can avoid major changes in the city landscape from removal of overmature trees. Maximum continuous benefit at minimum costs is assured if planning, maintenance, and forethought are applied to present-day planting of trees in our cities.

REFERENCES

Apple, J.D., and P.D. Manion. 1986. Increment core analysis of declining Norway maples, *Acer platanoides*. Urban Ecol. *9*: 309–321.

Bassett, J.R., and W.C. Lawrence. 1975. Status of street tree inventories in the U.S. J. Arboric. *1*:48–52.

Baxter, D.V. 1952. Pathology in forest practice, 2nd ed. John Wiley & Sons, Inc., New York. 601 pp.

Berrang, P., D.F. Karnosky, and B.J. Stanton. 1985. Environmental factors affecting tree health in New York City. J. Arboric. *11*:185–189.

Boyce, J.S. 1938. Forest pathology. McGraw-Hill Book Company, Inc., New York. 600 pp.

BURNS, B.S., and P.D. MANION. 1984. Spatial distribution of declining urban maples. Urban Ecol. *8*:127–137.

CHAPMAN, D.J. 1981. Tree species selection with an eye toward maintenance. J. Arboric. *7*:313–316.

CRAUL, P.J. 1985. A description of urban soils and their desired characteristics. J. Arboric. *11*:330–339.

DAWSON, J.O., and M.A. KHAWAJA. 1985. Changes in street-tree composition of two Urbana, Illinois neighborhoods after fifty years: 1932–1982. J. Arboric. *11*:344–348.

GETZ, D.A., A. KAROW, and J.J. KIELBASO. 1982. Inner city preferences for tree and urban forestry programs. J. Arboric. *8*:258–263.

NOWAK, D.J. 1986. Silvics of Norway maple (*Acer platanoides* L.). M.S. thesis, SUNY College of Environmental Science and Forestry, Syracuse, N.Y. 152 pp.

SANTAMOUR, F.S., JR. 1971. Trees for city planting: yesterday, today, and tomorrow. Arborist's News *36*:25, 27–28.

SANTAMOUR, F.S., JR., H.D. GERHOLD, and S. LITTLE, eds. 1976. Better trees for metropolitan landscapes. Proceedings of a symposium. USDA For. Serv. Northeast. For. Exp. Stn.

VALENTINE, F.A., R.D. WESTFALL, and P.D. MANION. 1978. Street tree assessment by a survey sampling procedure. J. Arboric. *4*:49–57.

23

PATHOLOGICAL CONSIDERATIONS OF INTENSIVELY MANAGED FOREST PLANTATIONS

- **HISTORICAL BASIS FOR APPROACHING DISEASE CONTROL THROUGH THE DOMINANT FACTOR**
- **PLANT MODIFICATION TO CONTROL DISEASES**
- **PATHOGEN MODIFICATION TO CONTROL DISEASES**
- **ENVIRONMENTAL MODIFICATION TO CONTROL DISEASES**
- **CONCLUSIONS**

Before discussing disease considerations in intensively managed plantations, it may be necessary to justify the need for such plantations. If one looks around, it appears that natural forests abound. Why then are intensively managed plantations being considered when there appears to be such surplus?

Actually much of what appears to be forest land is not really available to the timber industry. In many situations, other values placed on the forest far outweigh the commercial value of the timber.

Therefore, one reason for intensively managed plantations comes from the need to produce more wood fiber on fewer and fewer acres.

A second advantage of intensive management concerns the need to better control the economic investment of land and trees. The productivity of unmanaged forests is very low if productivity is expressed in terms of growth per acre per year. Low productivity was not a consideration when land was inexpensive and return on investment could be acquired through harvesting the crop of trees that already occupied the site. Forestry has shifted from harvesting to growing of trees, and the economic returns are therefore affected by the rate of fiber production.

Pathological considerations can be more appropriately justified with intensively managed plantations. The economic return is higher; therefore, money can be spent on protecting the crop and the investment. With natural forests it is usually more economically sound to let the diseases run their course and salvage-cut to extract whatever value can be gained from the forest.

Beech bark disease is a prime example of the economic dilemma of natural forests. Even if we had a control for this disease, it could not be economically justified. You may be amazed to learn that in 1978 a 10-in. (26-cm) beech had an approximate market value to the landowner of $0.50. How much can the forest owner spend to protect this type of investment? He or she will obviously salvage-cut as much timber as possible and use the rest for firewood.

Intensively managed forest plantations will supply the fiber needs of the future. In the South, such plantations already represent a significant proportion of the growing stock. New approaches to harvesting plantations will also become necessary. Equipment that harvests the whole tree, including the tap root, is now being used on a small scale for southern pines. Other approaches to intensive management utilize a modified silage chopper to harvest short-rotation tree crops much like corn. Sycamore and other hardwood species have been utilized for this type of management system on an experimental basis.

Rapid early growth of seedlings has always been a prime objective in establishing plantations, but little could be done to affect growth. Recent work with mycorrhizal inoculation of seedlings in the nursery suggests that it may be possible to affect the survival ability and initial growth of seedlings in plantations.

These and other intensive management activities are not that distant. We should not move in this direction without recognizing the pathological risks of intensive management. Only through maintaining a watchful eye on the developing plantation can we recognize the problems as they develop. We also have to develop an organized way of looking at disease control problems, because the investment in the plantation will not allow us to blindly apply control procedures.

In Chapter 1, the categories of plant disease were introduced and the methods for recognition for each were presented. The abiotic diseases are recognized primarily on the basis of the random distribution of diseased individuals, lack of consistent relationship with specific biotic agents, and lack of host specificity. The control of abiotic disease involves, first, recognition of the causal agent and, second, adjustment of future management activities through appropriate manipulation of either the plant or the pathogen.

Abiotic diseases are simple-interest diseases and therefore do not accelerate over time. Rapid increase in an abiotic disease may occur because of an unusually high level of the disease-inducing agent. Once the abiotic disease is recognized, it is very difficult to significantly affect the amount of future loss due to the disease. Therefore, control is usually directed at reducing losses in the next crop, by changing species or modifying the site.

Decline diseases are slow, progressive deterioration diseases of complex biotic and abiotic origin. As discussed in Chapter 18, declines are best managed by recognition of predisposing and inciting conditions and attempting to avoid or counteract them. Declines are also simple-interest diseases, best controlled through management activities for the next crop.

Biotic plant diseases are characterized by the involvement of a pathogen which is living and capable of reproducing on the host. Reproduction by the pathogen results in nonrandom distribution and intensification of the disease, often at a compound-interest rate. With biotic plant disease, control may be through management of the plant, the pathogen, or the environment. These management activities may be directed toward reducing losses in the present crop, future crops, or both.

I suggest that effective control of biotic-induced diseases be based on the principle that diseases do not generally result from the interaction among the plant, pathogen, and environment with equal importance to each, but that each disease has a dominant angle on the disease triangle. The most effective control involves manipulation of the dominant angle of the triangle.

HISTORICAL BASIS FOR APPROACHING DISEASE CONTROL THROUGH THE DOMINANT FACTOR

Investigations on the application of forest pathology to silviculture and management in the national forests began with the establishment of the U.S. Forest Service Regional Forest Research Stations in 1907. During the early part of this century, emphasis was placed on cutting practices as they affect disease. In 1916, Meinecke, in his monograph "Forest pathology in forest regulation," emphasized the disease-control benefits of sanitation and hygiene during cutting.

From the beginnings of pathology in Europe, two schools of thought developed. One group, pathogeneticists, of which Robert Hartig was a member, assumed that disease developed because plants inherited a susceptibility to disease. A second group, predispositionists, headed by Sarauer and Ward, assumed that disease developed because environmental factors predisposed plants to disease.

A third, but less-well-defined school of thought, with roots embedded in the overthrow of the spontaneous-generation concept, were concerned with the pathogen in disease. The pathogeneticists and predispositionists acknowledged the role of the pathogens but assumed a general or universal pathogen presence.

If we look at these three groups today, we do not see conflicting concepts of plant disease but our present concept of disease—plant, environment, and pathogen interacting to produce disease.

It is inappropriate to simplify these three schools into common, equally important factors in the development of all diseases. They originated as conflicting concepts because evidence for each could be observed in the real world.

In a practical sense, one of the three is usually the dominant factor in a given disease. The resource manager, attempting to minimize losses due to disease, should recognize the dominant or controlling factor in the diseases of concern and attempt to exploit that information to the fullest extent.

For economic and environmental reasons, it is most appropriate to integrate disease control into silvicultural practices. These practices will be discussed in relation to their effect on the plant, the pathogen, or the environmental aspects of plant disease.

PLANT MODIFICATION TO CONTROL DISEASES

Genetic variation in susceptibility can be directly utilized in tree improvement breeding programs or can be indirectly exploited in selective cutting practices. Chapter 20 discussed breeding for disease resistance. Even if genetically improved tree varieties are not yet commonly available, the resource manager can partially control diseases through plant modifications now. The probability of improvement of the next stand with selected quality seed trees is much better than with random regeneration from low-quality cull trees with large seed crops. Although large seed crops are often produced on trees that are stressed or deteriorating for various reasons, these trees obviously carry genes for susceptibility to problems and should be avoided. Logging practices of the past have generally removed the best trees and left the less desirable ones to regenerate many of the low-quality second-growth stands we have today. These types of logging practices should obviously be discontinued in favor of more careful selection of superior seed trees.

Fungicides applied to trees to prevent disease is another form of plant modification, but economic and environmental concerns restrict the use of this approach.

PATHOGEN MODIFICATION TO CONTROL DISEASES

Disease control, particularly of introduced diseases, has usually centered on the pathogen. But eradication of the pathogen is generally expensive and almost impossible to accomplish, even in a limited area.

Partial reduction in the pathogen population may or may not reduce the overall impact of the disease. A good example of disease impact reduction with partial pathogen reduction is the selective removal of trees with dwarf mistletoe infection in the upper crown. Inoculum produced in the upper crown is more effectively dispersed to intensify the disease than is inoculum in the lower crown.

Pathogens with small wind-disseminated spores are more difficult to control through pathogen manipulation. There is no reason to assume that selective removal of heart-rot- or root-rot-infected trees has any effect on the potential for disease in the surrounding stand. There is also no reason to assume that selective removal of canker-fungi-infected trees or branches has any major impact on diseases spread.

Canker fungi and decay fungi are generally limited by either host or environmental factors. So effective control of these is more appropriately applied to host improvement or environmental manipulation.

Clear-cutting, if done in large blocks, may reduce inoculum of disease agents, but the effects of clear-cutting on the environmental aspect of disease is probably more significant.

Fire is an effective control agent for dwarf mistletoe and brown spot needle blight of longleaf pine. Fire, like clear-cutting, also has environmental effects on disease.

Continuous cropping of the same species on the same site can be expected to increase disease problems. Most tree species develop in forest stands following successional sequences. Trees have evolved, with their pathogens, a balance of dynamically changing successions. The balance is shifted in favor of increasing the pathogen population if a stand is artificially maintained in the same species for more than one generation. These types of problems are well documented in agriculture, but are just beginning to show in forestry. A root rot pathogen such as *Heterobasidion annosum* is more of a problem in plantations than in naturally regenerated stands. Logging and fire control have resulted in extensive regeneration of Douglas-fir in the Pacific northwest. The second-growth Douglas-fir is being seriously damaged by *Phellinus weirii,* which produced only minor damage in the original stands.

Continuous cropping is not something that can be eliminated. Economic pressures for specific products and greater yields will force more extensive use of continuous cropping. It is important to recognize and anticipate the problems of this type of cultural practice, and weigh the advantages and disadvantages. If at all possible, one should consider normal succession patterns and adjust crops to more closely approximate the species changes encountered in natural stands.

Stumps produced in thinning or clear-cutting are ideally suited for infection by root decay fungi. Pathogens in the root system will continue to survive for a period, so the stump cut surface is much like a large wound in a living tree. Large numbers of wounds, whether resulting from cutting or from natural disasters, allow a major buildup in fungal pathogens and potential problems for regeneration of the surviving stand. Anything that favors rapid colonization and decomposition of stumps by saprobic decomposers should be considered control of these diseases through pathogen manipulation.

ENVIRONMENTAL MODIFICATION TO CONTROL DISEASES

Environmental modification, like fertilizing to improve growth or thinning to reduce competition, enhances the growth of trees and resistance to facultative parasitic microorganisms, but obligate pathogens, particularly rusts, are affected in just the opposite way and may actually increase more rapidly in vigorous trees.

Other types of environmental modification may be directed more at affecting the conditions for dispersal and infection by the pathogen. *Lophodermium* and other foliage diseases of Scots pine Christmas trees are more serious in plantations where tall weeds grow around the trees and along edges of the plantation adjacent to native forests. These foliage diseases are favored by prolonged moist conditions, which the shading of weeds and large trees provide. Weed control and cutting of large shading trees along the edge of the plantations is an environmental manipulation which affects disease development by affecting dispersal and infection of the pathogen.

The most significant environmental manipulation for control of tree diseases is the matching of species to sites. The selection of species to plant is more often determined by the availability of planting stock or desirability of a particular crop than by the ability of the species to grow well on the site. An even finer matching of trees to site goes beyond species suitability to using local seed sources. If one compares local seed sources with other seed sources of the same species, one usually finds that trees grown from local seed sources generally perform better than most other sources. These types of comparison, called provenance tests, have been made for a number of trees and the results are usually the same. Locally adapted plants are always among the best performers. Environmental stresses on non-locally adapted plants reduce tree vigor and, in many instances, predispose the tree to serious diseases.

CONCLUSIONS

Pathological understanding can contribute to improved intensive forest management. Neglect and misuse of pathological understanding has, in the past, cost a great deal of money. Both time and yield have been lost.

Control of tree diseases should not emphasize local suppression efforts. Control of diseases is more appropriately handled through effective planning and avoiding potential buildups. Just like modern fire control, there must be continuous assessment of potential problems and environmental conditions favoring buildup of pathogens. We must understand disease cycles and recognize where the cycle can be best manipulated.

At any point in time, we in forest pathology are concerned with the problems of the trees of various ages. There is not much we can do for the mature trees. We can prescribe cultural practices for the actively growing crops. But we can have our most significant impact on the future if we apply a little of our wealth of pathological understanding to planning, initiation, and management of the plantations we start today.

REFERENCES

BAXTER, D.V. 1952. Pathology in forest practice. John Wiley & Sons, Inc., New York. 601 pp.

BOYCE, J.S. 1938. Forest pathology. McGraw-Hill Book Company, New York. 600 pp.

CAMPBELL, W.A., and P. SPAULDING. 1942. Stand improvement of northern hardwoods in relation to diseases in the northeast. USDA For. Serv. Allegheny For. Exp. Stn. Occas. Pap. 5. 25 pp.

DINUS, R.J. 1974. Knowledge about natural ecosystems as a guide to disease control in managed forests. Proc. Am. Phytopathol. Soc. 1:184–190.

HANSBROUGH, J.R. 1965. Biological control of forest tree diseases. J. Wash. Acad. Sci. 55:41–44.

HEPTING, G.H. 1961. Forest pathology in forest management in the United States. *In* Recent advances in botany. University of Toronto Press, Toronto, pp. 1565–1569.

HEPTING, G.H. 1970. How forest disease and insect research is paying off: the case for forest pathology. J. For. *68*:78–81.

HUBERT, E.E. 1931. An outline of forest pathology. John Wiley & Sons, Inc., New York. 543 pp.

MEINECKE, E.P. 1916. Forest pathology in forest regulation. USDA Bull. 275, pp. 1–62.

SCHMIDT, R.A. 1978. Diseases in forest ecosystems: the importance of functional diversity. p. 287–315 *In* Plant disease: an advanced treatise, Vol. II, ed. J.G. Horsfall and E.B. Cowling. Academic Press, Inc., New York.

SHEA, K.R. 1971. Disease and insect activity in relation to intensive culture of forests. Proc. 15th Int. Union For. Res. Org. (IUFRO), pp. 109–119.

INDEX

397